Biopackaging

Biopackaging

Editor

Martin A. Masuelli

Área de Química Física
Departamento de Química
Facultad de Química, Bioquímica y Farmacia
Universidad Nacional de San Luis
Laboratorio de Membranas y Biomateriales—Instituto de
Física Aplicada—CONICET
Laboratorio de Investigación y Servicios de Química
Física (LISEQF-UNSL)
San Luis
Argentina

CRC Press
Taylor & Francis Group
Boca Raton London New York

CRC Press is an imprint of the
Taylor & Francis Group, an **informa** business

A SCIENCE PUBLISHERS BOOK

Cover credit

Images taken from the Thesis "Final Work of Food Engineering of Marisa Zanon, 2014. Universidad Nacional de San Luis, Argentina"

CRC Press
Taylor & Francis Group
6000 Broken Sound Parkway NW, Suite 300
Boca Raton, FL 33487-2742

First issued in paperback 2021

Library of Congress Cataloging-in-Publication Data

Names: Masuelli, Martin A. (Martin Alberto), editor.
Title: Biopackaging / editor, Martin A. Masuelli, Departamento de Quâimica, Facultad de Quâimica, Bioquâimica y Farmacia, Universidad Nacional de San Luis, Laboratorio de Membranas-Instituto de Fâisica Aplicada-CONICET, San Luis, Argentina.
Description: Boca Raton, FL : CRC Press, 2017. | "A science publishers book." | Includes bibliographical references and index.
Identifiers: LCCN 2017021425| ISBN 9781498749688 (hardback : alk. paper) | ISBN 9781498749695 (e-book : alk. paper)
Subjects: LCSH: Packaging--Materials. | Containers--Materials. | Packaging--Technological innovations. | Containers--Technological innovations. | Green technology.
Classification: LCC TS1982 .B56 2017 | DDC 658.5/64--dc23
LC record available at https://lccn.loc.gov/2017021425

Visit the Taylor & Francis Web site at
http://www.taylorandfrancis.com

and the CRC Press Web site at
http://www.crcpress.com

The 20th century was made with the synthetic polymers packaging, the 21st century will be called packaging with biopolymers or "Biopackaging".

Preface

The definition of biopackaging is related to natural and synthetic biodegradable polymers called biopolymers, these materials serve to protect and protect food, drugs, etc. In tandem with the development of synthetic commercial polymers, we might, for example, require that biopolymers use be limited to synthetic molecules fabricated from biological "monomer" units such as amino acids and monosaccharaides.

The packaging is a container or wrapping which contains products temporarily for the purpose of grouping units of a product with a view to its handling, transport and storage.

Other functions of the packaging are: to protect the contents, to facilitate the manipulation, to inform about its conditions of handling, legal requirements, composition, ingredients, etc. Within the commercial establishment, the packaging can help sell the merchandise through its graphic and structural design.

Bio-packaging can be defined simply as an extrinsic element of the product that protects it, where the fundamental characteristic is that its matrix is not polymeric, on the contrary, it is biopolymeric (proteins, polysaccharides, etc.).

The main theme of this book is concentrated on the biopackaging and the principles which underlie them. The book is directed to those who have relatively little background in the application of physical methods to the study of biological macromolecules used in biopackaging—the undergraduate who wishes to familiarize himself (herself) with the area or the researcher who is faced with proceeding in a new direction. Thus, although much of the material is presented at an elementary level, an effort has been made to reference the more important aspects and to present a current account of the status of biopolymer-packaging research. We are pleased to acknowledge the assistance of our colleagues in providing useful suggestions and material and the many investigators who provided us with photographs, original figures and tables, several of which have not previously been published. I am particularly grateful for the contribution of the authors of each chapter of this book for the diligent efforts, endless patience, help and joy brought to this project.

The chapters are:

Chapter 1. Edible Active Packaging for Food Application: Materials and Technology. Mengxing Li and Ran Ye. China.

Chapter 2. Active Bio-Packaging in the Food Industry. Ricardo Stefani. Brasil.

Chapter 3. Antimicrobial Active Packaging. Cintia B. Contreras, German Charles, Ricardo Toselli and Miriam C. Strumia. Argentina.

Chapter 4. Transport Phenomena in Biodegradable and Edible Films. Tomy J. Gutiérrez and Kelvia Álvarez. Venezuela.

Chapter 5. Formulation, Properties and Performance of Edible Films and Coatings from Marine Sources in Vegetable and Fruits. Armida Rodríguez-Félix, Tomás J. Madera-Santana and Yolanda Freile-Pelegrín. Mexico.

Chapter 6. Agroindustrial Biomass: Potential Materials for Production of Biopolymeric Films. Delia R. Tapia-Blácido, Bianca C. Maniglia, Milena Martelli-Tosi and Vinícius F. Passos. Brasil.

Chapter 7. Vegetable Nanocellulose in Food Packaging. C. Gómez H., A. Serpa, J. Velásquez-Cock, C. Castro, B. Gómez H., L. Vélez, P. Gañán and R. Zuluaga. Colombia.

Chapter 8. Xylan Polysaccharide Fabricated into Biopackaging Films. Xiaofeng Chen, Junli Ren, Chuanfu Liu, Feng Peng and Runcang Sun. United States of America.

Chapter 9. Reactive Extrusion for the Production of Starch-based Biopackaging. Tomy J. Gutiérrez, M. Paula Guarás and Vera A. Alvarez. Venezuela.

Chapter 10. Polyhydroxyalkanoates – A Prospective Food Packaging Material: Overview of the State of the Art, Recent Developments and Potentials. Katrin Jammernegg, Franz Stelzer and Stefan Spirk. Austria.

Chapter 11. Active Biopackaging Based on Proteins. Mariana Pereda, María R. Ansorena and Norma E. Marcovich. Argentina.

This book benefits teachers, professors, undergraduate students, graduate and postgraduate students, researchers and engineers with an interest in advanced technologies on biopackaging.

Martin A. Masuelli

Acknowledgements

We first would like to thanks all the authors contributing to the realization of this ebook: Mengxing Li & Ran Ye, China, Ricardo Stefani, Cintia B. Contreras, German Charles, Ricardo Toselli, Miriam C. Strumia, Tomy J. Gutiérrez, Kelvia Álvarez, Armida Rodríguez-Félix, Tomás J. Madera-Santana, Yolanda Freile-Pelegrín, Delia R. Tapia-Blácido, Bianca C. Maniglia, Milena Martelli-Tosi, Vinícius F. Passos, C. Gómez Hoyos, A. Serpa, J. Velásquez-Cock, C. Castro, B. Gómez H., L. Vélez, P. Gañán, R. Zuluaga, Xiaofeng Chen, Junli Ren, Chuanfu Liu, Feng Peng, Runcang Sun, M. Paula Guarás, Vera A. Alvarez, Katrin Jammernegg, Franz Stelzer, Stefan Spirk, Mariana Pereda, María R. Ansorena, Norma E. Marcovich.

We would like to thank the Área de Química Física – Departamento de Química – Facultad de Química Bioquímica y Farmacia – Universidad Nacional de San Luis – Instituto de Física Aplicada – CONICET, Argentina for giving us the time and opportunity for the writing and edition work of this ebook.

We are grateful to the editing team at CRC Press for their help, especially Raju Primlani and Amanda Parida.

I would like to thank Maria Guadalupe Garcia for her help in correcting the English in this work.

Finally, we wish to express our gratitude to my wife, Maria Gabriela and my daughter Gabriela Daiana, for their patience and support during the long hours of writing, reviewing, commenting, and editing of this book.

Contents

State of the Art

Martin A. Masuelli and Marisa Zanon

The current market approach to flexible packaging is technically structured, highly visual, easy to handle, functional, suitable for automatic machinery and creative forms. The current requirements of food producers are higher than some years ago, so it is a priority to compete with containers that successfully perform at least the following functions:

Containment: must provide mechanical strength and elongation, these properties often determine the amount of plastic material needed to form the wall of a container. They must be resistant to perforations and tearing, since many products have irregular geometries such as sharp edges or sharp tips so the material must be resistant to mechanical damage that could produce these forms, yielding elastically without breaking or deforming.

Protection: aspects such as seal tightness and barrier functions are included. All flexible containers must be closed, the vast majority of them being heat sealed. The integrity of the heat sealing is achieved by knowing the packaging material and the sealing conditions such as temperature, pressure, residence time of the material between the jaws, film tension and film thicknesses. When there is control over these parameters, drawbacks such as capillary leakage, seal unevenness, delaminations, burned material residues, adhesion degradation and low seal strength will be avoided. In terms of barrier functions, the industry provides packaging with low permeabilities to both gases and vapours as well as light and aroma. Gases such as nitrogen and carbon dioxide are used in technologies for packaging in a modified atmosphere and controlled atmosphere. The ideal mixture of these gases will depend on the type of product to be packaged, the effect sought and the properties of each gas.

Processing capacity: It is important to consider several aspects, such as those described below.

- Thermal properties of the package, that is, if it is able to withstand the direct application of heat during baking, sterilization or some other method of heat preservation. Resistance to low temperatures is also considered as some foods are kept refrigerated or frozen;

- Hot tack resistance;
- Mechanical properties of the package, which can be used in automatic forming-packaging machines, for which good elongation capacity and dimensional stability are required for adequate thermoforming;
- Coefficient of friction for its optimum slip and working speed in the forming-packaging machine.

Presentation of the product: the packaging is a very important element of attraction for the potential client. Therefore, the flexible packaging must have suitable optical properties such as gloss, transparency or opacity. In addition to supporting the impression with an attractive, colourful design and provide all the necessary information for the consumer (declaration of ingredients, nutritional contribution, date of elaboration and date of expiration, country of origin, etc.). It should also be easy to manipulate and open, so that it suits all types of consumer.

Packaging-content relationship: the migration of components and residues from packaging materials to food is a phenomenon of great practical relevance, originating the incorporation of unwanted substances that, sometimes, translate into alterations of the sensorial and nutritional characteristics of the food, or the incorporation of toxic substances that generate the rejection of the product by the consumer. They could also affect the physical-chemical and/or mechanical characteristics of the packaging. Packaging/food interaction phenomena could trigger gradual changes over time in food quality due to the migration of components from the polymer matrix or loss of substances and/or components through permeation, migration and sorption phenomena. Therefore, the components that make up a packaging material must be present on the positive lists of use of resins for direct contact with food and meet the limits on total and specific migration phenomena, which is specified in international legislation (Illanes 2004).

In recent years, innovation in food packaging has been directed towards the development of new technologies that bring an additional benefit to the traditional functions, among these emerging technologies are the so-called active and intelligent packaging, redefining some of the functions of packaging (Martínez and López 2011).

The concept of active packaging began with the change in protection functions when moving from passive to active packaging. Previously, the materials of the primary containers were considered passive; their function was limited to protection against oxygen and moisture through an inert barrier, that is, they offered a barrier to the outside environment to delay the adverse effects that the environment produced on the food over time. Its functions were to contain and protect the food; however, there are a lot of internal and external factors that inevitably alter packaged foods. For this reason new conservation technologies have been developed in which active or intelligent properties are introduced to the passive packaging. These new containers act in a coordinated way with the product and the environment to improve health and sensorial properties, extend the shelf life and thus improve the quality of the packaged food (Fernández 2012).

Active Packaging

Active containers act to correct the defects inherent in a passive package and are designed to intentionally incorporate components that release substances into the packaged food or in its environment or that absorb substances from the food or its environment. Active packaging technologies involve some physical, chemical and biological actions to modify the interactions between the packaging and the product to achieve the desired results. They are divided into three main categories according to the type of interaction: absorption systems, emission systems and other types of systems (Martínez and López 2011).

Smart Packaging

Smart containers are defined as those that incorporate some system that monitors and communicates useful information of the properties and/or state of the packaged food. Among them are devices with various applications and ways of capturing and displaying information. They can be classified in the following (Pineda 2012):

Time-temperature indicators: They aim to estimate the quality and integrity of the product. There are two types of devices: those that reflect the cumulative effect of time and temperature on the exposure of the product to temperatures above a critical level (known as TTIs) and those indicating whether the product has been subjected underneath or above a threshold value or TI.

Generally, we can find them in the form of labels, which acts by different physical-chemical properties or enzymatic reactions. In addition, we can find partial history indicators, which record the changes of temperature when exceeding a threshold temperature; and full story indicators, which record independently of the critical temperature.

This type of indicators are the most widespread, and many patents can be found in this respect, since they are simple, relatively cheap, easy to use and provide an accurate and in many cases irreversible response (Fernández 2012).

Microbiological quality indicators: They are based on the identification of different volatile metabolites generated by microbial growth in the food, such as carbon dioxide, acetaldehyde, ammonia, alcohols and fatty acids, as well as the variation of acidity (pH) due to microbial growth.

Indicators of oxygen, carbon dioxide or other gases in the food atmosphere: Among its main functions is to determine some type of leakage within the food packaging with controlled atmospheres, and also for the determination of the freshness of foods as in the case of the identification of ethylene in fruits.

Biodegradable Packaging

The indiscriminate use of synthetic packaging has generated serious ecological problems contributing to environmental pollution caused by solid waste with low degradability, which has led to the search for natural biopolymers. Taking advantage of natural resources as a source of conservation and recycling becomes an excellent choice and innovation in the development of new biodegradable products. Their total biodegradation in products such as CO_2, water and later, in organic fertilizer, is a great advantage over synthetic ones (Villada et al. 2007).

Progress in Biodegradable Packaging Films

The publications of the bio-films are framed in the first place in the development of new materials from polysaccharides of great abundance and timber-agroindustrial production (Selke 2006; John and Thomas 2008) as they are the cellulose acetate (Cerqueira et al. 2012), the starch (Mohan et al. 2013), as well as the use of plasticizers (Shi et al. 2013). Other available but expensive polysaccharides are alginates (Pathak et al. 2009; Liakos et al. 2013), chitosan (Li et al. 2013; Thakhiew et al. 2010), gellan gum (Rao et al. 2010) and guar gum (Yang and Paulson 2000; Saurabh et al. 2013). From an original point of view, the development of new films of biopolymers of various types such as arabinoxylans (Narasimha Murthy et al. 2004), soybean polysaccharides (Heikkinen et al. 2013), watercress polysaccharides (Jouki et al. 2013) and β-D-Glucans with arabinoxylans (Tajik et al. 2013). Some advances in these films will be described below.

In a study that evaluated the influence of glycerine and corn oil on the physicochemical properties of the galactomannan-based films of *Gleditsia triacanthos* and chitosan (Cerqueira et al. 2012), it was found that the presence of glycerol gave a more hydrophilic structure and the oil of corn, a decrease in the affinity of the film matrix to water. However, the two polysaccharides have different behaviour in terms of glass transition temperature, water vapour permeability and elongation at break which are related to the particularities of their structure. In the galactomannan, the groups –COOH and –OH are the specific sites for the adsorption of water, while in chitosan, the groups are –OH and –NH_2.

In investigating arabinoxylan (Tajik et al. 2013), it was found to have low tensile strength, high crystallinity, and high oxygen permeability, while branched arabinoxylan substantially decreases tensile strength and oxygen permeability.

As for soy polysaccharide (Ying et al. 2013), preliminary studies have shown that these films have good appearance and satisfactory mechanical properties, that is, they have excellent potential as a film-forming agent.

The films obtained from watercress seed (Jouki et al. 2013) showed that the water vapour permeability increases as the glycerol content increases from 25% to 50% w/w, resulting in improved films flexibility, significantly decreasing the tensile strength and obtaining higher elongation values. Colour measurement values showed that increasing the concentration of glycerol in the polymer matrix causes an increase in the opacity of the film.

References

Cerqueira, M., B. Souza, J. Teixeira and A. Vicente. 2012. Effect of glycerol and corn oil on physicochemical properties of polysaccharide films. A comparative study. Food Hydrocol. 27: 175–184.

Fernández, C. 2012. Envases activos e inteligentes: control de la calidad y seguridad del producto. Instituto tecnológico del embalaje, transporte y logística. España.

Heikkinen, S., K. Mikkonen, K. Pirkkalainen, R. Serimaa, C. Joly and M. Tenkanen. 2013. Specific enzymatic tailoring of wheat arabinoxylan reveals the role of substitution on xylan film properties. Carbohydr. Polym. 92: 733–740.

Illanes, J. 2004. Envases flexibles plásticos: usos y aplicación en la industria alimentaria. Universidad austral de Chile. Chile.

John, M. and S. Thomas. 2008. Biofibres and biocomposites. Carbohydr. Polym. 71: 343–364.

Jouki, M., N. Khazaei, M. Ghasemlou and M. Nezhad. 2013. Effect of glycerol concentration on edible film production from cress seed carbohydrate gum. Carbohydr. Polym. 96: 39–46.

Li, R., P. Hu, X. Ren, S. Worley and T. Huang. 2013. Antimicrobial N-alamine modified chitosan films. Carbohydr. Polym. 92: 534–539.

Liakos, I., L. Rizzello, I. Bayer, P. Pompa, R. Cingolani and A. Athanassiou. 2013. Controlled antiseptic release by alginate polymer films and beads. Carbohydr. Polym. 92: 176–183.

Martínez, Y. and A. López. 2011. envases activos con agentes antimicrobianos y su aplicación en los alimentos. Temas selectos de ingeniería en alimentos. 5-2: 1–12.

Mohan, T., R. Kargl, A. Doliskaa, H. Ehmann, V. Ribitsch and K. Stana-Kleinschek. 2013. Enzymatic digestion of partially and fully regenerated cellulose model films from trimethylsilyl cellulose. Carbohydr. Polym. 93: 191–198.

Narasimha Murthy, S. and S. Paranjothy Hiremath. 2004. Evaluation of carboxymethyl guar films for the formulation of transdermal therapeutic systems. Int. J. Pharm. 272: 11–18.

Pathak, T., J. Yun, S. Lee, D. Baek and K. Paeng. 2009. Effect of cross-linker and cross-linker concentration on porosity, surface morphology and thermal behavior of metal alginates prepared from algae (Undaria pinnatifida). Carbohydr. Polym. 78: 717–724.

Pineda, D. 2012. Envases inteligentes en la industria de alimentos. Ministerio de economía de El Salvador.

Rao, M.S., S.R. Kanatt, S.P. Chawla and A. Sharma. 2010. Chitosan and guar gum composite films: Preparation, physical, mechanical and antimicrobial properties. Carbohydr. Polym. 82: 1243–1247.

Saurabh, C., S. Gupta, J.S. Bahadur, M. Variyar and A. Sharma. 2013. Radiation dose dependent change in physiochemical, mechanical and barrier properties of guar gum based films. Carbohydr. Polym. 98: 1610–1617.

Selke, S. 2006. Plastics recycling and biodegradable plastics. pp. 8.1-8.109. *In*: Handbook of Plastics Technologies. McGraw-Hill, New York.

Shi, A., L. Wang, D. Li and B. Adhikari. 2013. Characterization of starch films containing starch nanoparticles Part 1: Physical and mechanical properties. Carbohydr. Polym. 96: 593–601.

Tajik, S., Y. Maghsoudlou, F. Khodaiyan, S. Jafari, M. Ghasemlou and M. Salami. 2013. Soluble soybean polysaccharide: A new carbohydrate to make a biodegradable film for sustainable green packaging. Carbohydr. Polym. 97: 817–824.

Thakhiew, W., S. Devahastin and S. Soponronnarit. 2010. Effects of drying methods and plasticizer concentration on some physical and mechanical properties of edible chitosan films. J. Food Eng. 99: 216–224.

Villada, H., H. Acosta and R. Velasco. 2007. Biopolímeros naturales usados en empaques biodegradables. Revista Temas Agrarios.

Yang, L. and A.T. Paulson. 2000. Effects of lipids on mechanical and moisture barrier properties of edible gellan film. Food Res. Int. 33: 571–578.

Ying, R., C. Rondeau-Mouro, C. Barron, F. Mabille, A. Perronnet and L. Saulnier. 2013. Hydration and mechanical properties of arabinoxylans and β-d-glucans films. Carbohydr. Polym. 96: 31–38.

1

Edible Active Packaging for Food Application: Materials and Technology

Mengxing Li[1] and Ran Ye[*,2]

[1] Biological Systems Engineering, University of Nebraska–Lincoln,
  Lincoln, NE, USA
[2] Roha USA, Global Center of Expertise, St Louis, MO, USA

Introduction

As shown in Robertson (2012), food packaging has several functions. It functions as a mechanical support to protect foods from the surroundings and to maintain the safety and quality of packaged food, which is what traditional food packaging is meant for. Food packaging can also restrict the exchange of gases and serve as a barrier for the transfer of aromatic compounds. Other major functions include containment, convenience, marketing and communication. A product in a container helps avoid spillage or disbursement. It usually contains mandatory law required information, namely weight, source, ingredients, and now, nutritional value and cautions for use. Marketing is the most important function at the point of purchase (Kotler et al. 2014). In view of the food processors, the requirements for packaging materials include processability, heat stability, recyclability and low cost (Quintavalla and Vicini 2002). More advanced functions include delayed oxidation and controlled respiration rate, microbial growth, moisture migration and solute transport (Dutta et al. 2009).

During the past decades, in response to consumer demands, industrial food production trends towards minimally processed, fresh, tasty and convenient food products are becoming more and more popular (Lagaron et al. 2004), which leads to the concept of edible active packaging. Meanwhile, the concept of edible active packaging falls entirely into the category of "green

---
*Corresponding author: ran.ye.email@gmail.com

packaging", which follows the suggestion of US environmental protection act (EPA). Formulations of edible packaging are exempt from chemically synthesized substances and are enriched with natural substances that bring health benefits and maintain nutritional and sensory characteristics (Falguera et al. 2011a). Therefore, edible packaging has become one of the major areas of research in food packaging in the US and Japan (Véronique 2008).

Conventional food preservation from microbial contamination includes thermal processing, drying, freezing, refrigeration, irradiation, modified atmosphere packaging and adding antimicrobial agents or salts (Quintavalla and Vicini 2002). Unfortunately, some of these techniques cannot be applied to minimal processed and ready-to-eat food products. Edible active packaging incorporated with antimicrobial, antioxidant agents can release the bioactive compounds slowly into the surface of the food, thus efficiently extending the shelf life of those food products.

Active packaging is defined as the packaging system that possesses attributes beyond basic functions, which are usually obtained by incorporating active compounds in the packaging system and/or applying actively functional polymers. From this definition, it can be understood that active packaging not only has a barrier function, but also plays an active role in food preservation and quality assurance during the storage and marketing process (Pereira de Abreu et al. 2012).

Edible packaging is defined as utilizing materials edible compounds such as proteins, polysaccharides, lipids and/or resins, and other edible components, derived from diverse renewable sources. Edible films are usually stand-alone structures that are preformed separately from the food and then placed on or between food components or sealed into edible pouches, whereas edible coatings are thin layers of edible materials formed directly onto the surface of the food products (Krochta 2002). Soft-gel capsules, hard-gel capsules, tablet coatings, and microcapsules made from edible materials could also be considered edible packaging (Janjarasskul and Krochta 2010). Edible packaging has been driven by both increasing consumer demand for safe, high-quality, convenient foods with long shelf lives and ecological consciousness of limited natural resources and the environmental impact of packaging waste. The role of active packaging in minimizing the supply chain was reviewed in Verghese et al. (2015); they concluded that packaging can reduce food waste in the food supply chain significantly. Zhang et al. (2015) performed a life cycle assessment of essential oil component-enabled packaging for fresh beef to look into the effect of active packaging on minimizing food losses. The sensitivity analysis showed that the breakeven point (that is, the threshold where the environmental impact becomes equal to the control case) of the balance can be achieved across four impact categories evaluated, including global warming, fossil energy demand, acidification potential and eutrophication potential.

Edible packaging mostly serves as a primary edible packaging together with nonedible packaging as secondary packaging to add additional protection from the atmosphere and prevents contamination from microorganisms or foreign particles. Edible packaging can convert multilayer packaging to a single-component package, leading to source reduction and recyclability improvement without compromising protective functions (Krochta 2002). The researchers in the food and pharmaceutical industries deem edible packaging as a useful alternative or addition to conventional packaging to reduce water and to create novel applications for improving product stability, quality, safety, variety and convenience for consumers (Janjarasskul and Krochta 2010). Edible active packaging is different from plastic active packaging in that the polymer made into packaging are edible, which is the main scope of this article. Edible materials used as the packaging materials make the active compounds more compatible and distributed homogeneously in the polymer matrices (Gómez-Estaca et al. 2014). Mellinas et al. (2016) reviewed the industrial applications of edible active packaging explicitly.

In this chapter, recent developments in methods of preparing the structural matrices of edible active packaging are summarized. The properties of active packaging are also emphasized. Lastly, the novel applications of edible active packaging are highlighted.

Materials

Edible packagings are categorised due to their structural matrix. Packaging materials applied in edible active packaging mainly fall into the categories of polysaccharides, proteins, lipids and composite. The use of only lipids in the packaging structural matrix is rare. Lipids are usually introduced to multicomponent or composite packaging, to improve the moisture barrier properties, but negatively affect the transparency of the films. For composite packaging, the mechanical strength and barrier properties are largely affected by the compatibility of the compounds used in the packaging. A plasticizer is necessary to be added into the formulation to get enough flexibility (Sobral et al. 2001). The amount of plasticizer could be up to 50% of the packaging weight; therefore in some literatures plasticizer belongs to the packaging materials. A general scheme for edible packaging based on the materials is summarized in the review of Sánchez-González et al. (2015). Specifically, Falguera et al. (2011b) reviewed the common hydrocolloids used as the structural matrix of edible packaging. One trend regarding the derivation of edible packaging from agricultural and industrial waste occurred and, Zia et al. (2016) comprehensively reviewed biopolymers derived from renewable resources for packaging applications.

Carbohydrates

Starch is one of the most common and important polysaccharides in the world. Sánchez-González et al. (2015) reviewed the strategies of preparing starch-based edible active packaging such as blending with antimicrobial polymers, active nanoparticles, and antimicrobial or antioxidant compounds. Different starch formulations may lead to edible active packaging with specific characteristics and properties. G Volpe et al. (2010) reviewed all the patents regarding the use of polysaccharides as biopolymers for food shelf life extension. Common polysaccharides used for edible packaging materials could be found in the review of Mellinas et al. (2016).

Novel polysaccharides materials such as pectin, chitosan and gum from new resources are attracting the increasing attention of researchers in recent years. Espitia et al. (2014) reviewed the physical-mechanical and antimicrobial properties of edible films from pectin; the esterification degree was found to directly affect gelation and the film forming ability. Chitosan is one of the most widely used biopolymers for preparation of edible active packaging. It showed particular physicochemical properties, such as biodegradability, biocompatibility with human tissues, null toxicity, and very importantly, antimicrobial and antifungal properties (Aider 2010). Aider (2010) and Dutta et al. (2009) reviewed the applications of chitosan for active antimicrobial films in the food industry. The disadvantage of chitosan-based film is the cost which is too high for chitosan to be applied in food packaging. Thus, chitosan combined with other cheap film forming biopolymers to prepare edible active packaging is proposed. Dutta et al. (2009) reviewed chitosan-starch film under different treatment methods including supercritical fluid drying, microwave and irradiation, optimisation of biocidic properties and how biocatalysts improved quality and shelf life of foods were widely discussed.

More and more polysaccharide polymers were found and/or applied to edible packaging. Mikkonen and Tenkanen (2012) applied xylan and mannans for food packaging. Xylan and mannan based films and coating show low oxygen and grease permeability and relatively high tensile strength. Cashew gum (Carneiro-da-Cunha et al. 2009), Brea gum (Spotti et al. 2016) and galactomannans (Cerqueira et al. 2009) were also newly found interesting polymer materials. Razavi et al. (2015) used sage seed gum for edible film preparation and the film showed low permeability to water vapor, acceptable mechanical properties, surface hydrophilicity, and transparency. Martínez-Abad et al. (2016) did a thorough study on enzymatic extraction and modification of lignocellulosic plant polysaccharides for packaging applications.

Proteins

Proteins are good candidates for edible active packaging due to the particular structure and mechanical stability of protein formed film. Proteins can be derived from animals (casein, whey protein, gelatin and egg albumin) or plant sources (corn, soybean, wheat, cottonseed, peanut and rice) (Mellinas et al. 2016).

Protein forming into film is caused by the denaturation of the protein initiated by heat, solvents or pH change (Janjarasskul and Krochta 2010). Edible packaging prepared from protein materials generally has good mechanical, optical and barrier properties. Specifically, the hydrophilicity of protein and polysaccharide-based edible films makes them excellent barriers to nonpolar aroma compounds. Flavor and aroma encapsulation by carbohydrate and protein emulsion-based films have been proposed (Hambleton et al. 2009). The disadvantage causing by hydrophilicity is that the packaging will absorb moisture when exposed to a high humidity environment, which will compromise the mechanical and barrier properties.

Modification of a certain functional group or the surface of the protein film is a good approach to overcome the disadvantage of the protein film. Enzymatic modification is newer and more advantageous than physical and chemical modification. Transglutaminase (TGase) is widely studied to enhance the cross-linking by catalysing ε-(γ-glytamyl)-lysyl cross-links between the glutamine and lysyl groups. Wang et al. (2015) applied TGase to induce cross-linking of gelatin-calcium carbonate composite films. Differential scanning calorimetry (DSC), the mechanical properties and the water vapor permeability studies revealed that TGase favored the strong intramolecular polymerisation of the peptides in gelatin. Di Pierro et al. (2013) prepared whey protein/pectin film at complexation pH (pH 5.1), they found that an increase of both tensile strength (2 fold) and elongation to break (10 fold) and a reduction of elasticity. Al-Saadi et al. (2014) compared the effect of heat and TGase modification on the solubility of goat milk casein based film; the film was less soluble when treated with TGase but more cross-linked when treated by TGase.

Composite

Carbohydrates and proteins are great materials due to their excellent mechanical and structural properties, but the critical issue is the poor barrier capacity against moisture invasion (Sobral et al. 2001). One way to overcome the disadvantage is to prepare a hydrocolloid film with a lipid layer or incorporate lipids into the hydrocolloid film formulation. Otoni et al. (2016) added micro/nano-droplets of Actem and Tween 60 into an isolated soy protein (ISP); the resulted film showed reduced strength and increased elongation. Peng et al. (2013) incorporated tea extracts into chitosan films and results showed that water vapor permeability decreased with the addition of tea extract. In the review of Mellinas et al. (2016) titled as "active edible films: current state and future trends", composite films was manufactured, based on alginate, chitosan, protein derived from agricultural and industrial residues and essential oils. Formulation of different ingredients provides opportunities to design specific biopackaging based on users' needs.

Processes

The processes for preparation of edible active packaging are conventionally divided into two categories: wet and dry processes. For all the different wet processes, casting is the step differentiating it from the dry processes, while extrusion is the step differentiating the dry processes from the wet processes. Electrospinning, as a novel process, is emphasized for its applications. For any process, the forming of edible active packaging includes a sufficiently cohesive and continuous structural matrix. During the process, usually the covalent bonds (e.g. disulfide bond), ionic bonds are formed and H-bonding occurs. Covalent bonds forming the polymer network are more stable than other bonds. The mechanisms of different processes are summarized as follows.

Casting

Solvent casting technique is the most commonly used technique when preparing edible active packaging. In the review of Mellinas et al. (2016), it is one of the steps (dispersion or gelatinization, homogenization (in the case of emulsions or mixtures), casting, and drying) in wet processing. Casting is a batch procedure used widely on a laboratory scale. Solvent casting is based on a conservation mechanism where hydrocolloids dispersed in an aqueous suspension are precipitated after drying (Kester and Fennema 1986). The solvent solution is heated to above the melting temperature, homogenized (for an emulsion system), and cast on a flat surface while cooling. The most common solvents for edible packaging are water and ethanol. Drying methods include air drying, hot surface drying, infrared and microwave techniques. Air drying is usually performed under room temperature, leading to a more homogeneous polymer network than a high temperature drying process. On the other hand, solvent casting is also a method to evaluate the film forming property of biomaterials. There are standard methods for predicting how coating from different formulation would perform on a certain type of food (Falguera et al. 2011b).

Extrusion

The extrusion process is preferred by the industry since most conventional packaging structures are manufactured completely or partially by extrusion processes. Compared to casting technology, extrusion is more severe concerning the parameters in the thermo-mechanical process. Whether or to what degree the active compounds or the packaging materials will be degraded during the process determine the packaging behavior (Gómez-Estaca et al. 2014). The extrusion process is based on the thermal properties of packaging materials, including phase transitions, glass transition and gelatinization characteristics. In the extrusion process, the packaging materials are plasticized and heated above the glass transition temperature

(T_g) with a low amount of water to construct a uniform melt by using heat, pressure and shear in the barrel of the extruder. During the extrusion process, the parameters such as temperature, feed rate, and screw configuration rate, product inlet/outlet pressures will affect the conformational changes, aggregation and chemical cross-linking (Janjarasskul and Krochta 2010; Liu et al. 2005). High temperature conditions are one of the reasons for antioxidant activity loss, demonstrating that casting processes produce materials with less loss of antioxidant in the film with respect to the nominal content due to milder processing and manufacture conditions (López-de-Dicastillo et al. (2010)).

Electrospinning

Electrospinning, as an electro-hydrodynamic process, applies electrostatic forces to produce electrically charged jets from viscoelastic polymer solutions which upon drying, by the evaporation of the solvent, give rise to ultra-thin structures (Li and Xia 2004), which is a new approach to construct nanostructured packaging materials (Fernández et al. 2008). Electrospinning is a simple, versatile and low cost process that is capable of producing fibers scaling from nanometers to micrometers. To achieve a good film, a large number of highly specific conditions including solution properties (polymer concentration, viscosity, electrical conductivity, surface tension and solvent volatility), environmental conditions (temperature, air velocity and humidity) and process conditions (voltage, spinning distance and flow rate) must be considered (Fabra et al. 2015). Fabra et al. (2015) investigated the effect of the film-processing conditions, relative humidity and aging on wheat gluten films coated with electrospun polyhydryalkanoate; the outer electrospun layer was in excellent adhesion with the inner wheat gluten layer and the mechanical and barrier properties were both improved for the multilayer compared to wheat gluten single layer film. Furthermore, the same authors developed layered active packaging systems using different hydrocolloid matrices (whey protein isolate (WPI), zein and soy protein isolate (SPI) as shell materials, alpha-tocopherol was encapsulated within these hydrocolloid materials). Their results revealed that the alpha-tocopherol was stable when the systems were applied to industrial sterilization process, and the alpha-tocopherol release profile showed a delayed release (Fabra et al. 2016). Neo et al. (2013) built a gallic acid loaded zein electrospun fiber mats as active packaging materials. They found that the fiber mats exhibited antimicrobial properties and gallic acid diffuses from the electrospun fibers in a Fickian diffusion manner.

Functions

Basic Functions

The most important criteria for edible packaging is a controllable mass transfer between food materials and the surrounding environments.

Permeability should be considered under the specific environmental conditions that the food product will encounter. Janjarasskul and Krochta (2010) mentioned that the edible packaging should be good barriers for moisture, oxygen, aroma and oil. They also reviewed the barrier properties of edible packaging made from different materials and compared the barrier properties of edible packaging with that of petroleum-based packaging materials. The requirement for barrier properties of edible packaging should inhibit moisture exchange, have a low oxygen permeability between finished food products and the surrounding materials and have a high aroma barrier efficiency. Thus, a migrating compound has low affinity to film materials and low diffusivity through the polymer matrix. Janjarasskul and Krochta (2010) also stated the functions of enhancers of food products for improvement of structure integrity and appearance.

Novel Functions and Applications

The biopolymer-based food packaging materials can serve as a carrier to deliver or control release active compounds (Janjarasskul and Krochta 2010). Formulations can be designed to carry desired food additives (including antioxidants, antimicrobials, pigments, flavors, spices, salts, nutrient, light absorbers) as well as pharmaceutical or nutraceutical ingredients in the form of hard capsules, softgel capsules, microcapsules, soluble films/strips, flexible pouches or coatings (Han and Gennadios 2005). In the reviews of Ozdemir and Floros (2004) and Fabra et al. (2015), the known packaging systems with novel properties were reviewed. Of all the strategies that have been devised, the properties can fall into two categories: either retention of desirable molecules (i.e., aldehydes, oxygen) or release of active substances (i.e., carbon dioxide, aromas). Incorporation of antimicrobials and antioxidants into the packaging materials have been widely studied, and the mechanism of how antimicrobial and antioxidant affect packages is required to be more explicitly elucidated.

Antimicrobial Property

Antimicrobial active packaging innovated the concept of active packaging and developed to reduce, inhibit or stop the growth of food-borne bacteria and organoleptic deterioration on food surfaces (Appendini and Hotchkiss 2002). Food surface is where the highest intensity of microbial contamination is found for fresh or minimally processed foods. There are three advantages. Firstly, avoid directly applying antimicrobials to the foods (Gómez-Estaca et al. 2014). Secondly, high concentrations of antimicrobials could be maintained on food surfaces using fewer amount of antimicrobials compared to directly applying antimicrobials to foods itself (Ramos et al. 2014). Thirdly, the antimicrobials will migrate from the packaging material to the food surface slowly and maintain a high concentration of antimicrobials at the food surface, compensating the continuous using up of antimicrobials during storage (Mastromatteo et al. 2010).

Traditional antimicrobials applied in antimicrobial active packaging include ethanol, ozone, peroxide, sorbitol, sulfur dioxide, antibiotics, silver-zeolite, organic salts (sorbates, benzoates, propionates, quaternary ammonium salts), enzyme (glucose oxidase), and bioactive polymer (chitosan) (Restuccia et al. 2010). Novel antimicrobial agents incorporated into active packaging form an area of great interest. Petide (bacteriocin) includes nisin, divergicin and pediocin (Tahiri et al. 2009). Spices and essential oils also show strong antimicrobial and antioxidant properties. Spices are rich in phenolic compounds such as flavonoids and phenolic acids. Essential oils are mostly plant extracts. Antimicrobial packaging is of particular interest to the red meat industry (Véronique 2008). The active packaging with antifungal properties was reviewed in the study of Aider (2010). Van Long et al. (2016) prepared lysozyme/low methyl pectin complexes for antimicrobial edible food packaging; they found that the complex formation considerably decreased the lysozyme antimicrobial activity, which was probably due to lysozyme substrate diffusional limitation and/or a decrease of the enzyme mobility.

Antioxidant Property

Lipid oxidation is happening together with microbial growth, which leads to a loss of both sensorial (flavor) and nutritional quality of foods and even the formation of toxic aldehydes. Sharing the same idea with antimicrobial packaging, the sustained release of antioxidant during food storage can keep the antioxidant working to the fullest. Typical food additives such as butylated hydroxyanisole (BHA), butylated hydroxytoluene (BHT), tertiary butylhydroquinone (TBHQ), ascorbic acid and tocopherols are commonly used antioxidant in food industry, which are also applied in active packaging (Restuccia et al. 2010). There were previous studies reporting antioxidants incorporated into plastic packaging. Wessling et al. (1998) reported BHT release from polyethylene films; Torres-Arreola et al. (2007) found delayed lipid oxidation and protein denaturation with BHT incorporated into low density polyethylene.

Antioxidants work in active packaging based on two modes of actions: the release of antioxidants into the food and the scavenging of undesirable compounds such as oxygen, radical oxidative species or metal ions from the headspace or from the food (Gómez-Estacaet al. 2014). Scavengers refers to the substances that react with, modify or trap substances involved in the oxidation process. Therefore the package should make the pro-oxidant substances accessible to the location where scavengers are incorporated (Gómez-Estaca et al. 2014).

For many antioxidants, the bioactive compounds themselves are antimicrobials as well. The bioactive compounds from plant extract, essential oils, herbs and spice extract and exhibit antimicrobial and antioxidant properties, are reviewed by Valdés et al. (2015).

There are a few issues concerning spices and essential oils directly incorporated into packaging. It may alter the sensory characteristics of protected food. Spices and essential oils have low water solubility. Instead of directly applying spices or essential oils into packaging materials, prepare inclusion complex of antimicrobials with cyclodextrin (Hill et al. 2013) or make micro- or nano-encapsulation of antimicrobial (Chen et al. 2015) and incorporate processed antimicrobial (inclusion or encapsulation of antimicrobial) into packaging matrices would overcome these disadvantages. Alternative methods include nanoemulsions through the ultrasonication (Otoni et al. 2014), encapsulation into nanoliposomes through the sonication of their aqueous dispersions, have also been reported. Moreover, these delivery system could act as carriers of bioactive compounds to be transported to target sites such as the intestine without losing their activity during their passage through the gastrointestinal tract (Korhonen 2002). In addition, it has been found that in emulsion-based film, the smaller the particle size or lipid globules and the more homogeneously distributed, the lower water vapor permeability is (Pérez-Gago and Krochta 2001). The edible coating system with delivery properties was termed as second generation of coating materials in Slaughter et al. (2012).

Other plant extracts, such as pomegranate and grape seed extract (Duran et al. 2016), green tea extract (Siripatrawan and Noipha 2012), brewery residual stream (Barbosa-Pereira et al. 2014) and essential oils (Patrignani et al. 2015) are good antioxidant agents that could be incorporated into edible active packaging. Meanwhile those antioxidant agents also show antimicrobial properties, which are probably due to their hydrophobic nature (Sanchez-Gonzalez et al. 2011).

Novel Technologies Applied in Food Packaging

Modification

Packaging modification is aimed at improving the mechanical or barrier properties. Cross-linking or by blending with other polymers or nano-sized particles can strengthen the packaging (Mikkonen and Tenkanen 2012). González et al. (2015) reviewed the modification methods of proteins for food packaging applications. Enzyme modification has been reviewed in the section "protein". Construction of multilayered packaging using lamination is another method to improve packaging performance. Nevertheless, the introduction of these modification increases the complexity of the film-making process. The compatibility of different packaging materials or different components is also needed to be considered.

Nanostructured Packaging

One dimensional nanostructured edible packaging has been a subject of great interest due to their unique properties and intriguing applications. Fabra et al. (2013) reviewed nanostructured biolayers in food packaging, including monolayer biocoatings, multi-layered biostructures, and layer-by-layer nanobioassemblies constructed by electrospinning. For these nanostructured biolayers, the most improved properties are barrier properties.

Nanofillers were added to reinforce the biopolymers to improve the properties of edible packaging and enhance the cost efficiency (Sorrentino et al. 2007). Traditional nanoparticles include clay, silica and talc (Rhim and Ng 2007) and novel nanoparticles include tripolyphosphate-chitosan (de Moura et al. 2009), microcrystalline cellulose (Bilbao-Sáinz et al. 2010) and silicon dioxide (Tang et al. 2009). These nanoparticles improve the moisture barrier properties and efficiently inhibit microbial growth.

Cruz-Romero et al. (2013) applied nano-sized solubilisates of benzoic acid and sorbic acid for antimicrobially active packaging and found that the nano-sized solubilisates had significantly higher antimicrobial properties than their non-nano equivalents. Cerqueira et al. (2013) constructed a delivery device consisting poly (N-isopropylacrylamide) nanohydrogels and polysaccharide-based films. Incorporation of the nanohydrogels did not change the main properties of the film while the film exhibited an increase of water affinity. The employment of nanocomposite to edible packaging materials promises to improve barrier and mechanical properties and facilitate effective incorporation of bioactive ingredients. Mihindukulasuriya and Lim (2014) gave an overall review of nanotechnology development in food packaging.

The safety concerns remain an issue if nanostructured packaging is widely applied in food industry. The nanoparticles may penetrate into cells and eventually remain in the human tissues. Currently the data about the toxicological effects of nanoparticles on human health are still limited; it is of great need to conduct risk and safety assessment before the commercialization of nanostructured food packaging.

Bioactive Delivery and Controlled Release

Bioactive delivery has been an old topic for functional foods. Bioactives are incorporated into food formulations and processed into food products. There are two limitations for direct incorporation of bioactives into food (Del Nobile and Conte 2013). First, once the bioactives are consumed in reactions, the functions or protections cease and the quality of the food deteriorates at an increasing rate. Second, bioactives cannot be released selectively on the target (food surface) where most microbial contamination occurs. Fernández et al. (2008) reviewed the methods (covalent bonding, adsorption and entrapment) to incorporate the biocatalysts into polymers traditionally used in food packaging applications. Under controlled release, bioactives (antimicrobial, antioxidant agents, enzymes, etc.) can be incorporated into the biopolymer, and either be repeatedly used within the packaging or, migrate to the food product surface in a controlled manner (Dutta et al. 2009; Véronique 2008). As a result, a predetermined concentration of compound is maintained in the food surface to achieve a desired shelf life (Mastromatteo et al. 2010).

The release profile and the corresponding release mechanism is most studied since they are of special significance in order to engineer or modify the system to large scale or different application. Neo et al. (2013) loaded 5%, 10% and 20% gallic acid on zein-based film, at the first 10 min of immersion in water, 58%, 60% and 78% of gallic acid was released, respectively. The different amount of incorporated gallic acid presents or near the fiber surface explained the variation in the gallic acid release rate. Fichian diffusion was found to be dominant release mechanism, which suggested that the release of gallic acid was partially attributed to the diffusion or permeation through the swollen zein sub-micron fibrous matrix, and partly through the water filled pores and channels in the matrix structure (Roy and Rohera 2002). Similar release behaviors were also found in the studies of Del Nobile et al. (2008), Ramos et al. (2014) and Manzanarez-López et al. (2011).

Most mathematical models built to describe the mass transfer of bioactive compound migration into the food simulant were based on Fich's second law in the assessment of compliance with specific migration limits and describing the effective transport of migrating species with time (Barbosa-Pereira et al. 2014). Change of the formation/rupture of hydrogen bonds, hydrophobic and/or electrostatic interactions within the polymer matrices would change the diffusion coefficient. The kinetic models shown in some research showed high initial release rates and further observation of a plateau at different times and temperatures (Graciano-Verdugo et al. 2010; Iñiguez-Franco et al. 2012; Ramos et al. 2014), which may be due to hydrogen bonding changes between packaging and bioactives.

Diffusion coefficient indicates the release behavior of bioactives embedded in edible packaging. Diffusion coefficient is influenced by many parameters. In the study of Imran et al. (2014), diffusion variations were solely influenced by biopolymer nature (affinity, charge, Tg) and structure (free volume), diffusion medium (polarity, pH) and temperature (4 °C, 40 °C). Recent studies of bioactive controlled release systems using edible active packaging are shown in Table 1.

Future Perspectives

Though extensive research has been conducted on active packaging for the past decades, there is still a lack of knowledge for the acceptability of edible active packaging to consumers and their effectiveness in packaging, or to their economic and environmental impact (Véronique 2008). Edible active packaging is based on a deliberate interaction between the migrating substances and the food/or the surrounding environments, there is a food safety concern (Restuccia et al. 2010). Food safety concerns include migration of substances from the environment to food through packaging or directly from packaging, insufficient labeling of the packaging and non-efficacious

Table 1. Recent studies of bioactives controlled release based on edible active packaging systems

Bioactives	Packaging	Release test medium	Release model	Release rate	References
Natamycin	Alginate/pectin	Water	Fick's second law	Varied from 3.2×10^{-9} to 9.2×10^{-12} cm^2/s	Bierhalz et al. 2012
Natamycin	Alginate/chitosan	Water	Fick's second law	Varied from 2.6×10^{-11} to 2.5×10^{-12} cm^2/s	da Silva et al. 2012
Thymol	Zein	Water	Fick's second law	Varied from 1.4×10^{-11} to 8.9×10^{-12} cm^2/s	Del Nobile et al. 2008
Nisin	Hydroxypropyl methylcellulose (HPMC), chitosan (CTS), sodium caseinate (SC) and polylactic acid (PLA)	95% ethanol	Fick's second law	Varied from 1.9×10^{-13} to 97×10^{-13} cm^2/s	Imran et al. 2014
Caffeine	Microfibrillated cellulose (MFC)	Water/ethanol 10 wt.%	Fick's second law	15 times longer duration than controls	Lavoine et al. 2014
Lysozyme	Modified zein	Water	NA	2.5 to 17 fold lower release rate than controls	Arcan and Yemenicioğlu 2013
Lysozyme	Modified sodium caseinate	Sodium acetate buffer	NA	Immobilized lysozyme within the protein matrix	Mendes de Souza et al. 2010
Gallic acid	Gallic acid loaded zein (Ze-GA) electrospun fibers	Water	Fickian diffusion (Higuchi model)	NA	Neo et al. 2013

NA: not applicable.

operation of the packaging, etc. (Shettleworth 2012). European Food Safety Agency set legal basis (Regulations 1935/2004/EC and 450/2009/EC) for the correct use, safety and marketing of active packaging. The life cycle impacts of edible active packaging across the food supply also needs more attention. Packaging developers must therefore consider the product and its packaging as a complete system, the impact of the system on the environment to optimize sustainability.

Edible packaging materials have been considered as attractive alternatives because of their unique properties, including the ability to protect foods with their barrier and mechanical properties, enhance sensory characteristics, control-release active ingredients, and control mass transfer between components of heterogeneous foods. As an emerging technology, this technology not only provides the functions of conventional packaging including extending the shelf life, maintain the sensory and nutritional quality, but also envision new possible application of functional foods. Processes and materials applied should be considered for each specific product and application. For the packaging materials, develop innovative biopolymers from agricultural commodities and/or food-waste products is fashionable (Valdés et al. 2014; Mellinas et al. 2016).

References

Aider, M. 2010. Chitosan application for active bio-based films production and potential in the food industry: A review. LWT-Food Sci. Technol. 43(6): 837–842.

Al-Saadi, J.S., K.A. Shaker and Z. Ustunol. 2014. Effect of heat and transglutaminase on solubility of goat milk protein-based films. Int. J. Dairy Technol. 67(3): 420–426.

Appendini, P. and J.H. Hotchkiss. 2002. Review of antimicrobial food packaging. Innov. Food Sci. Emerg. Technol. 3(2): 113–126.

Arcan, I. and A. Yemenicioğlu. 2013. Development of flexible zein–wax composite and zein–fatty acid blend films for controlled release of lysozyme. Food Res. Int. 51(1): 208–216.

Barbosa-Pereira, L., I. Angulo, J.M. Lagarón, P. Paseiro-Losada and J.M. Cruz. 2014. Development of new active packaging films containing bioactive nanocomposites. Innov. Food Sci. Emerg. Technol. 26: 310–318.

Bierhalz, A.C.K., M.A. da Silva and T.G. Kieckbusch. 2012. Natamycin release from alginate/pectin films for food packaging applications. J. Food Eng. 110(1): 18–25.

Bilbao-Sáinz, C., R.J. Avena-Bustillos, D.F. Wood, T.G. Williams and T.H. McHugh. 2010. Composite edible films based on hydroxypropyl methylcellulose reinforced with microcrystalline cellulose nanoparticles. J. Agric. Food. Chem. 58(6): 3753–3760.

Carneiro-da-Cunha, M.G., M.A. Cerqueira, B.W. Souza, M.P. Souza, J.A. Teixeira and A.A. Vicente. 2009. Physical properties of edible coatings and films made with a polysaccharide from Anacardium occidentale L. J. Food Eng. 95(3): 379–385.

Cerqueira, M.A., A.C. Pinheiro, B.W. Souza, Á.M. Lima, C. Ribeiro, C. Miranda, et al. 2009. Extraction, purification and characterization of galactomannans from non-traditional sources. Carbohydr. Polym. 75(3): 408–414.

Cerqueira, M.A., M.J. Costa, C. Fuciños, L. Pastrana and A.A. Vicente. 2013. Natamycin-loaded poly (n-isopropylacrylamide) nanohydrogels for smart edible packaging: Development and characterization. pp. 150–151. *In*: 1st International Conference in Polymers with special Focus in Early Stage Researchers, Barcelona, Spain.

Chen, H., Y. Zhang and Q. Zhong. 2015. Physical and antimicrobial properties of spray-dried zein–casein nanocapsules with co-encapsulated eugenol and thymol. J. Food Eng. 144: 93–102.

Cruz-Romero, M., T. Murphy, M. Morris, E. Cummins and J. Kerry. 2013. Antimicrobial activity of chitosan, organic acids and nano-sized solubilisates for potential use in smart antimicrobially-active packaging for potential food applications. Food Control 34(2): 393–397.

da Silva, M.A., A.C.K. Bierhalz and T.G. Kieckbusch. 2012. Modelling natamycin release from alginate/chitosan active films. Int. J. Food Sci. Tech. 47(4): 740–746.

de Moura, M.R., F.A. Aouada, R.J. Avena-Bustillos, T.H. McHugh, J.M. Krochta and L.H. Mattoso. 2009. Improved barrier and mechanical properties of novel hydroxypropyl methylcellulose edible films with chitosan/tripolyphosphate nanoparticles. J. Food Eng. 92(4): 448–453.

Del Nobile, M.A., A. Conte, A.L. Incoronato and O. Panza. 2008. Antimicrobial efficacy and release kinetics of thymol from zein films. J. Food Eng. 89(1): 57–63.

Del Nobile, M.A. and A. Conte. 2013. Bio-based packaging materials for controlled release of active compounds. pp. 91–107. *In*: Del Nobile, M.A. and A. Conte (eds.). Packaging for Food Preservation. Springer, New York.

Di Pierro, P., G. Rossi Marquez, L. Mariniello, A. Sorrentino, R. Villalonga and R. Porta. 2013. Effect of transglutaminase on the mechanical and barrier properties of whey protein/pectin films prepared at complexation pH. J. Agric. Food. Chem. 61(19): 4593–4598.

Duran, M., M.S. Aday, N.N.D. Zorba, R. Temizkan, M.B. Büyükcan and C. Caner. 2016. Potential of antimicrobial active packaging 'containing natamycin, nisin, pomegranate and grape seed extract in chitosan coating' to extend shelf life of fresh strawberry. Food Biop. Proc. 98: 354–363.

Dutta, P.K., S. Tripathi, G.K. Mehrotra and J. Dutta. 2009. Perspectives for chitosan based antimicrobial films in food applications. Food Chem. 114(4): 1173–1182.

Espitia, P.J.P., W.-X. Du, R.D.J. Avena-Bustillos, N.D.F.F. Soares and T.H. McHugh. 2014. Edible films from pectin: Physical-mechanical and antimicrobial properties—A review. Food Hydrocolloid. 35: 287–296.

Fabra, M.J., M.A. Busolo, A. Lopez-Rubio and J.M. Lagaron. 2013. Nanostructured biolayers in food packaging. Trends Food Sci. Technol. 31(1): 79–87.

Fabra, M.J., A. Lopez-Rubio and J.M. Lagaron. 2015. Effect of the film-processing conditions, relative humidity and ageing on wheat gluten films coated with electrospun polyhydryalkanoate. Food Hydrocolloid. 44: 292–299.

Fabra, M.J., A. López-Rubio and J.M. Lagaron. 2016. Use of the electrohydrodynamic process to develop active/bioactive bilayer films for food packaging applications. Food Hydrocolloid. 55: 11–18.

Falguera, V., J. Pagán and A. Ibarz. 2011a. Effect of UV irradiation on enzymatic activities and physicochemical properties of apple juices from different varieties. LWT-Food Sci. Technol. 44(1): 115–119.

Falguera, V., J.P. Quintero, A. Jiménez, J.A. Muñoz and A. Ibarz. 2011b. Edible films and coatings: Structures, active functions and trends in their use. Trends Food Sci. Technol. 22(6): 292–303.

Fernández, A., D. Cava, M.J. Ocio and J.M. Lagarón. 2008. Perspectives for biocatalysts in food packaging. Trends Food Sci. Technol. 19(4): 198–206.

Gómez-Estaca, J., C. López-de-Dicastillo, P. Hernández-Muñoz, R. Catalá and R. Gavara. 2014. Advances in antioxidant active food packaging. Trends Food Sci. Technol. 35(1): 42–51.

González, A., M.C. Strumia and C.I.A. Igarzabal. 2015. Modification strategies of proteins for food packaging applications. pp. 127. *In*: Cirillo, G., U.G. Spizzirri and F. Iemma (eds.). Functional Polymers in Food Science: From Technology to Biology, Volume 1: Food Packaging. John Wiley & Sons, Inc., Hoboken, NJ, USA.

Graciano-Verdugo, A.Z., H. Soto-Valdez, E. Peralta, P. Cruz-Zárate, A.R. Islas-Rubio, S. Sánchez-Valdes, et al. 2010. Migration of α-tocopherol from LDPE films to corn oil and its effect on the oxidative stability. Food Res. Int. 43(4): 1073–1078.

Hambleton, A., F. Debeaufort, A. Bonnotte and A. Voilley. 2009. Influence of alginate emulsion-based films structure on its barrier properties and on the protection of microencapsulated aroma compound. Food Hydrocolloid. 23(8): 2116–2124.

Han, J.H. and A. Gennadios. 2005. Edible films and coatings: A review. Innov. Food Packag. 239: 262.

Hill, L.E., C. Gomes and T.M. Taylor. 2013. Characterization of beta-cyclodextrin inclusion complexes containing essential oils (trans-cinnamaldehyde, eugenol, cinnamon bark, and clove bud extracts) for antimicrobial delivery applications. LWT-Food Sci. Technol. 51(1): 86–93.

Imran, M., A. Klouj, A.-M. Revol-Junelles and S. Desobry. 2014. Controlled release of nisin from HPMC, sodium caseinate, poly-lactic acid and chitosan for active packaging applications. J. Food Eng. 143: 178–185.

Iñiguez-Franco, F., H. Soto-Valdez, E. Peralta, J.F. Ayala-Zavala, R. Auras and N. Gámez-Meza. 2012. Antioxidant activity and diffusion of catechin and epicatechin from antioxidant active films made of poly (l-lactic acid). J. Agric. Food. Chem. 60(26): 6515–6523.

Janjarasskul, T. and J.M. Krochta. 2010. Edible packaging materials. Annu. Rev. Food Sci. Technol. 1: 415–448.

Kester, J. and O. Fennema. 1986. Edible films and coatings: A review. Food Technol. 40: 47–59.

Korhonen, H. 2002. Technology options for new nutritional concepts. Int. J. Dairy Technol. 55(2): 79–88.

Kotler, P., K.L. Keller, F. Ancarani and M. Costabile. 2014. Marketing Management 14/e. Pearson. Prentice Hall, Upper Saddle River, NJ, USA.

Krochta, J.M. 2002. Proteins as raw materials for films and coatings: Definitions, current status, and opportunities. pp. 1–41. *In*: Gennadios, A. (ed.). Protein-based Films and Coatings. CRC Press LLC, Boca Raton, FL, USA.

Lagaron, J., R. Catalá and R. Gavara. 2004. Structural characteristics defining high barrier properties in polymeric materials. Mater. Sci. Technol. 20(1): 1–7.

Lavoine, N., I. Desloges and J. Bras. 2014. Microfibrillated cellulose coatings as new release systems for active packaging. Carbohydr. Polym. 103: 528–537.

Li, D. and Y. Xia. 2004. Electrospinning of nanofibers: Reinventing the wheel? Adv. Mater. 16(14): 1151–1170.

Liu, L., J.F. Kerry and J.P. Kerry. 2005. Selection of optimum extrusion technology parameters in the manufacture of edible/biodegradable packaging films derived from food-based polymers. J. Food Agric. Environ. 3(3/4): 51.

López-de-Dicastillo, C., J.M. Alonso, R. Catalá, R. Gavara and P. Hernández-Muñoz. 2010. Improving the antioxidant protection of packaged food by incorporating natural flavonoids into ethylene– vinyl alcohol copolymer (EVOH) films. J. Agric. Food. Chem. 58(20): 10958–10964.

Manzanarez-López, F., H. Soto-Valdez, R. Auras and E. Peralta. 2011. Release of α-tocopherol from poly (lactic acid) films, and its effect on the oxidative stability of soybean oil. J. Food Eng. 104(4): 508–517.

Martínez-Abad, A., A.C. Ruthes and F. Vilaplana. 2016. Enzymatic-assisted extraction and modification of lignocellulosic plant polysaccharides for packaging applications. J. Appl. Polym. Sci. 133(2): 1:15.

Mastromatteo, M., M. Mastromatteo, A. Conte and M.A. Del Nobile. 2010. Advances in controlled release devices for food packaging applications. Trends Food Sci. Technol. 21(12): 591–598.

Mellinas, C., A. Valdés, M. Ramos, N. Burgos, M.D.C. Garrigós and A. Jiménez. 2016. Active edible films: Current state and future trends. J. Appl. Polym. Sci. 133(2): 1:15.

Mendes de Souza, P., A. Fernández, G. López-Carballo, R. Gavara and P. Hernández-Muñoz. 2010. Modified sodium caseinate films as releasing carriers of lysozyme. Food Hydrocolloid. 24(4): 300–306.

Mihindukulasuriya, S. and L.-T. Lim. 2014. Nanotechnology development in food packaging: A review. Trends Food Sci. Technol. 40(2): 149–167.

Mikkonen, K.S. and M. Tenkanen. 2012. Sustainable food-packaging materials based on future biorefinery products: Xylans and mannans. Trends Food Sci. Technol. 28(2): 90–102.

Neo, Y.P., S. Swift, S. Ray, M. Gizdavic-Nikolaidis, J. Jin and C.O. Perera. 2013. Evaluation of gallic acid loaded zein sub-micron electrospun fibre mats as novel active packaging materials. Food Chem. 141(3): 3192–3200.

Otoni, C.G., S.F. Pontes, E.A. Medeiros and N.D.F. Soares. 2014. Edible films from methylcellulose and nanoemulsions of clove bud (Syzygium aromaticum) and oregano (Origanum vulgare) essential oils as shelf life extenders for sliced bread. J. Agric. Food. Chem. 62(22): 5214–5219.

Otoni, C.G., R.J. Avena-Bustillos, C.W. Olsen, C. Bilbao-Sáinz and T.H. McHugh. 2016. Mechanical and water barrier properties of isolated soy protein composite edible films as affected by carvacrol and cinnamaldehyde micro and nanoemulsions. Food Hydrocolloid. 57: 72–79.

Ozdemir, M. and J.D. Floros. 2004. Active food packaging technologies. J. Agric. Food. Chem. 44(3): 185–193.

Patrignani, F., L. Siroli, D.I. Serrazanetti, F. Gardini and R. Lanciotti. 2015. Innovative strategies based on the use of essential oils and their components to improve safety, shelf-life and quality of minimally processed fruits and vegetables. Trends Food Sci. Technol. 46(2): 311–319.

Peng, Y., Y. Wu and Y. Li. 2013. Development of tea extracts and chitosan composite films for active packaging materials. Int. J. Biol. Macromol. 59: 282–289.

Pereira de Abreu, D., J. Cruz and P. Paseiro Losada. 2012. Active and intelligent packaging for the food industry. Food Rev. Int. 28(2): 146–187.

Pérez-Gago, M.B. and J.M. Krochta. 2001. Lipid particle size effect on water vapor permeability and mechanical properties of whey protein/beeswax emulsion films. J. Agric. Food. Chem. 49(2): 996–1002.

Quintavalla, S. and L. Vicini. 2002. Antimicrobial food packaging in meat industry. Meat Sci. 62(3): 373–380.

Ramos, M., A. Beltrán, M. Peltzer, A.J.M. Valente amd M.d.C. Garrigós. 2014. Release and antioxidant activity of carvacrol and thymol from polypropylene active packaging films. LWT-Food Sci. Technol. 58(2): 470–477.

Razavi, S.M.A., A. Mohammad Amini and Y. Zahedi. 2015. Characterisation of a new biodegradable edible film based on sage seed gum: Influence of plasticiser type and concentration. Food Hydrocolloid. 43: 290–298.

Restuccia, D., U.G. Spizzirri, O.I. Parisi, G. Cirillo, M. Curcio, F. Iemma, et al. 2010. New EU regulation aspects and global market of active and intelligent packaging for food industry applications. Food Control 21(11): 1425–1435.

Rhim, J.-W. and P.K. Ng. 2007. Natural biopolymer-based nanocomposite films for packaging applications. Crit. Rev. Food Sci. Nutr. 47(4): 411–433.

Robertson, G.L. 2012. Food Packaging: Principles and Practice. CRC Press, Boca Raton, FL, USA.

Roy, D.S. and B.D. Rohera. 2002. Comparative evaluation of rate of hydration and matrix erosion of HEC and HPC and study of drug release from their matrices. Eur. J. Pharm. Sci. 16(3): 193–199.

Sanchez-Gonzalez, L., M. Vargas, C. Gonzalez-Martínez, A. Chiralt and M. Chafer. 2011. Use of essential oils in bioactive edible coatings: A review. Food Eng. Rev. 3(1): 1–16.

Sánchez-González, L., E. Arab-Tehrany, M. Cháfer, C. González-Martínez and A. Chiralt. 2015. Active edible and biodegradable starch films. Polysaccharides: Bioact. Biotechnol. 717–734.

Shettleworth, S.J. 2012. Do animals have insight, and what is insight anyway? Can. J. Exp. Psychol. 66(4): 217–226.

Siripatrawan, U. and S. Noipha. 2012. Active film from chitosan incorporating green tea extract for shelf life extension of pork sausages. Food Hydrocolloid. 27(1):102–108.

Slaughter, A., X. Daniel, V. Flors, E. Luna, B. Hohn and B. Mauch-Mani. 2012. Descendants of primed Arabidopsis plants exhibit resistance to biotic stress. Plant physiol. 158(2): 835–843.

Sobral, P.D.A., F. Menegalli, M. Hubinger and M. Roques. 2001. Mechanical, water vapor barrier and thermal properties of gelatin based edible films. Food Hydrocolloid. 15(4): 423–432.

Sorrentino, A., G. Gorrasi and V. Vittoria. 2007. Potential perspectives of bio-nanocomposites for food packaging applications. Trends Food Sci. Technol. 18(2): 84–95.

Spotti, M.L., J.P. Cecchini, M.J. Spotti and C.R. Carrara. 2016. Brea Gum (from Cercidium praecox) as a structural support for emulsion-based edible films. LWT-Food Sci. Technol. 68: 127–134.

Tahiri, I., M. Desbiens, C. Lacroix, E. Kheadr and I. Fliss. 2009. Growth of Carnobacterium divergens M35 and production of divergicin M35 in snow crab by-product, a natural-grade medium. LWT-Food Sci. Technol. 42(2): 624–632.

Tang, H., H. Xiong, S. Tang and P. Zou. 2009. A starch-based biodegradable film modified by nano silicon dioxide. J. Appl. Polym. Sci. 113(1): 34–40.

Torres-Arreola, W., H. Soto-Valdez, E. Peralta, J.L. Cárdenas-López and J.M. Ezquerra-Brauer. 2007. Effect of a low-density polyethylene film containing butylated hydroxytoluene on lipid oxidation and protein quality of Sierra fish (Scomberomorus sierra) muscle during frozen storage. J. Agric. Food Chem. 55(15): 6140–6146.

Valdés, A., A.C. Mellinas, M. Ramos, M.C. Garrigós and A. Jiménez. 2014. Natural additives and agricultural wastes in biopolymer formulations for food packaging. Frontiers Chem. 2: 6.

Valdés, A., A. Mellinas, M. Ramos, N. Burgos, A. Jiménez and M. Garrigós. 2015. Use of herbs, spices and their bioactive compounds in active food packaging. RSC Adv. 5(50): 40324–40335.

Van Long, N.N., C. Joly and P. Dantigny. 2016. Active packaging with antifungal activities. Int. J. Food Microbiol. 220: 73–90.

Verghese, K., H. Lewis, S. Lockrey and H. Williams. 2015. Packaging's role in minimizing food loss and waste across the supply chain. Packag. Technol. Sci. 2015: 603–620.

Véronique, C. 2008. Bioactive packaging technologies for extended shelf life of meat-based products. Meat Sci 78(1): 90–103.

Volpe, M.G., M. Malinconico, E. Varricchio and M. Paolucci. 2010. Polysaccharides as biopolymers for food shelf-life extension: Recent patents. Rec. Pat. Food Nutr. Agri. 2(2): 129–139.

Wang, Y., A. Liu, R. Ye, W. Wang and X. Li. 2015. Transglutaminase-induced crosslinking of gelatin–calcium carbonate composite films. Food Chem. 166: 414–422.

Wessling, C., T. Nielsen, A. Leufven and M. Jägerstad. 1998. Mobility of -tocopherol and BHT in LDPE in contact with fatty food simulants. Food Addit. Contam. 15(6): 709–715.

Zhang, H., M. Hortal, A. Dobon, J.M. Bermudez and M. Lara-Lledo. 2015. The effect of active packaging on minimizing food losses: Life cycle assessment (LCA) of essential oil component-enabled packaging for fresh beef. Packag Technol. Sci. 28(7): 761–774.

Zia, K.M., A. Noreen, M. Zuber, S. Tabasum and M. Mujahid. 2016. Recent developments and future prospects on bio-based polyesters derived from renewable resources: A review. Int. J. Biol. Macromol. 82: 1028–1040.

2

Active Bio-Packaging in the Food Industry

Ricardo Stefani*

Laboratório de Estudos de Materiais (LEMAT), Universidade Federal de Mato Grosso. Campus Araguaia, Av. Senador Valdon Varjão, 6390, Campus UFMT, Barra do Garças – MT. Brazil 78600-000

Introduction

Due to consumer and industry demand for better quality food, in recent years active and smart food packaging technology has emerged (Aider 2010; Reis et al. 2015; Brizio and Prentice 2014). The main goal of an active or smart package is to control and monitor the quality of food products in real time. In this context, the development of bio-based packages for food is in active development (Cian et al. 2014; Lorenzo et al. 2014; Wen et al. 2016). Bio-based packaging for food is mostly made from naturally occurring renewable polymers or natural polymers blended with artificial biodegradable polymers that have a good film and coating-forming properties.

In the literature, the reported active films are mostly carbohydrate-based polymers such as chitosan (Aider 2010; Silva-Pereira et al. 2015; Abugoch et al. 2016; Souza et al. 2015), starch (Arismendi et al. 2013; Lu 2009; Vásconez et al. 2009; Silva-Pereira et al. 2015), cellulose (Sonkaew et al. 2012; El-Sayed et al. 2011; Geng and Xiao 2009), pectin (Kaur et al. 2014) and blends of other natural or synthetic polymers. In addition to carbohydrates, other biodegradable polymers such as polylactic acid (PLA) (Samsudin et al. 2014), proteins (Abugoch et al. 2016; González and Igarzabal 2013; Xiping et al. 2014; Liu et al. 2015; Chen et al. 2010), alginate (Belščak-Cvitanović et al. 2011; Vu and Won 2013) and other polymers (Nath et al. 2015) are used for active and smart bio-packaging in food.

*Corresponding author: rstefani@ufmt.br

The use of these polymers in food packaging has a clear environmental advantage, since they are renewable and biodegradable, thus reducing environmental pollution. Moreover, active compounds can be incorporated into the polymeric matrix, which is useful to turn the biodegradable polymeric film into a low cost active or smart polymer. This approach has been used to develop antioxidant (Cian et al. 2014; de Abreu et al. 2011; López-de-Dicastillo et al. 2012; Li et al. 2014), antimicrobial (Muriel-Galet et al. 2015; Dutta et al. 2009; Wen et al. 2016), time-temperature indicator (TTI) (Kim et al. 2012; Kim et al. 2013; Brizio and Prentice 2014), pH indicator (Wang et al. 2004; Brizio and Prentice 2014; Kim et al. 2012; Silva-Pereira et al. 2015; Salinas et al. 2014) as well as oxygen barrier monitoring (Marek et al. 2013; Laufer et al. 2012) films.

Thus, bio-packaging for food can be a versatile and low-cost approach to control and preserve the quality of a food product as it travels from the manufacturer to the final consumer. Finally, biopolymer films have numerous potential applications within the food industry. This chapter examines some of these applications and assesses the challenges associated with the development and use of active and smart bio-based packaging in the food industry.

Preparation of Films for Bio-based Food Packaging

Active and intelligent films for food packaging are typically prepared using the casting technique (Arzate-Vázquez et al. 2012), which can produce films efficiently at a low cost. In this method, hydrocolloid suspensions are made by suspending the required amount of biopolymer in distilled or deionized water, which is then poured onto acrylic or glass plates and dried until the weight is constant. To be successfully applied in the development of active and smart films, this technique requires polymers that have high stability and the ability to form networks of structured and thermally stable copolymers. As casting does not require the use of complicated laboratory instruments, it represents a versatile, inexpensive, rapid and simple method of preparing films. This approach has been extensively covered in the existing literature (El-Sayed et al. 2011; Bitencourt et al. 2014; Nisa et al. 2015; Li et al. 2014; Hosseini et al. 2016; Yang et al. 2016; Pereira et al. 2014) and represents the method of choice for the production of new and improved active and smart polymer films.

In addition to casting, films for active food packaging have been prepared by other techniques. Curcumin has been successfully encapsulated in a chitosan-PVA-silver nanocomposite film (Vimala 2011), and layer-by-layer thin films with applications in active food packaging such as oxygen barrier (Jang et al. 2008; Laufer et al. 2012), antioxidant (Shutava and Lvov 2012; Shutava et al. 2009; Deladino et al. 2013), and antimicrobial packaging have been reported (Hosseini et al. 2016; Brasil et al. 2012; Hosseini et al. 2009; Neethirajan and Jayas 2010). Electron spin resonance has also been applied in the development of active films for food packaging (Wen et al. 2016; Fabra et al. 2013) with antioxidant and antibacterial properties.

Characterization of Bio-packaging Films

Bio-packaging films can be prepared using many types of biopolymers. The physical, chemical, physicochemical, and functional properties of the films depend on the selection of biopolymers (Ramos et al. 2014; Pereira de Abreu et al. 2012). Furthermore, a suitable biopolymer for packaging applications should have good thermal stability, water vapour permeability (WVP), and mechanical resistance. Thus, the produced films should be characterized by their spectroscopy, morphology, thermal stability, mechanical strength, and behaviour against humidity.

Spectroscopic Characterization

Spectroscopic characterization of the films is important to determine their composition. The chemical structure of polymeric films is characterized using UV-visible (UV/Vis) (Vikova 2011; Nafisi et al. 2011), X-ray photoelectron (Huo et al. 2016), FT-IR (Hosseini et al. 2013; Zhang et al. 2015), and Raman spectroscopy (Dashtdar et al. 2013). These techniques provide information about the functional groups that are present in the chemical structure of the film and the manner in which they interact with each other.

Morphological Characterization

Polymeric film morphological characterization can identify defects in the microstructure of the polymeric matrix that can affect the film properties. For morphological characterization of films, atomic force microscopy (AFM), scanning electronic microscopy (SEM), and X-ray diffraction (XRD) are widely applied (Lewandowska 2012; Hosseini et al. 2016; Zhou et al. 2012; Zhu et al. 2012). Microscopic analysis provides useful composition and topology information about the microstructure and surface characteristics, and can be useful in studies that aim to examine the structural integrity of the polymeric matrix.

Thermal Characterization

Differential scanning calorimetry (DSC) and thermogravimetric analysis (TG) are both thermoanalytical techniques that provide information about the thermal stability of the films and how their components interact with one another (Winkler et al. 2014). Since these interactions can result in changes to the melting point of each compound (Kanimozhi et al. 2016), they provide a clear picture of how these interactions affect the properties of the films (Silva-Pereira et al. 2015; Huo et al. 2016; Tripathi et al. 2009).

Mechanical Properties

In food packaging applications, it is important that biopolymer films demonstrate excellent mechanical properties. A biopolymer film that is suitable for food packaging applications will typically demonstrate mechanical properties that are comparable to that of the polymers that are in commercial use. These mechanical properties are mostly measured using the method described in ASTM D1708-10 (Pereira Jr et al. 2014; Abugoch et al. 2016; Liu et al. 2014), which is suitable for determining the mechanical properties of plastics or the traction of films with thicknesses ranging from 0.0025 mm to 2.5 mm, which is sufficient for use in food packaging. Other techniques, such as dynamic mechanical thermal analysis (DMTA), have also been reported in the literature (Ma et al. 2008; Reis et al. 2015).

Characterization of Water and Moisture Absorption

WVP measurements assess how fast the polymeric matrix absorbs moisture and thus swells or reduces in size. WVP provides a good indication of how fast the polymeric matrix will absorb water (Garcia et al. 2006).

The swelling index, which is a measure of the mass of water absorbed by the polymeric matrix, is also widely reported in the literature and is often gravimetrically determined (Kanimozhi et al. 2016; Mangala et al. 2003). The standard test method that is most commonly applied to measure WVP is ASTM E96M-14.

Applications of Bio-based Packaging in the Food Industry

It is known that active and smart packaging technologies have effects on the shelf life of food products (Barbosa-Pereira et al. 2014; Lorenzo et al. 2014; Benito-Peña et al. 2016). Active packaging represents one method by which food waste can be reduced (Marsh and Bugusu 2007). It is recognized that incorporating antioxidant and antimicrobial compounds into packaging films can improve the shelf life of products and decrease oxidation in foods, thus antimicrobial and antioxidant food preservation methods are well established. Release of these active agents can be controlled over an extended period to maintain the quality of the food product, without the need for the direct addition of any substances to the food.

Among the technologies for active and smart packaging, the ones concerning carbon dioxide absorbers (Nopwinyuwong et al. 2010; Xu et al. 2013; Mills 2005), foodborne pathogen detectors (Jokerst et al. 2012; Hossain et al. 2012; Pires et al. 2011; Ebrahimi and Schönherr 2014; Bento et al. 2015), and food spoilage monitoring based on pH changes need further development (Heising et al. 2014; Yoshida et al. 2014; Brizio and Prentice 2014). In this section, state-of-the-art applications of active bio-based packaging for food and perspectives for future development will be discussed.

Antioxidant Food Packaging

Synthetic antioxidants such as BHT, BHA and organophosphate have been used in industry, and there is a current trend in using natural antioxidants, since some synthetic antioxidants have demonstrated physiological effects (Yehye et al. 2015). Natural antioxidants exhibit beneficial effects when used in food packaging (Gómez-Estaca et al. 2014; Han et al. 2015; Belščak-Cvitanović et al. 2011; Sonkaew et al. 2012). The most common naturally occurring antioxidants incorporated into active food packaging films include polyphenols (Shutava et al. 2009; Belščak-Cvitanović et al. 2011; Liu et al. 2015; Konecsni et al. 2012; Muriel-Galet et al. 2015), essential oils (Wen et al. 2016; Hosseini et al. 2009; Teixeira et al. 2014; Llana-Ruíz-Cabello et al. 2015), and tocopherols (Barbosa-Pereira et al. 2014; Chen et al. 2012), which are directly absorbed in the active film or encapsulated in the form of particles.

Despite the significant amount of scientific work on antioxidant bio-packaging films that has been produced in recent years, few of these films have been commercialized. This is because films based on biopolymers are very hydrophilic and absorb high contents of water (Vu and Won 2013),that is, they swell considerably.

This problem is not restricted to antioxidant films but is widespread among bio-based films. To overcome this issue, most of the antioxidant active packaging that has been commercialized incorporates independent devices, such as a sachet or seals that contain an oxygen scavenger, such as ferrous oxide (Moon and Shibamoto 2009), polyphenols and phenols (de Abreu et al. 2011; Benito-Peña et al. 2016; Woranuch et al. 2015), or carotenes (Fabra et al. 2015). However, it is clear that further research on antioxidant films is necessary before they can be used commercially.

Antimicrobial Food Packaging

Bacterial contamination is a threat of considerable proportion in both medicine and the food industry, being one of the leading causes of infections and foodborne diseases. Thus, to avoid illness and bacterial food spoilage, fast and reliable methods of bacteria control are of great importance because microbiological contamination, as well as resistant bacterial infections, remains a critical issue in both developed and developing nations. The existence of inexpensive, accurate, simple and portable methods to control bacterial growth could be a valuable resource to monitor food spoilage and prevent illness caused by bacteria.

Therefore, the development of active antimicrobial food packaging is a key requirement for food safety. Currently, typical and state-of-the-art methods employed for bacteria control in food include the direct use of antimicrobial additives in food, such as sodium and phosphorous-based additives (Gutiérrez 2013). Unfortunately, many of these additives can cause undesirable side effects such as allergies (Gutiérrez 2013; Hosseini et al. 2013). Thus, many efforts have been made to overcome this limitation, including the development of antimicrobial active packaging. As this type of application has garnered a great deal of attention in the last several years (Brasil et al. 2012; Ibarguren et al. 2015; Rodríguez et al. 2008; Martin et al. 2013; Bonilla et al. 2014; Gorrasi et al. 2016), there is a growing demand for systems that can indirectly control the food microbial flora.

Essential oils are composed mainly of mono- and sesqui-terpenes that are known for their antimicrobial activity (Bizzo et al. 2009), and they are incorporated into biopolymeric films to form active packages. Essential oils have been integrated into films as pure substances, pure essential oils or a mixture of different essential oils. Clove, garlic and origanum essential oils have been incorporated into fish protein to form active antimicrobial films (Teixeira et al. 2014). Origanum essential oil has been incorporated into a polylactic acid (PLA) with polybutylene succinate (PBS) blended film to form an active film with cymene, terpinene and thymol as major active compounds (Llana-Ruíz-Cabello et al. 2015). A new antimicrobial active packaging for white bread as a target food was tested and was found to prevent spoilage by *Rhizopus stolonifer*; the active component of this film is cinnamon, which is obtained from cinnamon essential oil (Wen et al. 2016).

In addition to essential oils, some flavonoids and phenolic compounds also have antimicrobial activity (Cushnie and Lamb 2005), and as flavonoids also have antioxidant activity (Burda and Oleszek 2001; Ray et al. 2007) they have been incorporated into biopolymeric films to create active films with both antimicrobial and antioxidant properties. It is reported that low-cost active films produced based on food-grade gelatine with the addition of flavonoid ester prunin laurate as antimicrobial compounds (Ibarguren et al. 2015) are active against *Listeria monocytogenes*, *Staphylococcus aureus*, and *Bacillus cereus*, providing an alternative for preservation of foods susceptible to contamination by these pathogens.

A biodegradable starch-PVA film impregnated with catechin was developed and tested on red meat (Wu et al. 2010), and the produced film had both antimicrobial activity against *Escherichia coli* and *Listeria* and antioxidant properties. Both these properties are highly desirable for the packaging of meat products prior to consumption.

Active Packaging for Detecting Foodborne Bacteria

Many species of bacteria, such as *E. coli*, *Pseudomonas putida*, *Pseudomonas aeruginosa*, *Staphylococcus* sp., and *Salmonella* sp., are responsible for foodborne infections and diseases (Miranda et al. 2011). Thus, to avoid food-related illnesses and bacterial food spoilage, fast and reliable methods of bacteria detection are of great importance. The existence of inexpensive, accurate, simple and portable methods to detect bacteria could be a valuable resource to detect food spoilage caused by bacteria. Therefore, the rapid and sensitive detection of pathogens is a key requirement in food safety diseases (Miranda et al. 2011).

Currently, typical and state-of-the-art methods employed for bacteria detection and identification include separation, culturing and counting of bacterial cells diseases (Miranda et al. 2011). Despite those systems being rapid and reliable, the majority of them still require some laboratory instrumentation such as UV/Vis or fluorescence spectrophotometers, imaging processing software, or even chromatography techniques for the detection of pathogens, which requires trained personnel. Although these methods are suitable for food safety control in industry and agencies, they are not appropriate for use in the supermarket or consumers' home. This limitation surely limits real-time applications, such as intelligent food packaging.

As real-time applications of food bacteria detection have garnered a great deal of attention in the last few years (Kim et al. 2012; Pereira de Abreu et al. 2012; Jokerst et al. 2012; Wan et al. 2014; Li et al. 2013), there is a growing demand for systems that can visually indicate the presence of contamination and spoilage in food products caused by improper storage conditions or bacteria. Therefore, a bacterial detection sensor that is capable of showing an intense colour change that is visible to the naked eyes allows better visualization of the presence of contamination that can cause infections and food spoilage, without the need of expensive instrumentation or specialized handling. Such systems could indicate to the consumers the existence of pathogens in a safe and intuitive way. The most simple and low-cost way to achieve this goal is through the use of a chromogenic reagent, which changes its colour due to the presence of bacteria.

The use of chromogenic reagents to identify bacteria is well established, being widely used in classical detection methods (Syed et al. 1984). Due to its sensitivity and simplicity, it has emerged in recent years as a cheap and reliable approach to the development of new sensors for bacterial detection (Carey et al. 2011; Cellier et al. 2014a; Cellier et al. 2014b). The principle behind modern chromogenic reagents relies on the colour change caused by reactions of a specific bacteria-released enzyme with a given substrate (Carey et al. 2011; Cellier et al. 2014a; Cellier et al. 2014b) or due to pH changes in the medium caused by bacteria-released substances (Kim et al. 2012). Since many bacteria can induce pH changes, detection of these changes is a straightforward and effective way to detect pathogen presence, and this is usually achieved by the use of pH sensitive dyes (Gram et al. 2002; Salinas et al. 2014; Li et al. 2013). Therefore, indicator dyes combined with biopolymers are promising sensors that can visually indicate changes in food that are induced by bacterial contamination.

Active Packaging for Detecting pH Changes in Food

Since a great deal of food spoilage caused by bacterial or chemical degradation can induce pH changes in food, detecting these changes is a straightforward and effective way to identify food that is unsuitable for consumption. This detection is usually achieved by using pH-sensitive dyes incorporated into a polymeric matrix. Biopolymer films associate with natural pH-indicator dyes to obtain smart polymers. These smart polymers can be applied as part of an intelligent food package to indicate food deterioration (Silva-Pereira et al. 2015; Kuswandi et al. 2011; Pacquit et al. 2006; Salinas et al. 2012; Nopwinyuwong et al. 2010; Kim et al. 2013).

Meat products such as beef, pork, chicken, and fish release trimethylamines during spoilage caused by microorganisms. Hence, smart packages sensitive to trimethylamine concentrations were developed (Xiao-Wei et al. 2016; Shukla et al. 2015; Salinas et al. 2014). These packages are based on a polymeric matrix and a sensitive dye, which can indicate the presence or absence of trimethylamines through colour changes. For meat products, the smart package can be arranged as a single sensor or as a sensor array, which has sensitivity as low as 10 ppm (Xiao-Wei et al. 2016). Another method for monitoring meat freshness is based on temperature changes during storage since improper storage temperatures can lead the product to spoil. This type of sensor is known as a TTI and it records changes in temperature that occur during storage. A TTI is a simple and low-cost device that can monitor and indicate the overall temperature history based on a colorimetric change (Kim et al. 2013). Several types of TTI have been developed, including enzymatic–, pH-indicator–, and microbial-growth–based TTIs (Kim et al. 2012; Kim et al. 2012; Wu et al. 2013; Kerry et al. 2006; Lu et al. 2013; Heising et al. 2014).

Most of these TTIs are based on biopolymers, natural dyes, and even bacteria-formed biofilms (Kim et al. 2012; Zhang et al. 2014; Bento et al. 2015; Golasz et al. 2013). Study results show that indicator dyes combined with biopolymers are promising sensors that can visually indicate changes that are induced by chemical reactions.

Conclusions

Academics and the food industry have been searching for technologies and methods to produce high-quality products that self-monitor their quality. Hence, smart and active packaging systems are promising technologies that aim to support the growing demand for the control of food quality as it makes its way from the manufacturer to the final consumer. Furthermore, the development of new technologies and smart materials that are simple and economically viable is important for the food industry in order to develop new bio-active self-monitoring materials for the conservation of food products.

References

Abugoch, L., N. Caro, E. Medina, M. Díaz-dosque, L. Luis and C. Tapia. 2016. Food hydrocolloids novel active packaging based on films of chitosan and chitosan/quinoa protein printed with chitosan-tripolyphosphate-thymol nanoparticles via thermal ink-jet printing. Food Hydrocoll. 52: 520–532.

Aider, M. 2010. Chitosan application for active bio-based films production and potential in the food industry: Review. LWT - Food Sci. Technol. 43: 837–842.

Arismendi, C., S. Chillo, A. Conte, M.A. Del Nobile, S. Flores and L.N. Gerschenson. 2013. Optimization of physical properties of xanthan gum/tapioca starch edible matrices containing potassium sorbate and evaluation of its antimicrobial effectiveness. LWT - Food Sci. Technol. 53: 290–296.

Arzate-Vázquez, I., J.J. Chanona-Pérez, G. Calderón-Domínguez, E. Terres-Rojas, V. Garibay-Feblesc, A. Martínez-Rivas, et al. 2012. Microstructural characterization of chitosan and alginate films by microscopy techniques and texture image analysis. Carbohydr. Polym. 87: 289–299.

Barbosa-Pereira, L., J.M. Cruz, R. Sendón, A.R.B. Quirós, A. Ares, M. Castro-López, et al. 2014. Development of antioxidant active films containing tocopherols to extend the shelf life of fish. Food Control. 31: 236–246.

Belščak-Cvitanović, A., R. Stojanović, V. Manojlović, D. Komes, I.J. Cindrić, V. Nedović, et al. 2011. Encapsulation of polyphenolic antioxidants from medicinal plant extracts in alginate–chitosan system enhanced with ascorbic acid by electrostatic extrusion. Food Res. Int. 44: 1094–1101.

Benito-Peña, E., V. González-Vallejo, A. Rico-Yuste, L. Barbosa-Pereira, J.M. Cruz, A. Bilbao, et al. 2016. Molecularly imprinted hydrogels as functional active packaging materials. Food Chem. 190: 487–494.

Bento, L., M. Silva-Pereira, K. Chaves and R. Stefani. 2015. Development and evaluation of a smart packaging for the monitoring of ricotta cheese spoilage. MOJ Food Process. Technol. 1: 3–5.

Bitencourt, C.M., C.S. Fávaro-Trindade, P.J.A. Sobral and R.A. Carvalho. 2014. Gelatin-based films additivated with curcuma ethanol extract: Antioxidant activity and physical properties of films. Food Hydrocoll. 40: 145–152.

Bizzo, H.R., A.M.C. Hovell and C.M. Rezende. 2009. Óleos essenciais no Brasil: aspectos gerais, desenvolvimento e perspectivas. Quim. Nova. 32: 588–594.

Bonilla, J., E. Fortunati, L. Atarés, A. Chiralt and J.M. Kenny. 2014. Physical, structural and antimicrobial properties of poly vinyl alcohol–chitosan biodegradable films. Food Hydrocoll. 35: 463–470.

Brasil, I.M., C. Gomes, A. Puerta-Gomez, M.E. Castell-Perez and R.G. Moreira. 2012. Polysaccharide-based multilayered antimicrobial edible coating enhances quality of fresh-cut papaya. LWT - Food Sci. Technol. 47: 39–45.

Brizio, A.P.D.R. and C. Prentice. 2014. Use of smart photochromic indicator for dynamic monitoring of the shelf life of chilled chicken based products. Meat Sci. 96: 1219–1226.

Burda, S. and W. Oleszek. 2001. Antioxidant and antiradical activities of flavonoids. J. Agric. Food Chem. 49: 2774–2779.

Carey, J.R., K.S. Suslick, K.I. Hulkower, J.A. Imlay, K.R.C. Imlay, C.K. Ingison, et al. 2011. Rapid identification of bacteria with a disposable colorimetric sensing array. J. Am. Chem. Soc. 133: 7571–7576.

Cellier, M., A.L. James, S. Orenga, J.D. Perry, A.K. Rasul, S.N. Robinson, et al. 2014a. Novel chromogenic aminopeptidase substrates for the detection and identification of clinically important microorganisms. Bioorg. Med. Chem. 22: 5249–5269.

Cellier, M., E. Fazackerley, A.L. James, S. Orenga, J.D. Perry, G. Turnbull, et al. 2014b. Synthesis of 2-arylbenzothiazole derivatives and their application in bacterial detection. Bioorg. Med. Chem. 22: 1250–1261.

Chen, X., D.S. Lee, X. Zhu and K.L. Yam. 2012. Release kinetics of tocopherol and quercetin from binary antioxidant controlled-release packaging films. J. Agric. Food Chem. 60: 3492–3497.

Chen, Y.-C., S.-H. Yu, G.-J. Tsai, D.-W. Tang, F.-L. Mi and Y.-P. Peng. 2010. Novel technology for the preparation of self-assembled catechin/gelatin nanoparticles and their characterization. J. Agric. Food Chem. 58: 6728–6734.

Cian, R.E., P.R. Salgado, S.R. Drago, R.J. González and A.N. Mauri. 2014. Development of naturally activated edible films with antioxidant properties prepared from red seaweed *Porphyra columbina* biopolymers. Food Chem.146: 6–14.

Cushnie, T.P.T. and A.J. Lamb. 2005. Antimicrobial activity of flavonoids. Int. J. Antimicrob. Agents. 26: 343–356.

Dashtdar, H., M.R. Murali, A.A. Abbas, A.M. Suhaeb, L. Selvaratnam, L.X. Tay, et al. 2013. PVA-chitosan composite hydrogel versus alginate beads as a potential mesenchymal stem cell carrier for the treatment of focal cartilage defects. Knee Surg. Sports Traumatol. Arthrosc. 23: 1368-1377.

de Abreu, D.A.P., P.P. Losada, J. Maroto and J.M. Cruz. 2011. Natural antioxidant active packaging film and its effect on lipid damage in frozen blue shark (*Prionace glauca*). Innov. Food Sci. Emerg. Technol. 12: 50–55.

Deladino, L., A.S. Navarro and M.N. Martino. 2013. Carrier systems for yerba mate extract (*Ilex paraguariensis*) to enrich instant soups. Release mechanisms under different pH conditions. LWT - Food Sci. Technol. 53: 163–169.

Dutta, P.K., S. Tripathi, G.K. Mehrotra and J. Dutta. 2009. Perspectives for chitosan based antimicrobial films in food applications. Food Chem. 114:1173–1182.

El-Sayed, S., K.H. Mahmoud, A.A. Fatah and A. Hassen. 2011. DSC, TGA and dielectric properties of carboxymethyl cellulose/polyvinyl alcohol blends. Phys. B Condens. Matter. 406: 4068–4076.

Fabra, M.J., M.A. Busolo, A. Lopez-Rubio and J.M. Lagaron. 2013. Nanostructured biolayers in food packaging. Trends Food Sci. Technol. 31: 79–87.

Fabra, M.J., A. Lopez-Rubio, E. Sentandreu and J.M. Lagaron. 2015. Development of multilayer corn starch-based food packaging structures containing beta-carotene by means of the electro-hydrodynamic processing. Starch-Stärke. 67: 1–8.

Garcia, M.A., A. Pinotti and N.E. Zaritzky. 2006. Physicochemical, water vapor barrier and mechanical properties of corn starch and chitosan composite films. Starch-Stärke. 58: 453–463.

Geng, N.N. and C.M. Xiao. 2009. Facile preparation of methylcellulose/poly(vinyl alcohol) physical complex hydrogels with tunable thermosensitivity. Chinese Chem. Lett. 20: 111–114.

Golasz, L., J. Silva and S. Silva. 2013. Film with anthocyanins as an indicator of chilled pork deterioration. Ciência e Tecnol. Aliment. 2012: 155–162.

Gómez-Estaca, J., C. López-de-Dicastillo, P. Hernández-Muñoz, R. Catalá and R. Gavara. 2014. Advances in antioxidant active food packaging. Trends Food Sci. Technol. 35: 42–51.

González, A. and C.I.A. Igarzabal. 2013. Soy protein – Poly (lactic acid) bilayer films as biodegradable material for active food packaging. Food Hydrocoll. 33: 289–296.

Gorrasi, G., V. Bugatti, L. Tammaro, L. Vertuccio, G. Vigliotta and V. Vittoria. 2016. Active coating for storage of Mozzarella cheese packaged under thermal abuse. Food Control. 64: 10–16.

Gram, L., L. Ravn, M. Rasch, J.B. Bruhn, A.B. Christensen and M. Givskov. 2002. Food spoilage--interactions between food spoilage bacteria. Int. J. Food Microbiol. 78: 79–97.

Gutiérrez, O.M. 2013. Sodium- and phosphorus-based food additives: Persistent but surmountable hurdles in the management of nutrition in chronic kidney disease. Adv. Chronic Kidney Dis. 20: 150–156.

Han, T., L. Lu and C. Ge. 2015. Development and properties of high density polyethylene (HDPE) and ethylene-vinyl acetate copolymer (EVA) blend antioxidant active packaging films containing quercetin. Packag. Technol. Sci. 28: 415–423.

Heising, J.K., P.V. Bartels, M.A.J.S. van Boekel and M. Dekker. 2014. Non-destructive sensing of the freshness of packed cod fish using conductivity and pH electrodes. J. Food Eng. 124: 80–85.

Hossain, S.M.Z., C. Ozimok, C. Sicard, S.D. Aguirre, M.M. Ali, Y. Li, et al. 2012. Multiplexed paper test strip for quantitative bacterial detection. Anal. Bioanal. Chem. 403: 1567–1576.

Hosseini, M.H., S.H. Razavi and M.A. Mousavi. 2009. Antimicrobial, physical and mechanical properties of chitosan-based films incorporated with thyme, clove and cinnamon essential oils. J. Food Process. Preserv. 33:727–743.

Hosseini, S.F., M. Zandi, M. Rezaei and F. Farahmandghavi. 2013. Two-step method for encapsulation of oregano essential oil in chitosan nanoparticles: Preparation, characterization and in vitro release study. Carbohydr. Polym. 95: 50–56.

Hosseini, S.F., M. Rezaei, M. Zandi and F. Farahmandghavi. 2016. Development of bioactive fish gelatin/chitosan nanoparticles composite films with antimicrobial properties. Food Chem. 194: 1266–1274.

Huo, W., G. Xie, W. Zhang, W. Wang, J. Shan, H. Liu, et al. 2016. Preparation of a novel chitosan-microcapsules/starch blend film and the study of its drug-release mechanism. Int. J. Biol. Macromol. 87: 114–122.

Ibarguren, C., G. Céliz, A.S. Díaz, M.A. Bertuzzi, M. Daz and M.C. Audisio. 2015. Gelatine based films added with bacteriocins and a flavonoid ester active against food-borne pathogens. Innov. Food Sci. Emerg. Technol. 28: 66–72.

Jang, W.-S., I. Rawson and J.C. Grunlan. 2008. Layer-by-layer assembly of thin film oxygen barrier. Thin Solid Films. 516: 4819–4825.

Jokerst, J.C., J.A. Adkins, B. Bisha, M.M. Mentele, L.D. Goodridge and C.S. Henry. 2012. Development of a paper-based analytical device for colorimetric detection of select foodborne pathogens. Anal. Chem. 84: 2900–2907.

Kanimozhi, K., S.K. Basha and V.S. Kumari. 2016. Processing and characterization of chitosan/PVA and methylcellulose porous scaffolds for tissue engineering. Mater. Sci. Eng. C. 61: 484–491.

Kaur, I., R. Gupta, A. Lakhanpal and A. Kumar. 2014. Investigation of immobilization and hydrolytic properties of pectinase onto chitosan-PVA copolymer. Int. J. Adv. Chem. 2: 117–123.

Kerry, J.P., M.N. O'Grady and S.A. Hogan. 2006. Past, current and potential utilisation of active and intelligent packaging systems for meat and muscle-based products: A review. Meat Sci. 74: 113–130.

Kim, E., D.Y. Choi, H.C. Kim, K. Kim and S.J. Lee. 2013. Calibrations between the variables of microbial TTI response and ground pork qualities. Meat Sci. 95: 362–367.

Kim, K., E. Kim and S.J. Lee. 2012. New enzymatic time-temperature integrator (TTI) that uses laccase. J. Food Eng.113: 118–123.

Kim, M.J., S.W. Jung, H.R. Park and S.J. Lee. 2012. Selection of an optimum pH-indicator for developing lactic acid bacteria-based time–temperature integrators (TTI). J. Food Eng. 113: 471–478.

Konecsni, K., N.H. Low and M.T. Nickerson. 2012. Chitosan–tripolyphosphate submicron particles as the carrier of entrapped rutin. Food Chem. 134: 1775–1779.

Kuswandi, B., Y. Wicaksono, A. Abdullah, L.Y. Heng and M. Ahmad. 2011. Smart packaging: Sensors for monitoring of food quality and safety. Sens. Instrum. Food Qual. Saf. 5: 137–146.

Laufer, G., C. Kirkland, M.A. Priolo and J.C. Grunlan. 2012. High oxygen barrier, clay and chitosan-based multilayer thin films: An environmentally friendly foil replacement. Green Mater. 1: 4–10.

Lewandowska, K. 2012. Surface studies of microcrystalline chitosan/poly(vinyl alcohol) mixtures. Appl. Surf. Sci. 263: 115–123.

Li, J., L.-J. Wu, S.-S. Guo, H.-E. Fu, G.-N. Chen and H.-H. Yang. 2013. Simple colorimetric bacterial detection and high-throughput drug screening based on a graphene-enzyme complex. Nanoscale. 5: 619–623.

Li, J.-H., J. Miao, J.-L. Wu, S.-F. Chen and Q.-Q. Zhang. 2014. Preparation and characterization of active gelatin-based films incorporated with natural antioxidants. Food Hydrocoll. 37: 166–173.

Liu, F., J. Antoniou, Y. Li, J. Yi, W. Yokoyama, J. Ma, et al. 2015. Preparation of gelatin films incorporated with tea polyphenol nanoparticles for enhancing controlled-release antioxidant properties. J. Agric. Food Chem. 63: 3987–3995.

Liu, R., X. Xu, X. Zhuang and B. Cheng. 2014. Solution blowing of chitosan/PVA hydrogel nanofiber mats. Carbohydr. Polym. 101: 1116–1121.

Llana-Ruíz-Cabello, M., S. Pichardo, N.T. Jiménez-Morillo, J.M. Bermúdez, S. Aucejo, F.J. González-Vila, et al. 2015. Molecular characterization of a bio-based active packaging containing *Origanum vulgare* L. essential oil using pyrolysis gas chromatography/mass spectrometry (Py-GC/MS). J. Sci. Food Agric. 96: 3207–3212.

López-de-Dicastillo, C., J. Gómez-Estaca, R. Catalá, R. Gavara and P. Hernández-Muñoz. 2012. Active antioxidant packaging films: Development and effect on lipid stability of brined sardines. Food Chem. 131: 1376–1384.

Lorenzo, J.M., R. Batlle and M. Gómez. 2014. Extension of the shelf-life of foal meat with two antioxidant active packaging systems. LWT - Food Sci. Technol. 59: 181–188.

Lu, B.L., W. Zheng, Z. Lv and Y. Tang. 2013. Development and application of time – temperature Indicators used on food during the cold chain logistics. Packag. Technol. Sci. 26: 80–90.

Lu, D.R. 2009. Starch-based completely biodegradable polymer materials. eXPRESS Polym. Lett. 3: 366–375.

Ma, X., P.R. Chang and J. Yu. 2008. Properties of biodegradable thermoplastic pea starch/carboxymethyl cellulose and pea starch/microcrystalline cellulose composites. Carbohydr. Polym. 72: 369–375.

Mangala, E., T.S. Kumar, S. Baskar and K.P. Rao. 2003. Development of chitosan/ poly (vinyl alcohol) blend membranes as burn dressings. Trends Biomater. Artif. Organs.17: 34–40.

Marek, P., J.J. Velasco-Veléz, T. Haas, T. Doll and G. Sadowski. 2013. Time-monitoring sensor based on oxygen diffusion in an indicator/polymer matrix. Sensors Actuators, B. Chem. 178: 254–262.

Marsh, K. and B. Bugusu. 2007. Food packaging--roles, materials, and environmental issues. J. Food Sci. 72: R39–R55.

Martin, J.G.P., E. Porto, S.M. de Alencar, E.M. da Glória, C.B. Corrêa and I.S.R. Cabral. 2013. Antimicrobial activity of yerba mate (*Ilex paraguariensis* St. Hil.) against food pathogens. Rev. Argent. Microbiol. 45: 93–98.

Mills, A. 2005. Oxygen indicators and intelligent inks for packaging food. Chem. Soc. Rev. 34: 1003–1011.

Miranda, O.R., X. Li, L. Garcia-Gonzalez, Z. Zhu, B. Yan, U.H.F. Bunz, et al. 2011. Colorimetric bacteria sensing using a supramolecular enzyme–nanoparticle biosensor. J. Am. Chem. Soc. 133: 9650–9653.

Moon, J.-K. and T. Shibamoto. 2009. Antioxidant assays for plant and food components. J. Agric. Food Chem. 57: 1655–1666.

Muriel-Galet, V., M.J. Cran, S.W. Bigger, P. Hernández-Muñoz and R. Gavara. 2015. Antioxidant and antimicrobial properties of ethylene vinyl alcohol copolymer films based on the release of oregano essential oil and green tea extract components. J. Food Eng. 149: 9–16.

Nafisi, S., G.B. Sadeghi and A. PanahYab. 2011. Interaction of aspirin and vitamin C with bovine serum albumin. J. Photochem. Photobiol. B. 105: 198–202.

Nath, S.D., C. Abueva, B. Kim and B.T. Lee. 2015. Chitosan-hyaluronic acid polyelectrolyte complex scaffold crosslinked with genipin for immobilization and controlled release of BMP-2. Carbohydr. Polym. 115: 160–169.

Neethirajan, S. and D.S. Jayas. 2010. Nanotechnology for the food and bioprocessing industries. Food Bioprocess Technol. 4: 39–47.

Nisa, I., B.A. Ashwar, A. Shah, A. Gani, A. Gani and F.A. Masoodi. 2015. Development of potato starch based active packaging films loaded with antioxidants and its effect on shelf life of beef. J. Food Sci. Technol. 52: 7245–7253.

Nopwinyuwong, A., S. Trevanich and P. Suppakul. 2010. Development of a novel colorimetric indicator label for monitoring freshness of intermediate-moisture dessert spoilage. Talanta. 81: 1126–1132.

Pacquit, A., K.T. Lau, H. McLaughlin, J. Frisby, B. Quilty and D. Diamond. 2006. Development of a volatile amine sensor for the monitoring of fish spoilage. Talanta. 69: 515–520.

Pereira de Abreu, D.A., J.M. Cruz and P. Paseiro Losada. 2012. Active and intelligent packaging for the food industry. Food Rev. 28: 146–187.

Pereira, V.A., I.N.Q. de Arruda and R. Stefani. 2014. Active chitosan/PVA films with anthocyanins from *Brassica oleraceae* (Red Cabbage) as time–temperature indicators for application in intelligent food packaging. Food Hydrocoll. 43: 1–9.

Pires, A.C.D.S., N.D.F.F. Soares, L.H.M. da Silva, M.D.C.H. da Silva, M.V. De Almeida, M. Le Hyaric, et al. 2011. A colorimetric biosensor for the detection of foodborne bacteria. Sensors Actuators, B Chem. 153: 17–23.

Ramos, M., A. Beltrán, M. Peltzer, A.J.M. Valente and M.D.C. Garrigós. 2014. Release and antioxidant activity of carvacrol and thymol from polypropylene active packaging films. LWT - Food Sci. Technol. 58: 470–477.

Ray, S., C. Sengupta and K. Roy. 2007. QSAR modeling of antiradical and antioxidant activities of flavonoids using electrotopological state (E-State) atom parameters. Cent. Eur. J. Chem. 5: 1094–1113.

Reis, L.C.B., C.O. de Souza, J.B.A. da Silva, A.C. Martins, I.L. Nunes and J.I. Druzian. 2015. Active biocomposites of cassava starch: The effect of yerba mate extract and mango pulp as antioxidant additives on the properties and the stability of a packaged product. Food Bioprod. Process. 94: 382–391.

Rodríguez, A., C. Nerín and R. Batlle. 2008. New cinnamon-based active paper packaging against *Rhizopus stolonifer* food spoilage. J. Agric. Food Chem. 56: 6364–6369.

Sadat Ebrahimi, M.M. and H. Schönherr. 2014. Enzyme-sensing chitosan hydrogels. Langmuir. 30: 7842–7850.

Salinas, Y., J.V. Ros-Lis, J.-L. Vivancos, R. Martínez-Máñez, M.D. Marcos, S. Aucejo, et al. 2012. Monitoring of chicken meat freshness by means of a colorimetric sensor array. Analyst. 137: 3635–3643.

Salinas, Y., J.V. Ros-Lis, J.L. Vivancos, R. Martínez-Máñez, S. Aucejo, N. Herranz, et al. 2014. A chromogenic sensor array for boiled marinated turkey freshness monitoring. Sensors Actuators, B Chem. 190: 326–333.

Samsudin, H., H. Soto-Valdez and R. Auras. 2014. Poly(lactic acid) film incorporated with marigold flower extract (*Tagetes erecta*) intended for fatty-food application. Food Control. 46: 55–66.

Shukla, V., G. Kandeepan and M.R. Vishnuraj. 2015. Development of on-package indicator sensor for real-time monitoring of buffalo meat quality during refrigeration storage. Food Anal. Methods. 8: 1591–1597.

Shutava, T.G., S.S. Balkundi and Y.M. Lvov. 2009. (-)-Epigallocatechin gallate/gelatin layer-by-layer assembled films and microcapsules. J. Colloid Interface Sci. 330: 276–283.

Shutava, T.G. and Y.M. Lvov. 2012. Encapsulation of natural polyphenols with antioxidant properties in polyelectrolyte capsules and nanoparticles. pp. 215–235. *In*: Diederich, M. and K. Noworyta (eds.). Natural Compounds as Inducers of Cell Death, 1st ed. Springer, Dordrecht, Netherlands.

Silva-Pereira, M.C., J.A. Teixeira, V.A. Pereira-Júnior and R. Stefani. 2015. Chitosan/ corn starch blend films with extract from *Brassica oleraceae* (red cabbage) as a visual indicator of fish deterioration. LWT - Food Sci. Technol. 61: 258–262.

Sonkaew P., A. Sane and P. Suppakul. 2012. Antioxidant activities of curcumin and ascorbyl dipalmitate nanoparticles and their activities after incorporation into cellulose-based packaging films. J. Agric. Food Chem. 60: 5388–5399.

Souza, M.P., A.F.M. Vaz, H.D. Silva, M.A. Cerqueira, A.A. Vicente and M.G. Carneiro-da-Cunha. 2015. Development and characterization of an active chitosan-based film containing quercetin. Food Bioprocess Technol. 8: 2183–2191.

Syed, S., F. Gusberti and N. Lang. 1984. Diagnostic potential of chromogenic substrates for rapid detection of bacterial enzymatic activity in healthy and disease associated periodontal plaques. J. Periodontal Res. 19: 618–621.

Teixeira, B., A. Marques, C. Pires, C. Ramos, I. Batista, J.A. Saraiva, et al. 2014. Characterization of fish protein films incorporated with essential oils of clove, garlic and origanum: Physical, antioxidant and antibacterial properties. LWT - Food Sci. Technol. 59: 533–539.

Tram, K., P. Kanda, B.J. Salena, S. Huan and Y. Li. 2014. Translating bacterial detection by DNAzymes into a litmus test. Angew. Chem. Int. Ed. Engl. 53: 12799–12802.

Tripathi, S., G.K. Mehrotra and P.K. Dutta. 2009. Physicochemical and bioactivity of cross-linked chitosan-PVA film for food packaging applications. Int. J. Biol. Macromol. 45: 372–376.

Vásconez, M.B., S.K. Flores, C.A. Campos, J. Alvarado and L.N. Gerschenson. 2009. Antimicrobial activity and physical properties of chitosan–tapioca starch based edible films and coatings. Food Res. Int. 42: 762–769.

Vikova, M. 2011. Alternative UV sensors based on color-changeable pigments. Adv. Chem. Eng. Sci. 1: 224–230.

Vimala, K. 2011. Fabrication of curcumin encapsulated chitosan-PVA silver nanocomposite films for improved antimicrobial activity. J. Biomater. Nanobiotechnol. 2: 55–64.

Vu, C.H.T. and K. Won. 2013. Novel water-resistant UV-activated oxygen indicator for intelligent food packaging. Food Chem. 140: 52–56.

Wan, Y., Y. Sun, P. Qi, P. Wang and D. Zhang. 2014. Quaternized magnetic nanoparticles-fluorescent polymer system for detection and identification of bacteria. Biosens. Bioelectron. 55: 289–293.

Wang, S., A.K. Singh, D. Senapati, A. Neely, H. Yu and P.C. Ray. 2010. Rapid colorimetric identification and targeted photothermal lysis of *Salmonella* bacteria by using bioconjugated oval-shaped gold nanoparticles. Chem. A Eur. J. 16: 5600–5606.

Wang, T., M. Turhan and S. Gunasekaran. 2004. Selected properties of pH-sensitive, biodegradable chitosan–poly(vinyl alcohol) hydrogel. Polym. Int. 53: 911–918.

Wen, P., D.-H. Zhu, K. Feng, F.-J. Liu, W.-Y. Lou, N. Li, et al. 2016. Fabrication of electrospun polylactic acid nanofilm incorporating cinnamon essential oil/β-cyclodextrin inclusion complex for antimicrobial packaging. Food Chem. 196: 996–1004.

Winkler, H., W. Vorwerg and R. Rihm. 2014. Thermal and mechanical properties of fatty acid starch esters. Carbohydr. Polym. 102: 941–949.

Woranuch, S., R. Yoksan and M. Akashi. 2015. Ferulic acid-coupled chitosan: Thermal stability and utilization as an antioxidant for biodegradable active packaging film. Carbohydr. Polym. 115: 744–751.

Wu, D., Y. Wang, J. Chen, X. Ye, Q. Wu, D. Liu, et al. 2013. Preliminary study on time–temperature indicator (TTI) system based on urease. Food Control. 34: 230–234.

Wu, J.-G., P.-J. Wang and S.C. Chen. 2010. Antioxidant and antimicrobial effectiveness of catechin-impregnated PVA-starch film on red meat. J. Food Qual. 33: 780–801.

Xiao-Wei, H., L. Zhi-Hua, Z. Xiao-Bo, S. Ji-Yong, Z. Han-Ping, Z. Jie-Wen, et al. 2016. Detection of meat-borne trimethylamine based on nanoporous colorimetric sensor arrays. Food Chem. 197: 930–936.

Xiping, G.A.O., T. Keyong, L.I.U. Jie, Z. Xuejing and Z. Yuqing. 2014. Compatibility and properties of biodegradable blend films with gelatin and poly(vinyl alcohol). J. Wuhan Univ. Technol. Sci. Ed. 29: 351–356.

Xu, Q., S. Lee, Y. Cho, M.H. Kim, J. Bouffard and J. Yoon. 2013. Polydiacetylene-based colorimetric and fluorescent chemosensor for the detection of carbon dioxide. J. Am. Chem. Soc. 135: 17751–17754.

Yang, H.-J., J.-H. Lee, M. Won and K.B. Song. 2016. Antioxidant activities of distiller dried grains with solubles as protein films containing tea extracts and their application in the packaging of pork meat. Food Chem. 196: 174–179.

Yehye, W.A., N.A. Rahman, A. Ariffin, S.B. Abd Hamid, A.A. Alhadi, F.A. Kadir, et al. 2015. Understanding the chemistry behind the antioxidant activities of butylated hydroxytoluene (BHT): A review. Eur. J. Med. Chem. 101: 295–312.

Yoshida, C.M.P., V.B.V. Maciel, M.E.D. Mendonça and T.T. Franco. 2014. Chitosan biobased and intelligent films: Monitoring pH variations. LWT - Food Sci. Technol. 55: 83–89.

Zhang, L., R. Li, F. Dong, A. Tian, Z. Li and Y. Dai. 2015. Physical, mechanical and antimicrobial properties of starch films incorporated with ε-poly-L-lysine. Food Chem. 166: 107–114.

Zhang, X., S. Lu and X. Chen. 2014. A visual pH sensing film using natural dyes from *Bauhinia blakeana* Dunn. Sensors Actuators, B Chem. 198: 268–273.

Zhou, H., X. Sun, L. Zhang, P. Zhang, J. Li and Y.-N. Liu. 2012. Fabrication of biopolymeric complex coacervation core micelles for efficient tea polyphenol delivery via a green process. Langmuir. 28: 14553–14561.

Zhu, H.-Y., Y.-Q. Fu, R. Jiang, J. Yao, L. Xiao and G.-M. Zeng. 2012. Novel magnetic chitosan/poly(vinyl alcohol) hydrogel beads: Preparation, characterization and application for adsorption of dye from aqueous solution. Bioresour. Technol. 105: 24–30.

3

Antimicrobial Active Packaging

Cintia B. Contreras[1], German Charles[2], Ricardo Toselli[2] and Miriam C. Strumia[1,*]

[1] Instituto de Investigación de Procesos de Ingeniería y Química Aplicada (IPQA), 2CONICET. Facultad de Ciencias Químicas, Universidad Nacional de Córdoba, Argentina, Medina Allende y Haya de la Torre, Ciudad Universitaria, Córdoba (X5000HUA), Argentina

[2] Centro de Química Aplicada (CEQUIMAP), Facultad de Ciencias Químicas, Universidad Nacional de Córdoba, Argentina, Medina Allende y Haya de la Torre, Ciudad Universitaria, Córdoba (X5000HUA), Argentina

Introduction

For centuries, the overriding need to contain and protect products of different nature has led to the use of different types of packaging technologies. Today, this phenomenon is particularly important when it comes to food packaging.

The FDA reported that globally, around 1.3 billion tons of food are disposed off each year due to problems associated with production, distribution and storage stages. In Latin America, this loss represents approximately 200 kg per individual annually, while in more developed countries it reaches 300 kg per year. In view of this scenario, packaging plays a key role in the food production and storage (Dobrucka 2013).

Synthetic polymers, specifically thermoplastics, are the main packaging materials due to their versatile solutions to a variety of needs. They exhibit highly advantageous characteristics, such as transparency, good presentation, transfer resistance, heat sealability, and good strength to weight ratio. Further, plastics are generally inexpensive and efficient, showing mechanical properties such as stress and tensile strength, being good barriers to oxygen and heat (Hu and Wang 2016). Although all these advantages have already been recognized, conventional packaging polymers have begun to be questioned due to growing environmental concerns.

*Corresponding author: mcs@fcq.unc.edu.ar

Most packaging materials are petroleum-based, thus derived from nonrenewable resources. Moreover, synthetic plastics are not biodegradable. Given that around 30 to 40% of municipal solid waste comes from packaging, the large amount of solid waste produced poses an environmental problem that needs to be urgently addressed.

Currently, many studies focus on providing a solution to these problems, trying to find a possible re-use and valorization of waste without harming the environment. For this reason, several researchers have studied the possibility of replacing traditional petroleum-derived plastics with low-cost biodegradable materials showing similar properties (De Azeredo et al. 2014; Fabra et al. 2013; Javidi et al. 2016). They can minimize the environmental impact caused by the overuse of conventional polymeric materials and therefore reduce increased production of plastics.

Biodegradable materials are polymers from renewable raw materials such as proteins and polysaccharides extracted from plant and animal products and by-products derived from agriculture, marine and microbial sources. Hence, many biopolymers including chitosan, starch, clay and pectin have been used in food packaging. These materials can be degraded by environmental conditions, for example, soil exposed optimum moisture, oxygen and microorganisms, giving simple substances, such as water, carbon dioxide and biomass.

Similar to conventional packaging, bio-based materials must have a number of important conditions, including containment and protection of food quality, serving as selective barriers for moisture transfer, oxygen uptake, lipid oxidation and loss of volatile aromas and flavours keeping food taste and quality.

The materials from natural polymers are associated with poor properties, often having inferior properties compared to those of commodity polymers. Unlike most synthetic plastics used as packaging materials, most bioplastics currently available do not meet the essential requirements of food packaging, especially in terms of barrier and mechanical properties.

By applying a chemical modification technique, we can improve these properties and achieve combinations of properties required for specific applications. The chemical structure of biopolymers opens up possibilities for reactive modification. Modification strategies of the starting material must improve water resistance, barrier gas and mechanical properties to avoid loss due to breakage or rupture of the material, in addition to increasing stability storage conditions. Therefore, copolymerization, grafting, transesterification and coupling reactions have been successfully used to yield polymers with improved properties (Strumia et al.; chapter 4 in Thakur 2015). The preparation of blends is another technique which allows for considerably enhanced impact resistance and barrier properties.

The trend to improve the stability and mechanical properties of bio-based packaging is shown in the growing number of works annually reported in the literature on the production and characterization of novel biopolymers and biocomposites, and the development of processing strategies designed to overcome the inherent limitations of these bio-based materials (Lagaron et al. 2016).

Biopolyesters, proteins and polysaccharides used in packaging applications are the biopolymers most widely studied. Among the most commonly used polysaccharide polymers we can find cellulose, chitosan and starch, which have gained significant attention lately. However, further studies on the structure-property relationship of these materials are required to exploit their full potential as food packaging (Kumar et al. 2010).

Thermoplastic sustainable biopolymers extensively researched include starch, polyhydroxy alkanoate (PHA) and polylactic acid (PLA). Specifically, starch and PLA biopolymers are the most important families of biodegradable materials, as they present an interesting balance of properties and have become commercially available, being produced on a large industrial scale. They are of particular interest in food packaging due to their excellent transparency and relatively good water resistance (Javidi et al. 2016).

Biopolymers and their mixtures are successfully applied in various industrial areas resulting in different consumer goods. They can be used directly in food alone or as multilayers. In the first case, edible packaging materials are intended to be an integral part of food and to be eaten with the products, thus they are also inherently biodegradable. Edible films are stand-alone structures, preformed separately from food and then applied to the food surface or between food components, or sealed into edible pouches. Gel capsules, microcapsules and tablet coatings made from edible materials could also be considered as edible packaging. Although edible packaging is not expected to replace conventional plastic packaging, it has a major role to play in the full development of renewable packaging (De Azeredo et al. 2014; González et al. 2015).

The formation of a multilayer applied to packaging is another interesting and innovative strategy to achieve enhanced barrier properties. It is based on layers and nanostructured layers made of biopolymers, which are able to retain transparency and enhance barrier properties without significant changes in mechanical performance. These technologies include monolayer biocoatings, multi-layered biodegradable polymers and layer-by-layer bioassemblies, through a novel methodology based on high voltage spinning to generate interlayers able to, on the one hand, improve the barrier properties of the biopolymers and, on the other, tie together the different biopolymer layers serving as a natural adhesive (Fabra et al. 2013).

Requirements of a Food Package

Environmentalists require that materials meet criteria in the manufacture package. In additions, package as a final product shall need also comply with mandatory requirements for use in food of different nature and shape.

Food is not an inalterable product that keeps indefinitely its physical, chemical and microbiological characteristics. In effect, food deteriorates over time by the action of living organisms, the physical and chemical environment or by the biological activity of food itself.

Food packaging plays a key role in the conservation, distribution and marketing of food. The functions traditionally assigned to packaging include containing, protecting, informing and attracting marketing; however, this conception has been modified over the years, in response to the major changes in marketing and consumer's lifestyle. This dynamic context has forced the advent of new forms of packaging and preservation.

Regardless of the material (synthetic or natural polymer) used for the preparation of a package, it must fulfill specific functions (Vanderroost et al. 2014), which acquire different levels of importance depending on the food product:

Containment: Proper containment of food products during storage, transportation and commercialization is perhaps one of the fundamental aspects. Poor containment conditions are directly related to food waste, economic loss and negative environmental effects. The efficiency of the package in containing food products is directly related to packaging material. The importance and hierarchy of containment are evident throughout the supply chain of products.

Each material has unique properties and applications for food packaging, that is, it is defined for each product. For example, specific materials should be select for liquid or solid foods, dry powders, grains, etc.

Protection: The degree of protection required by food products is key to the selection of packaging material and design. Properly selected packaging material is beneficial for extending food shelf life.

One of the most important requirements in the performance and design of food package is its ability to avoid "permeation" of different substances, or "barrier" property. These substances may cause deterioration or damage in the quality parameters of packaged products, such as, loss or gain of water vapour, oxygen rancidity or transfer of aromas (Marsh and Bugusu 2007; Richard et al. 2003).

Packaging made by glass and metal shows the best barrier properties, even though its use is limited due to issues of cost energy, weight, transportation and environmental impact. On the other hand, cellulose or paper packaging generally has poor barrier properties; consequently, its application to food products is particularly limited.

Plastic material, however, offers a number of advantages in terms of cost, energy requirement, versatility and recycling, which differ from those of glass, metal and paper. Generally, it has good barrier properties to small molecules; although, its gas permeability can be critical. To remedy this, the gas barrier can be controlled by modifying the material or by combining it with other classes of plastic or nonplastic compounds.

Communication: The benefits of a product are measured from the first contact with its packaging. This means that packages have the ability to become the sole differentiating agent among products with similar characteristics, and therefore consumers make choices on the basis of the benefits offered by a particular product.

Food package is used to inform consumers of the product characteristics. For example, package labeling satisfies legal requirements for product identification, nutritional value, ingredient declaration, net weight and manufacturer information. Additionally, the package supplies important information of the product such as cooking instructions, brand identification and pricing. All of these improvements may impact waste disposal (Marsh and Bugusu 2007).

Packaging becomes the medium through which a product transcends the functional barrier to convey a message linked to the senses, desires and emotions. Consequently, innovative or distinctive packaging can boost sales in a competitive environment. The package may be designed to enhance the product image and/or differentiate the product from that of its competition; this is a central function, that is, the marketing function (Jansson-Boyd 2010).

Convenience: Changes in consumer culture and lifestyle of modern societies have caused packaging industries to adapt their products to comply with these changes and satisfy the demands of consumers. Thus, a variety of designs has been incorporated in food packages in an effort to increase food acceptance. These designs include innovation in package opening, product dispensing, product visibility, package resealing and ultimate preparation of products before consumption (microwavability), ease of access, handling and disposal (Marsh and Bugusu 2007).

When choosing a product, package design and acceptance are key. For example, closing appliances are central in order to guarantee product quality.

Package construction plays a significant role in determining the shelf life of food products. The right selection of packaging material and technology ensures product quality and freshness during distribution and storage (Marsh and Bugusu 2007).

Active Packaging

In the last decades, however, one of the most innovative developments in food packaging has been 'active' packaging, based on deliberate interactions with food or food environment. The purpose of 'active packaging' involves extending food shelf-life and maintenance or even improvement of its quality. This kind of packaging interacts with the product or headspace between package and food system, to obtain a desired outcome. The packaging material performs an additional function beyond containment and basic protection; it remains an area of active research with full potential for commercial applications. Active packaging is used not only in food packaging but also in the pharmaceutical and consumer goods industry, with a common goal of improving shelf life, safety or quality of packaged goods (Lim 2011; Silvestre et al. 2011).

In the last two decades, food packaging innovation activities have gradually expanded toward the development of intelligent packaging. It contains a component that enables the monitoring of the condition of packaged food or the environment surrounding the food during transport and storage. Therefore, intelligent packaging is a system that provides the user with correct information on food conditions and packaging integrity in real time (Vanderroost et al. 2014).

Paradoxically, whereas the concept of active and intelligent packaging is now considered as 'modern', such concept belongs to ancestral traditions in all tropical areas of the world. In the regions of Africa, Asia and South America, vegetal leaves were, and still are, profusely and traditionally used for food packaging with important markets devoted to leave commercialization. Beyond their use as simple 'barriers', numerous vegetal leaves are used due to their ability to transfer aromatic, coloring, enzyme (e.g. papain) or antimicrobial substances (e.g. essential oils) to food. Vegetal leaf packaging interacts with the food to modify its texture or organoleptic properties or to slow down microbial spoilage. It is also used due to its ability to change color with temperature and/or time, thus playing the role of cooking or freshness indicators. Up to four different types of leaves are used as successive layers, each one having a highly specific function. Leaves have also been used in Mediterranean regions of Europe, for example, in traditional cheese to allow a good maturation process (Dainelli et al. 2008).

In active packaging systems, the protection function of a package is enhanced by incorporating active compounds including antimicrobials, preservatives, oxygen absorbers, water vapour absorbers and ethylene removers (Brody et al. 2008).

Active packaging can be classified into two main types (Fig. 1): (1) non-migratory active packaging acting without intentional migration and (2) active releasing packaging allowing controlled migration of non-volatile agents or emission of volatile compounds in the atmosphere surrounding food (Bastarrachea et al. 2015; Goddard and Hotchkiss 2007). Nowadays, a large number of research studies focus on non-migratory active packaging for eliciting a desirable response from food systems without the active component migrating from packaging into food. The most well-known examples of non-migratory active packaging are moisture absorbers, mostly based on the adsorption of water by a zeolite, cellulose and their derivatives, among others. The tendency in the market of moisture-absorbing systems is to introduce absorbing substances into the packaging material in order to make the active system invisible for the consumer, as in absorbing trays for fresh meat or fish.

Microbiological contamination and subsequent growth of microorganisms represent the main cause of food spoilage, decreasing food

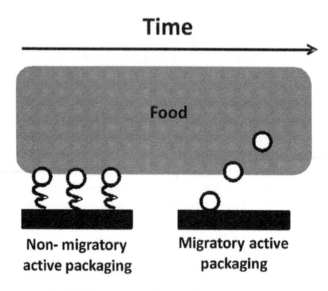

Fig. 1. Migratory and non-migratory packaging.

lifetime and compromising food security. Traditional methods to inhibit microbial growth include heat treatment, pilling, radiation, pre-cooking, controlled atmosphere and direct addition of antimicrobial agents, which can lead to a variety of drawbacks. Unfortunately, these conservation techniques cannot be equally applied to all kinds of food.

The antimicrobial packaging technology is a form of active packaging whose main purpose is to increase the shelf life of packaged foods, avoiding spoilage by the action of microorganisms. Design or model of antimicrobial packaging can vary significantly depending on the use to which they are intended. Therefore, to carry out its development, the following factors must be taken into account (Han 2013):

- Chemical nature of the package, the processes associated with its production and residual antimicrobial activity: The choice of antimicrobial agents usually depends on their limitations as compared to the different conditions related to processing package (extruded, laminated, printing, etc.) and their compatibility with the material (difference in polarity, solubility, etc.). One example involves the need to carry out the extrudate of low density polyethylene (LDPE) additivated with potassium sorbate at low temperatures, since the activity of the preservative can be compromised at elevated temperatures (Appendini and Hotchkiss 2002). Knowing the antimicrobial activity of the package after being processed is essential for the choice of antimicrobial agent, and therefore, for the design of active packaging.
- Characteristics of antimicrobial substance and food: Food components can significantly affect the effectiveness of antimicrobial substances.

The physicochemical characteristics of food can have a direct impact on its activity; for example, the pH of food can influence the ionization of the active site of the antimicrobial agent and alter its biological activity. Another factor that can substantially modify the activity and chemical stability of an active agent is the water activity of the packaged product. Each food has a particular micro flora, which must be taken into account when selecting and incorporating the antimicrobial agent (Quintavalla and Vicini 2002).

- Storage temperature: The temperature which preserves food can have an effect on antimicrobial activity, in addition to modifying the degree of transfer of the active agent toward food. Usually, by increasing the storage temperature, the migration of active agent is increased. Temperature can also affect the residual antimicrobial activity of the compound, so it is important to know the storage conditions of a particular product to predict the efficiency of the active compounds used.
- Physical properties of packaging materials in antimicrobial packaging: The performance and characteristics of packaging materials should be kept after the addition or incorporation of active substances. This is perhaps one of the most important objectives in antimicrobial packaging design.

Packaging systems for containing and protecting food are usually formed by the product, the container and the space between them. Particularly in antimicrobial packages, the active agent may be incorporated in packaging inedible parts such as films, bilayers, coatings or in the space between the material and the product through sachets or pads. Edible coatings, moreover, are also a form of antimicrobial packaging that has been recently developed to prevent deterioration mainly of fruits (Dobrucka 2013; Han 2005). The antimicrobial packaging can take several forms (Fig. 2), including:

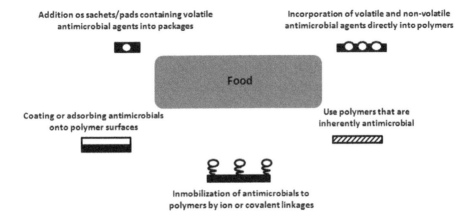

Addition os sachets/pads containing volatile antimicrobial agents into packages

Incorporation of volatile and non-volatile antimicrobial agents directly into polymers

Food

Coating or adsorbing antimicrobials onto polymer surfaces

Use polymers that are inherently antimicrobial

Inmobilization of antimicrobials to polymers by ion or covalent linkages

Fig. 2. Different ways of including the antimicrobial agent in active packaging.

(a) Addition of Sachets/Pads Containing Volatile Antimicrobial Agents into Packages

The most successful commercial application of antimicrobial packaging has been sachets that are enclosed loose or attached to the interior of a package. These systems use volatile substances that inhibit microbial growth by being released into the food environment. Three forms have predominated: oxygen absorbers, moisture absorbers and ethanol vapour generators. Oxygen and moisture absorbers are used primarily in bakery, pasta, produce and meat packaging to prevent oxidation and water condensation. Although oxygen absorbers may not be intended to be antimicrobial, a reduction in oxygen inhibits the growth of aerobes, particularly moulds. Moisture absorbers can indirectly reduce microbial growth.

(b) Incorporation of Volatile and Non-volatile Antimicrobial Agents Directly into Polymers

Incorporation of bioactive agents, including antimicrobials, into polymers has been commercially applied to drug and pesticide delivery, household goods, textiles, surgical implants and other biomedical devices. In these systems, antimicrobial compounds such as bactericides, enzymes, chelators, metal ions or organic acids are added to the polymer during the process of extruded plastic films (Hanušová et al. 2010). An example of such systems is the incorporation of sorbic acid into PVA films to be used in Gouda cheese (Hauser and Wunderlich 2011; Silveira et al. 2007). A further example is the extrudate linalool and methylchavicol (GRAS substances) with PE pellets used as antimicrobial packaging (Suppakul et al. 2011). However, one of the disadvantages involved relates to the fact that, due to the conditions employed in such processes (high temperatures, compression, etc.), the antimicrobial activity of the incorporated active agents may be diminished.

In several studies, PLA were used as polymer matrices to produce active food packaging films with the combination of natural extracts or essential oils which could come from plants (Llana-Ruiz-Cabello et al. 2015; Tawakkal et al. 2016). One example is the contribution of Javidi et al. (Javidi et al. 2016) to develop active bio-packaging films using hydrophobic compounds such as oregano essential oil, which improved the mechanical and physical properties and induced antimicrobial function in PLA films. The efficacy of PLA/essential oil film was demonstrated in rainbow trout delaying the growth of microorganisms; this antimicrobial activity remained throughout the period studied and contributed to extending the shelf-life of refrigerated trout (Javidi et al. 2016).

(c) Coating or Adsorbing Antimicrobials onto Polymer Surfaces

The coatings and films based on alginate, carrageenan, gelatin, zein, gluten, whey, carnauba, beeswax and fatty acids, proteins and lipids show advantages such as biodegradability, biocompatibility, good look and high barrier to oxygen and physical stress (Fakhouri et al. 2015; González and Alvarez Igarzabal 2013; Santacruz et al. 2015). Early developments in antimicrobial packaging incorporated fungicides into waxes to coat fruits and vegetables and shrink films coated with quaternary ammonium salts to wrap potatoes. Other developments included coating wrap cheese with sorbic acid (Hauser and Wunderlich 2011).

(d) Immobilization of Antimicrobials to Polymers by Electrostatic or Covalent Linkages

This type of immobilization of active agents is presented as an alternative to their incorporation into the polymer when extrudating or casting. It allows including on surfaces those antimicrobials that are not capable of supporting the mechanical conditions and high temperature of extrusion process, or which may be altered due to the presence of certain solvents used in the preparations of the polymeric films by casting (Appendini and Hotchkiss 2002). This way of incorporating the active agent has the advantage of reducing the amount of the compound added to the polymer matrix, minimizing the loss of antimicrobial activity and helping to maintain the concentration of the active compound in the food surface at appropriate levels to ensure its inhibitory effect. A few examples of ionic and covalent immobilization of antimicrobials onto polymers or other materials have been published. This type of immobilization requires the presence of functional groups on both the antimicrobial and the polymer. Examples of antimicrobials with functional groups are peptides, enzymes, polyamines and organic acids. In addition to functional antimicrobials and polymer supports, immobilization may require the use of 'spacer' molecules that link the polymer surface to the bioactive agent.

(e) Use of Inherently Antimicrobial Polymers

Some polymers are inherently antimicrobial, such as chitosan, which has been used as a coating and appears to protect fresh vegetables and fruit from fungal degradation. Chitosan is soluble in diluted aqueous acidic solutions due to the protonation of their amino groups. The cationic character confers unique properties to the polymer, such as antimicrobial activity and the ability to carry and slow-release functional ingredients. Charge density depends on the degree of deacetylation as well as pH. Quaternization of the nitrogen atoms of amino groups has been a usual chitosan modification, whose objective is to introduce permanent positive charges along polymer chains, providing the molecule with a cationic character independent of the aqueous medium pH (Aranaz et al. 2016).

Chitosan films have proved to be effective in extending the shelf life of fruit since they have retarded microbial growth on fruit surfaces. The polycationic structure of chitosan probably interacts with the predominantly anionic components (lipopolysaccharides, proteins) of microbial cell membranes, especially Gram-negative bacteria (De Azeredo et al. 2014; Helander et al. 2001).

Antimicrobial packaging has the ability to kill or inhibit the growth of microorganisms in order to extend the shelf life of perishable food while safeguarding its microbiological safety. This is achieved by creating unfavourable conditions in the environment of the microorganism, subtracting any essential component for development or simply by contacting a specific antimicrobial agent (Han 2013).

Knowing the microorganism that inhibits the antimicrobial selected, the material from which packaging is made and the food composition is central to consider the design of an antimicrobial package and determine its effectiveness and potential application.

All antimicrobial agents incorporated into the packaging material must be selected according to their action spectrum, mode of action and chemical composition, considering the growth rate of microorganisms to inhibit. The union of the active agents to the polymeric material must allow conserving the mode of action of the compound, since it is related to their antimicrobial activity. The additives present in the polymer, such as plasticizers, antistatic, lubricants or fillers should not interact unfavourably with the antimicrobial agent, since this interaction can adversely affect antimicrobial activity (Costa et al. 2011; Quintavalla and Vicini 2002).

Techniques for Measuring Antimicrobial Packaging Capacity

There are few standardized methods to measure the antimicrobial activity of polymeric materials. The best known is the methodology JIS Z 2801 which has been adopted as an International Organization for Standardization (ISO) procedure, ISO 22196. This standard determines the antibacterial activity and efficacy of antimicrobial agents in plastic surfaces and other nonporous surfaces. Antibacterial activity is measured by quantifying the survival of bacterial cells which have been held in close contact for 24 h at 35 °C with a surface that contains an antibacterial agent. The antibacterial effect is measured by comparing the survival of bacteria on a treated material with that achieved on an untreated material. The general procedure is shown in Fig. 3 (Japanese Standards Association 2000).

Alternately, to evaluate the effectiveness of an antimicrobial packaging, most studies have been carried out using culture media, which provide favourable conditions for the development of microorganisms and also for the actions of antimicrobial packaging conditions.

Among the group of microbiological tests in culture media, the most popular is the diffusion assay plate *in vitro*, which demonstrates the effectiveness of antimicrobial polymers. This is based on the material ability to suppress or inhibit the growth of selected microorganisms when

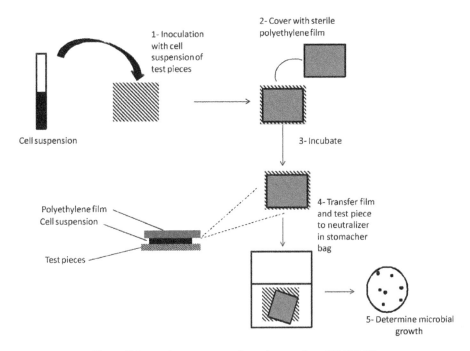

Fig. 3. Schematic representation of procedure JIS Z 2801.

the polymer is placed in direct contact with a solid agar surface previously inoculated with the target microorganism. This method attempts to simulate food wrapping and may represent what will happen when the material comes into contact with contaminated food surface and carry out the migration of the bioactive agent. In this methodology, incorporating appropriate reference samples and keeping a strict control of the repeatability of the test are very/ particularly important.

Other methodologies used include: determination of the minimum inhibitory concentration (MIC) and dynamic tests or shake flasks. On MIC assays, the antimicrobial potency of the material can be determined quantitatively and this can be compared to the corresponding free bioactive agent or starting material. The assay basically consists of planting, in different tubes, an inoculum of the target microorganism, together with the material to be tested containing different ratios of the antimicrobial agent. The MIC is the concentration at which total inhibition of microbial growth is observed.

The dynamic tests provide more detailed information on the antimicrobial polymer kinetics and growth of microorganisms. For this, liquid media (buffer, culture media or food) are inoculated with the target microorganism and placed in contact with the polymer to be evaluated. The system is slightly shaken for a given time and stored under prescribed conditions, which may represent the possible application of material. At different times samples of the liquid medium and seeded agar plate are taken to determine the decline in the population of microorganisms based on time (Appendini and Hotchkiss 2002).

Usually, the growth of microorganisms in the culture media is not the same as that on the surface of food, since in the later, the absence of certain nutrients and stress conditions suffered by microorganisms often affects development. Furthermore, the interaction between antimicrobial agents, microorganisms and component substances of the food often alters the effectiveness of bioactive systems in relation to that observed in the tests performed in culture media. For these reasons, it is essential that experiments determine the effectiveness of antimicrobial packaging in real food (Han 2005; Ollé Resa et al. 2014). Some antecedents where the capacity of antimicrobial active packaging has been evaluated on real food matrices include: measurement of antimicrobial efficiency of the coextruded polyamide/polyethylene film coated with a lacquer containing nisin and natamycin in two types of traditional cheese (Hanušová et al. 2010). Nisin was also applied to LDPE films in smoked salmon and hot dogs (Han 2013). Hauser and Wunderlich tested the antimicrobial efficacy of films coated with a lacquer containing sorbic acid, applying the test strains *Escherichia coli*, *Listeria monocytogenes* and *Saccharomyces cerevisiae* according to the Standard Method JIS Z 2801:2000. Storage tests were performed with Gouda cheese and inoculated pork loin (Hauser and Wunderlich 2011). Biopolymers of starch and chitosan with grape seed extracts were tested against *B. thermosphacta* and *L. monocytogenes* in pork loins and mortadella sausage (Realini and Marcos 2014).

Nanotechnology Biopackaging

Nanotechnology is a powerful, highly interdisciplinary tool for the development of innovative products. In view of its growing interest worldwide, nanotechnology is expected to provide a major boost to the development of new advanced systems to meet the needs of consumers. This type of technology involves exploitation of materials with one or more dimensions lower than 100 nm. Typical nanomaterials can be classified into three main types: particles, platelets and fibres. Due to of their nanosize, these materials have a very large surface/volume ratio that increases functional activity. The mixture of compatible polymers with nanomaterials can drastically improve the properties of the resulting material, such as enhanced mechanical strength, thermal stability, electrical conductivity and barrier food. Therefore, nanomaterials are promising for the development of advanced structures for application in the preparation of active and intelligent packaging.

As mentioned above, the greatest shortcoming in biopolymer materials is their weakest barrier and mechanical properties. This deficit in the biopolymer prevents the use of many packaging applications in food. Nanotechnology has proved to be a useful tool to improve these disadvantages by optimizing the preparation of nanocomposites. Nanocomposite preparation refers to that which is comprised as a single or a mixture of polymers with at least one organic or inorganic filler having a dimension lower than 100 nm. For example, montmorillonite (MMT) was added to the alginate biopolymer in various loading contents (1, 3 and 5 wt%). The addition of MMT decreased the water solubility (WS) of the film and caused an marked reduction in the water vapour permeability of alginate-based nanocomposites (Abdollahi et al. 2013).

However, when producing biopolymer-based nanocomposites, special attention must be directed to the use of nanoreinforcements to ensure that the final material is biodegradable and solely based on renewable resources (Petersson and Oksman 2006).

Some authors reported beneficial effects by adding chitin whiskers to biopolymer films. The chitin whiskers added to soy protein isolate (SPI) by Lu et al. (2004) greatly improved the tensile properties and water resistance of the matrix. Sriupayo et al. (2005) reported that whiskers improved the water resistance of chitosan films and enhanced their tensile strength up to a content of 2.96%, but impaired strength at higher contents. Similarly, Chang et al. reported that chitin nanoparticles were uniformly dispersed and showed a good interaction with a starch matrix at low loading levels (up to 5%), improving its mechanical, thermal and barrier properties. Such effects were ascribed to close interactions between the nanoparticles and the matrix, due to their chemical similarities. However, aggregation of nanoparticles occurred at higher loading levels, impairing the performance of the matrix (Chang et al. 2010).

On the other hand, chitosan nanoparticles rather than chitin are added more frequently as antimicrobials. In addition, they proved to be effective as nanoreinforcement to bio-based films. The incorporation of chitosan tripolyphosphate (CS-TPP) nanoparticles significantly improved the mechanical and barrier properties of the hydroxypropyl methylcellulose (HPMC) films (de Moura et al. 2009). The authors attributed such effects to the nanoparticles filling discontinuities in the HPMC matrix. However, higher nanoparticle loads (8% w/w) resulted in their aggregation in the nanocomposites, impairing the physical properties of the materials (De Azeredo et al. 2014).

The biodegradable starch/clay nanocomposite films using SiC were developed and used as food packaging material (Avella et al. 2005). These bionanocomposites showed a decrease in oxygen permeability with enhancement in thermal and chemical resistant properties (Dash and Swain 2013).

Cellulose whiskers are an example of a new generation of natural reinforcing materials in composites that can be produced from native cellulose (the most abundant biopolymer) by strong acid hydrolysis. These nanoparticles have gained considerable interest within the biopolymer community due to their cellulose renewable nature, abundance and good mechanical properties and to the large specific surface area of the nanowhiskers and their compatibility with biopolymers (Siró and Plackett 2010). There are interesting examples of the advantages of cellulose used as reinforcers. One includes the work of Hu and Wang (Hu and Wang 2016) reporting the blend of polyvinyl alcohol (PVA) and MCC bionanocomposites. PVA is among the most commonly used synthetic polymers due to its favorable properties such as nontoxicity, biocompatibility and film-forming ability. It is an environmentally friendly polymer easily biodegraded by adapted microorganisms in CO_2 and water, despite the fact that high hydrophilicity makes it highly sensitive to moisture attack. Otherwise, the absence of antibacterial properties makes it unable to provide a barrier against bacteria. This interesting approach involved the combination of microcrystalline cellulose (MCC) grafted with 3-chloro-2-hydroxypropyltrimethylammonium chloride (CHPTAC) for preparing quaternized cellulose blended with a polyvinyl alcohol (PVA) matrix to produce composite films via co-regeneration from the alkaline solution. These films showed improved surface roughness, hydrophobicity and water swelling ratio, and specifically, antibacterial activity, indicating that the biocompatible blend film will become an exceptional alternative in the functional bio-material field (Hu and Wang 2016).

Other green pathways for fabricating the nanocomposite renewable natural polymer and inorganic moieties involve a good option for replacing some petroleum-based polymers in packaging and other applications. Eco-friendly nanocomposite films with enhanced functional performance were developed by solution casting from the combination of nanoscale hydroxyapatite synthesized from eggshell waste and protein-based polymer extracted from defatted soybean residues. Significant improvements in tensile modulus and strength were achieved (Rahman et al. 2016).

Other bio-nanocomposite films based on soy protein isolate (SPI) and montmorillonite (MMT) were prepared using melt extrusion. A significant improvement was seen in their mechanical (tensile strength and percent elongation at break) and dynamic mechanical properties (glass transition temperature and storage modulus), thermal stability and water vapour permeability (Kumar et al. 2010).

Nanocomposite materials have been investigated for antimicrobial activity such as growth inhibitors, antimicrobial agents, antimicrobial carriers or antimicrobial packaging films. The main potential food applications for antimicrobial nanosystem films include meat, fish, poultry, bread, cheese, fruit and vegetables (Rhim et al. 2013).

Recently there has been increasing/particular/considerable interest in this type of active packaging, employing antimicrobial agents as carbon nanotubes or metal nanoparticles. Silver, gold and zinc nanomaterials have proved antimicrobial activity, being most commonly used in this type of materials; silver is especially found in many commercial applications, given its high stability and efficiency compared with those of about 150 different species of bacteria (Kumar and Münstedt 2005). These types of systems act by direct contact with the product or by their slow migration and subsequent action (Silvestre et al. 2011).

Metal and metal oxide nanoparticles are the nanosystems most widely used to develop antimicrobial active packaging. These particles function in direct contact; however, they can also migrate slowly and react preferentially with organic compounds in the food.

One of the nanocomposites most widely studied involves those prepared with silver nanoparticles. The use of silver (Ag) in antimicrobial packaging offers numerous advantages including high thermal stability and ease of incorporation into or onto numerous materials. This component has been classified as GRAS (Generally Recognized as Safe) by the Food and Drug Administration (FDA) to be used as a food preservative; however, the European Food Safety Authority (EFSA) has classified it as one from a list of additives with general restrictions. Silver antimicrobial nanocomposite films have been used for packaging with horticultural products, such as Chinese jujube, lettuce, fresh fruit salad, asparagus and fresh orange juice (Costa et al. 2011). Several studies have demonstrated the antimicrobial activity of silver nanoparticles included in chitosan, in PVA with cellulose nanocrystals or polyethylene (PE) films that are effective against both Gram-positive and Gram-negative bacteria, fungi, protozoa and certain viruses (C. Costa et al. 2011; Costa et al. 2012; Cruz-Romero et al. 2013; Gammariello et al. 2011; Mihindukulasuriya and Lim 2014).

Gammariello and co-workers evaluated the effects of an alginate-based coating containing Ag-MMT nanoparticles on the shelf-life of *Fior di latte* cheese with modified atmosphere packaging (Gammariello et al. 2011).

Titanium dioxide (TiO_2), zinc oxide (ZnO), silicon oxide (SiO_2) and magnesium oxide (MgO) are the oxide nanoparticles most widely studied for their ability to be UV blockers and photocatalytic disinfecting agents (Fujishima et al. 2000). It is well known that photocatalyst titanium dioxide (TiO_2), a wide band gap (3.2 eV) semiconductor under UV illumination, generates energy-rich electron hole pairs that can be transferred to the surface of TiO_2 and promote reactivity with the surface-absorbed molecules leading to the production of active radicals. These active radicals oxidize C-H bonds of polyunsaturated phospholipids in cell membranes of microorganisms resulting in their degradation. The photocatalyst reaction of TiO_2 leads to generation of hydroxyl radicals ($\cdot$OH) and reactive oxygen species (ROS) on the TiO_2 surface, leading to oxidation of polyunsaturated phospholipids in the cell membranes of microorganisms. As a consequence, the microorganism is inactivated. TiO_2 has been approved by the FDA as a non-toxic material for human food, drug and cosmetic and use in the food industry. Currently, there is considerable interest in the self-disinfecting property of TiO_2 for meeting hygienic design requirements in food processing and packaging surfaces. TiO$_2$ photocatalyst has been used to degrade organic pollutants and inactivate a wide spectrum of microorganisms. Some interesting studies have demonstrated the efficiency of TiO$_2$ nanoparticles. These results are of industrial relevance since an effective technique was put forward for the development a novel photocatalyst thin film as an active packaging system (Bodaghi 2013).

The antimicrobial properties of nano-ZnO and MgO have been reported (Silvestre et al. 2011). These nanoparticles are expected to provide a more affordable and safe food packaging solution in the future. It was demonstrated that ZnO nanoparticles exhibit better antibacterial activity when particle size decreases (Yamamoto 2001) and their activity does not require the presence of UV light but the presence of visible light (Jones et al. 2008). This class of nanoparticles has been incorporated in a number of different polymers (Azeredo 2013; Silvestre et al. 2011). Premanathan et al. (2011) reported antimicrobial activity of ZnO nanoparticles against *Listeria monocytogenes*, *Salmonella enteritidis* and *E. Coli*, indicating that it was essential for nanoparticles to contact or penetrate into microbial cells. They were bacteriostatic or bactericidal depending on the concentration. Composite chitosan-ZnO nanofibres were obtained using PVA as a support. The presence of nano-ZnO in the nanofibers was demonstrated to contribute to the antimicrobial activity of chitosan. The co-antimicrobial effect of chitosan and nano-ZnO was attributed to a sequence of effects, including: adherence of nanofibers to cell membranes through electrostatic attraction, leading to changes in microbial cell membranes (denaturation of proteins and permeability changes); disruption of cell membranes by ROS produced by nano-ZnO, causing leakage of intercellular contents; interference in cell metabolism by nano-ZnO and chitosan combination with DNA and RNA, blocking genome replication (Cruz-Romero et al. 2013).

On the other hand, the bactericidal mechanism of MgO nanoparticles was related to the production of high concentrations of superoxide anions (O$_2$-) on MgO surface, which can react with the carbonyl groups of peptide linkages in bacterial cell walls and destroy them.

Chitosan-magnesium oxide-based nanocomposite film containing clove essential oil was prepared by solution cast method. In this study, Sanuja et al. (Sanuja et al. 2014) showed that the incorporation of MgO in these nanocomposites increased mechanical property, film thickness, film opacity, and decreased swelling, barrier and solubility properties. Further antibacterial activity by disc diffusion method was followed against Staphylococcus aureus which was also found to have a higher inhibition effect for clove oil and nanoparticle incorporated chitosan film. Thus it will be an ideal material for active food packaging applications.

The effectiveness of MgO nanowires as antimicrobial agents has been demonstrated by Al-Hazmi et al. (2012). They showed increasing activity against *E. coli* and *Bacillus sp.*, with increasing concentration of MgO nanowires (Al-Hazmi et al. 2012).

Biopolymer nanoparticles or whiskers are usually used for their intrinsic antimicrobial properties, as chitosan or nanocarriers of antimicrobial compounds. Chitosan nanoparticles (CSNP) and their derivatives have far-reaching antimicrobial effects. The most accepted antimicrobial mechanism for chitosan relates to interactions of cationic chitosan with anionic cell membranes, increasing membrane permeability and eventually resulting in rupture and leakage of the intracellular material (Zhao et al. 2011).

Packaging and Society

Packaging for foodstuff is susceptible to continuous changes in the form of marketing, globalization and consumer's lifestyle. This dynamic environment implies that packaging technologies are affected by a large number of standards, guidelines and codes which must be complied with. As a result, society receives a set of benefits that seek to improve its quality of life. Some of the benefits attributed to packaging include:

- Prevention or decrease of product damage and food spoilage, leading to reduction of solid food waste.
- Decrease or elimination of risk of tampering and adulteration.
- Presentation of food in a hygienic and often aesthetically attractive way.
- Communication of important food information which helps consumers to make informed purchase choices.
- Promotion of goods in a competitive marketplace and increase in consumer's choice.
- Energy saving through the use of ambient packs that do not require refrigeration or frozen distribution and storage.

However, both society and industry packaging are aware of the controversy that involves the growth of new packaging technologies and the increase in production volume of food containers. This controversy mostly includes environmental issues and concerns for food security (Jansson-Boyd 2010; Koutsimanis et al. 2012; Richard et al. 2003).

Today there is a wide range of packaging with very different compounds and characteristics which are trying to meet the demand of the large number of food products on the market. Therefore it is necessary to select, for each use, the package and the most appropriate technology on the basis of different parameters. In this context, the task of carrying out innovations in food packaging emerges as a priority, which entails new packaging paradigms aimed at satisfying demands generated by products, consumers, markets, regulations, competition and the environment (Fig.4). These innovations in packaging technologies include: controlled atmosphere and modified packaging and, recently, biodegradable packaging or biopackaging for sensible food, and active and intelligent packaging.

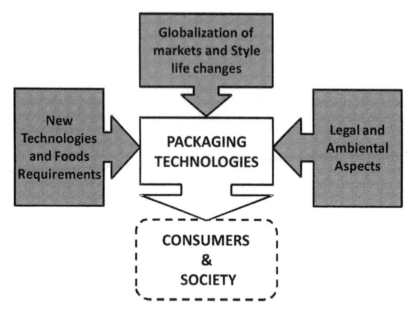

Fig. 4. Context for the development of new technologies for food packaging.

Conclusion

The biopackaging field has adopted technologies to improve the performance and/or add functionalities to biopolymer-based materials; hence, their range of application can be widened, and they can thus become more competitive in the polymer market. Moreover, nanotechnology has demonstrated a great potential to expand the use of biodegradable polymers, since the preparation of bionanocomposites has led to improvements in the overall performance of biopolymers, making them more competitive in a market dominated by non-biodegradable materials.

However, the creation of an "ideal" packaging with active and intelligent properties still needs further research and development to address several challenges. There are still some problems that need to be solved where knowledge about the structure/property relationship of different polymer platforms could be extended. A further important issue is to minimize the impact of environmental conditions on material performance. Finally, biomaterials must be compatible with their counterparts from petrochemicals in terms of price, quality and variety of use.

Through all civilizations, the antimicrobial function of a container has been a need in the area of food. It is known that the addition of antimicrobial compounds to films can effectively inhibit or delay the growth of microorganisms that may be present on the surface of food products. Some antibacterial compounds including antibiotics, essential oils and inorganic nanoparticles could be directly incorporated or covalently bonded into the packaging material. The second option is more convenient since it guarantees consumers that the antimicrobial agent will not be incorporated into the food. However, the toxic substance will go into the environment from the waste film and cause serious secondary pollution. Hence, all antibacterial systems need to be combined with other characteristics, such as environmental friendliness, low toxicity or non-toxicity, cost efficiency and easy incorporation.

References

Abdollahi, M., M. Alboofetileh, M. Rezaei and R. Behrooz. 2013. Food hydrocolloids comparing physico-mechanical and thermal properties of alginate nanocomposite films reinforced with organic and/or inorganic nanofillers. 32: 416–424.

Al-Hazmi, F., F. Alnowaiser, A.A. Al-Ghamdi, Attieh A. Al-Ghamdi, M.M. Aly, R.M. Al-Tuwirqi, et al. 2012. A new large – scale synthesis of magnesium oxide nanowires: Structural and antibacterial properties. Superlattices Microstruct. 52: 200–209.

Appendini, P. and J.H. Hotchkiss. 2002. Review of antimicrobial food packaging. Innov. Food Sci. Emerg. Technol. 3: 113–126.

Aranaz, I., R. Harris, F. Navarro-García, A. Heras and N. Acosta. 2016. Chitosan based films as supports for dual antimicrobial release. Carbohydr. Polym. 146: 402–410.

Avella, M., J.J. DeVlieger, M.E. Errico, S. Fischer, P. Vacca and M.V. Grazia. 2005. Biodegradable starch/clay nanocomposite films for food packaging applications. Food Chem. 93: 467–474.

Azeredo, H.M.C. 2013. Antimicrobial nanostructures in food packaging. Trends Food Sci. Technol. 30: 56–69.

Bastarrachea, L., D. Wong, M. Roman, Z. Lin and J. Goddard. 2015. Active packaging coatings. Coatings 5: 771–791.

Bodaghi, H. 2013. Evaluation of the photocataytic antimicrobial effects of a TiO_2 nanocomposite food packaging film by *in vitro* and *in vivo* tests. Food Sci. Technol. 50: 702–706.

Brody, A.L., B. Bugusu, J.H. Han, C.K. Sand and T.H. McHugh. 2008. Innovative food packaging solutions. J. Food Sci. 73: 107–116. doi:10.1111/j.1750-3841.2008.00933.x

Chang, P.R., J. Ruijuan, Y. Jiugao and M. Xiaofei. 2010. Starch-based composites reinforced with novel chitin nanoparticles. Carbohydr. Polym. 80: 420–425.

Coles, R., D. Mc Dowell and M. Krwan. 2003. Food Packaging Technology. CRC Press, London.

Costa, C., A. Conte, G.G. Buonocore and M.A. Del Nobile. 2011. Antimicrobial silver-montmorillonite nanoparticles to prolong the shelf life of fresh fruit salad. Int. J. Food Microbiol. 148: 164–167.

Costa, C., A. Conte, G.G. Buonocore, M. Lavorgna and M.A. Del Nobile. 2012. Calcium-alginate coating loaded with silver-montmorillonite nanoparticles to prolong the shelf-life of fresh-cut carrots. Food Res. Int. 48: 164–169.

Costa, F., I.F. Carvalho, R.C. Montelaro, P. Gomes and M.C.L. Martins. 2011. Covalent immobilization of antimicrobial peptides (AMPs) onto biomaterial surfaces. Acta Biomater. 7: 1431–1440.

Cruz-Romero, M.C., T. Murphy, M. Morris, E. Cummins and J.P. Kerry. 2013. Antimicrobial activity of chitosan, organic acids and nano-sized solubilisates for potential use in smart antimicrobially-active packaging for potential food applications. Food Control 34: 393–397.

Dainelli, D., N. Gontard, D. Spyropoulos, E. Zondervan-van den Beuken and P. Tobback. 2008. Active and intelligent food packaging: Legal aspects and safety concerns. Trends Food Sci. Technol. 19: S103–S112.

Dash, S. and S.K. Swain. 2013. Synthesis of thermal and chemical resistant oxygen barrier starch with reinforcement of nano silicon carbide. Carbohydr. Polym. 97: 758–763.

De Azeredo, H.M.C., M. Rosa, M. Souza Filho and K.W. Waldron. 2014. The use of biomass for packaging films and coatings. pp. 819–874. *In*: Waldron, K. (ed.). Advances in Biorefineries. Woodhead Publishing Limited, Langford Lane. doi:10.1533/9780857097385.2.819

De Moura, M.R., F.A. Aouada, R.J. Avena-Bustillos, T.H. McHugh, J.M. Krochta and L.H.C. Mattoso. 2009. Improved barrier and mechanical properties of novel hydroxypropyl methylcellulose edible films with chitosan/tripolyphosphate nanoparticles. J. Food Eng. 92: 448–453.

Dobrucka, R. 2013. LogForum. Sci. J. Logist. 9: 103–110. URL: http://www.logforum.net/vol9/issue2/no4

Fabra, M.J., M.A. Busolo, A. Lopez-Rubio and J.M. Lagaron. 2013. Nanostructured biolayers in food packaging. Trends Food Sci. Technol. 31: 79–87.

Fakhouri, F.M., S.M. Martelli, T. Caon, J.I. Velasco and L.H. Innocentiniei. 2015. Edible films and coatings based on starch/gelatin: Film properties and effect of coatings on quality of refrigerated Red Crimson grapes. Postharvest Biol. Technol. 109: 57–64.

Fujishima, A., T.N. Rao and D.A. Tryk. 2000. Titanium dioxide photocatalysis. J. Photochem. Photobiol. C. Photochem. Rev. 1: 1–21.

Gammariello, D., A. Conte, G.G. Buonocore and M.A. Del Nobile. 2011. Bio-based nanocomposite coating to preserve quality of Fior di latte cheese. J. Dairy Sci. 94: 5298–5304.

Goddard, J.M. and J.H. Hotchkiss. 2007. Polymer surface modification for the attachment of bioactive compounds. Prog. Polym. Sci. 32: 698–725.

González, A. and C.I. Alvarez Igarzabal. 2013. Soy protein—poly(lactic acid) bilayer films as biodegradable material for active food packaging. Food Hydrocoll. 33: 289–296.

González, A., M. Strumia and C. Alvarez Igarzabal. 2015. Modification strategies of proteins for food packaging applications. pp. 127–147. *In*: Cirillo, Giuseppe, Umile Gianfranco Spizzirri and Francesca Iemma (eds.). Functional Polymers in Food Science: From Technology to Biology, Volume 1: Food Packaging. John Wiley & Sons, Inc., Beverly, MA.

Han, J. 2005. Innovations in Food Packaging. Elsevier Science & Technology Books, Toronto.

Han, Jung H. 2013. Plastic films in food packaging. pp. 151–180. *In*: Ebnesajjad, S. (ed.). Plastic Films in Food Packaging. Elsevier Inc., Walham, MA. doi:10.1016/B978-1-4557-3112-1.00010-7

Hanušová, K., M. Šťastná, L. Votavová, K. Klaudisová, J. Dobiáš, M. Voldřich, et al. 2010. Polymer films releasing nisin and/or natamycin from polyvinyldichloride lacquer coating: Nisin and natamycin migration, efficiency in cheese packaging. J. Food Eng. 99: 491–496.

Hauser, C. and J. Wunderlich. 2011. Antimicrobial packaging films with a sorbic acid based coating. Procedia Food Sci. 1: 197–202.

Helander, I.M., E.L. Nurmiaho-Lassila, R. Ahvenainen, J. Rhoades and S. Roller. 2001. Chitosan disrupts the barrier properties of the outer membrane of gram-negative bacteria. Int. J. Food Microbiol. 71: 235–244.

Hu, D. and L. Wang. 2016. Physical and antibacterial properties of polyvinyl alcohol films reinforced with quaternized cellulose. J. Appl. Polym. Sci. 133: 1–8.

Jansson-Boyd, C.V. 2010. Consumer Psychology. McGraw Hill: Open University Press.

Japanese Standards Association, 2000. JIS Z 2801: 2000.

Javidi, Z., S.F. Hosseini and M. Rezaei. 2016. Development of flexible bactericidal films based on poly(lactic acid) and essential oil and its effectiveness to reduce microbial growth of refrigerated rainbow trout. LWT – Food Sci. Technol. 72: 251–260.

Jones, N., B. Ray, K.T. Ranjit and A.C. Manna. 2008. Antibacterial activity of ZnO nanoparticle suspensions on a broad spectrum of microorganisms. FEMS Microbiol. Lett. 279: 71–76.

Koutsimanis, G., K. Getter, B. Behe, J. Harte and E. Almenar. 2012. Influences of packaging attributes on consumer purchase decisions for fresh produce. Appetite 59: 270–280.

Kumar, P., K.P. Sandeep, S. Alavi, V.D. Truong and R.E. Gorga. 2010. Preparation and characterization of bio-nanocomposite films based on soy protein isolate and montmorillonite using melt extrusion. J. Food Eng. 100: 480–489.

Kumar, R. and H. Münstedt. 2005. Silver ion release from antimicrobial polyamide/silver composites. Biomaterials 26: 2081–2088.

Lagaron, J.M., A. Lopez-Rubio and M. Jose Fabra. 2016. Bio-based packaging. J. Appl. Polym. Sci. 133: 1–4.

Lim, L.T. 2011. Active and intelligent packaging materials. pp. 629–644. *In*: Moo-Young, M. (ed.). Comprehensive Biotechnology, 2nd Ed. Elsevier, Amsterdam.

Llana-Ruiz-Cabello, M., S. Pichardo, A. Baños and C. Núñez. 2015. Characterisation and evaluation of PLA films containing an extract of Allium spp. to be used in the packaging of ready-to-eat salads under controlled atmospheres. Food Sci. Technol. 64: 1354–1361.

Marsh, K. and B. Bugusu. 2007. Food packaging: Roles, materials, and environmental issues. J. Food Sci. 72: 39–55.

Mihindukulasuriya, S.D.F. and L.-T. Lim. 2014. Nanotechnology development in food packaging: A review. Trends Food Sci. Technol. 40: 149–167.

Ollé Resa, C.P., R.J. Jagus and L.N. Gerschenson. 2014. Natamycin efficiency for controlling yeast growth in models systems and on cheese surfaces. Food Control 35: 101–108.

Petersson, L. and K. Oksman. 2006. Biopolymer based nanocomposites: Comparing layered silicates and microcrystalline cellulose as nanoreinforcement. Compos. Sci. Technol. 66: 2187–2196.

Premanathan, M., K. Karthikeyan, K. Jeyasubramanian and G. Manivannan. 2011. Selective toxicity of ZnO nanoparticles toward Gram-positive bacteria and cancer cells by apoptosis through lipid peroxidation. Nanomedicine 7: 184–192.

Quintavalla, S. and L. Vicini. 2002. Antimicrobial food packaging in meat industry. Meat Sci. 62: 373–380.

Rahman, M.M., A.N. Netravali, B.J. Tiimob, V. Apalangya and V.K. Rangari. 2016. Bio-inspired "green" nanocomposite using hydroxyapatite synthesized from eggshell waste and soy protein. J. Appl. Polym. Sci. 133: 1–10.

Realini, C.E. and B. Marcos. 2014. Active and intelligent packaging systems for a modern society. Meat Sci. 98: 404–419.

Rhim, J., H. Park and C. Ha. 2013. Progress in polymer science bio-nanocomposites for food packaging applications. Prog. Polym. 38: 1629–1652.

Richard, C., M. Derek and K. Mark. 2003. Food Packaging Technology, 1st ed. Blackwell Publishing Ltd, Boca Raton, FL, USA.

Santacruz, S., C. Rivadeneira and M. Castro. 2015. Edible films based on starch and chitosan: Effect of starch source and concentration, plasticizer, surfactant's hydrophobic tail and mechanical treatment. Food Hydrocoll. 49: 89–94.

Sanuja, S., A. Agalya and M.J. Umapathy. 2014. Studies on magnesium oxide reinforced chitosan bionanocomposite incorporated with clove oil for active food packaging application. Int. J. Polym. Mater. Polym. Biomater. 63: 733–740.

Silveira, M.F., N.F.F. Soares, R.M. Geraldine, N.J. Andrade, D. Botrel and M.P.J. Gonçalves. 2007. Active film incorporated with sorbic acid on pastry dough conservation. Food Control 18: 1063–1067.

Silvestre, C., D. Duraccio and S. Cimmino. 2011. Food packaging based on polymer nanomaterials. Prog. Polym. Sci. 36: 1766–1782.

Siró, I. and D. Plackett. 2010. Biocomposite cellulose-alginate films: Promising packaging materials. Food Chem. 151: 343–351.

Sriupayo, J., P. Supaphol, J. Blackwell and R. Rujiravanit. 2005. Preparation and characterization of α-chitin whisker-reinforced chitosan nanocomposite films with or without heat treatment. Carbohydr. Polym. 62: 130–136.

Suppakul, P., K. Sonneveld, S.W. Bigger and J. Miltz. 2011. Diffusion of linalool and methylchavicol from polyethylene-based antimicrobial packaging films. LWT – Food Sci. Technol. 44: 1888–1893.

Tawakkal, S., M.J. Cran and S.W. Bigger. 2016. Release of thymol from poly(lactic acid)-based antimicrobial films containing kenaf fibres as natural filler. Food Sci. Technol. 66: 629–637.

Thakur, V. 2015. Surface modification of biopolymers. pp. 100–102. *In*: Thakur, V.K. and A.S. Singha (eds.). Surface Modification of Biopolymers. Wiley, Singapore.

Vanderroost, M., P. Ragaert, F. Devlieghere and B. De Meulenaer. 2014. Intelligent food packaging: The next generation. Trends Food Sci. Technol. 39: 47–62. doi:10.1016/j.tifs.2014.06.009

Yamamoto, O. 2001. Influence of particle size on the antibacterial activity of zinc oxide. Int. J. Inorg. Mater. 3: 643–646.

Zhao, L.M., L.E. Shi, Z.L. Zhang, J.M. Chen, D.D. Shi, J. Yang, et al. 2011. Preparation and application of chitosan nanoparticles and nanofibers. Brazilian J. Chem. Eng. 28: 353–362.

Transport Phenomena in Biodegradable and Edible Films

Tomy J. Gutiérrez [1,2,3]* and Kelvia Álvarez[1,2]

[1] Department of Analytical Chemistry, Faculty of Pharmacy, Central University of Venezuela (UCV), PO Box 40109, Caracas 1040-A, Venezuela

[2] Institute of Food Science and Technology (ICTA), Faculty of Sciences, Central University of Venezuela (UCV), PO Box 47097, Caracas 1041-A, Venezuela

[3] Thermoplastic Composite Materials (CoMP) Group, Institute of Research in Materials Science and Technology (INTEMA), Faculty of Engineering, National University of Mar del Plata (UNMdP) and National Council of Scientific and Technical Research (CONICET), Colón 10850, Mar del Plata 7600, Buenos Aires, Argentina

Introduction

Edible coatings and films have many desirable properties which have led to their use in the food industry. For example, they prevent and slowdown the exchange of moisture, gases and food components, serving as a means for incorporating additives such as antioxidants, antimicrobials and nutritionally valuable additives such as vitamins and minerals, which can improve the food quality and safety, thus extending the life of food (Skurtys et al. 2010).

This behavior improves the quality of food as a result of the mass transfer phenomena taking place within the film structure from polymeric materials of biological origin such as proteins, starch, alginate, chitosan, cellulose and its derivatives (Peelman et al. 2013).

The application of these materials for packaging creates a more complex ternary system (environment-material-food), in which the phenomena of mass transfer is shown as gas permeability, sorption of food components for biopolymer films and/or the release or transfer of components of biopolymers in foodstuffs.

*Corresponding author: tomy.gutierrez@ciens.ucv.ve; tomy_gutierrez@yahoo.es

Fundamentals of Transport Mechanisms

Mass transport phenomenon occurs due to the natural tendency of systems to reach equilibrium. Edible films or coatings, in most cases are exposed to various different conditions such as moisture and pressure, which can modify the structure of these materials, resulting in the modification of the chemical balances that coexist within the polymer matrix. For example, when a food is coated with a biopolymeric material at least four conditions can be identified (Fig. 1): 1) the external atmosphere (EA), 2) the film (P), 3) the internal atmosphere (IA) and 4) the food that may have one or more phases (F), trying to come into balance with each other.

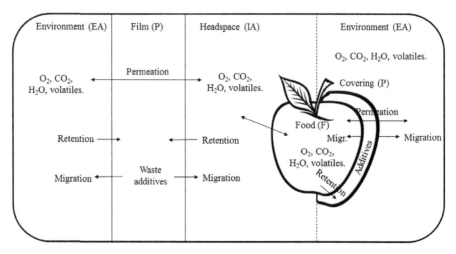

Fig. 1. Outline of mass transfer processes in edible films (left) and coatings (right).

Each phase is composed of numerous components, many of which have high mobility and low molecular weight such as oxygen, carbon dioxide, water, food flavor components or odors, and coating or film constituents. Other components may be considered non-mobile such as macromolecules that make up the food matrix or coating. Mobile components move within each phase and are transferred between the phases through the interfaces, until they occupy all phases, achieving a chemical balance. For this reason, transport phenomena can be considered complex consisting of multiple variables.

Chemical Balance: Solubility and Partition Coefficient

The chemical balance in accordance with the laws of thermodynamics is achieved when the dynamic potentials of each component ($"i"$, μ_i^α) in each of phases making up system (α = EA, IA, P or F) are equal (Hernández and Gavara 1999).

$$\mu_i^F = \mu_i^{EA} = \mu_i^P = \mu_i^{IA} \tag{1}$$

However, in these systems it is impractical to describe the balance in terms of the chemical potential of the components. As an alternative, it is preferable that they are expressed as ratios between concentrations and/or partial pressures of a compound between two phases. The choice depends mainly on the type of phase with which the exchange occurs. In the mass exchange between the film or coating and a gaseous phase (environment), the equilibrium can be expressed as the ratio of the concentration of compound in the condensed phase (film) and the partial pressure in air. This balance is shown experimentally by adsorption isotherms. Depending on the profile of the isotherm, it can be described by some different models, among others, the equations of Lanmuir, Flory-Huggins, B.E.T., G.A.B, Freundlich or Henry's Law. Henry's Law, is the simplest model and describes the concentration of a gaseous compound in a condensed medium ($c_i^{F \text{ or } P}$), which is directly proportional to the partial pressure of said compound in the gaseous medium to which the condensed medium is exposed ($p_i^{IA \text{ or } EA}$), where the constant of proportionality is the solubility coefficient (S_i):

$$C_i^{FoP} = S_i \cdot p_i^{IAoEA} \tag{2}$$

When S_i, is considered constant for a compound, only then will the polymer depend on temperature (T) according to the Van't Hoff Law:

$$S_i = S_i^o \cdot \exp\left(\frac{-\Delta H_S}{RT}\right) \tag{3}$$

where ΔH_s is the solubilization enthalpy, the constant S_i^o is the value of S at 0 K and R is the gas constant.

The solubility coefficient is widely used in the description of gas retention in synthetic polymers, since Henry's Law is satisfied in a reasonable way for a range of gas pressures. Also, it applies to the general description of the absorption of very volatile organic compounds in polymers, although in these systems Henry's Law usually shows deviations as result of the interactions between polymers and the penetrant. In hydrophilic polymers, the water sorption isotherm often deviates significantly from linearity, which generally has a profile type II according to the IUPAC classification (Fig. 2) (Perry and Green 1984). Generally, edible films and coatings have a strong hydrophilicity and often also show hysteresis phenomena in them; so the description of the isotherms can be of type II or type IV. It is worth noting that it is not usual to describe the sorption of vapours, especially water, through the solubility coefficient.

When the balance between two condensed phases is established, the partition constant or partition coefficient (K) describes the distribution of compound between both phases, for example, film (P) and food (F):

$$K_i = \frac{c_i^P}{c_i^F} \tag{4}$$

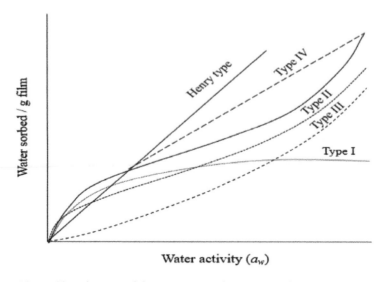

Fig. 2. Classification of the sorption isotherms according to IUPAC.

The partition coefficient also depends on the temperature according to the Van't Hoff Law (Eq. 3). Where K_i o S_i, allow to estimate the magnitude of the mass transfer process in the hypothetical case that the system reaches equilibrium. Similarly, high values of these constants indicate that the compounds are retained primarily in the film or coating. In the same way, the speed of advance towards equilibrium depends on the kinetics of the process.

Kinetic Aspects: Diffusion and Permeability Coefficient

Within each of the phases making up system, the substances present move at different rates. In gaseous phases, or in liquid medium fluids, the diffusion kinetics is so fast, that solutes are found to be homogeneously distributed. In solid mediums or viscous liquids, the substances diffuse more slowly and in general are considered to follow the Fick Laws (Crank 1975), where the penetration flow is J for films or coatings in the direction of its thickness (z), is proportional to the concentration gradient along the thickness ($\partial c_i / \partial z$):

$$J_i = \frac{q_i}{A.t} = -D_i \frac{\partial c_i^P}{\partial z} \; ; \; \frac{\partial c_i^P}{\partial t} = D_i \left(\frac{\partial^2 c_i^P}{\partial z^2} \right) \tag{5}$$

The constant of proportionality is the diffusion coefficient (D_i) which is generally considered constant for a solute and solid matrices, and is dependent on the temperature according to the Arrhenius Law:

$$D_i = D_i^0 . \exp \left(\frac{-E_D}{RT} \right) \tag{6}$$

where E_D is the activation energy for the diffusion process. However, it has been observed that the D value also depends on the thermo-mechanical conditions under which the film was obtained. High values of D indicate rapid diffusion rates.

Clearly, both the extent and the speed, or the thermodynamic and the kinetic aspects together, affect the mass transfer processes and must be taken into account (Salame 1989).

Moreover, the applications of edible films and coatings are designed to slow the natural processes between the environment and food. Film systems are opposed to balance. For this reason, the process kinetics acts as a limiting factor, establishing a new equilibrium at the interfaces food/film (or coating) and film/environment. Under these conditions, the process reaches a stationary state in which the concentration gradient in the film ($\partial c_i/\partial z$) is constant and can be re-written using Fick's first Law (Eq. 5) as follows:

$$J_i = \frac{q_i}{A.t} = -D_i \frac{\partial c_i^P}{\partial z} = -D \frac{c_i - c_e}{l} \tag{7}$$

where c_i and c_e are the concentrations of the substance transported at the outer end and inside the film and ℓ, is its thickness. The application of the Henry's Law with the above equation can be expressed as a function of partial pressures of the gas in the outer and interior environments, p_i and p_e:

$$J_i = \frac{qi}{A.t} = -D_i^P \frac{c_i - c_e}{\ell} = -D_i^P \frac{s_i p_i - s_i p_e}{\ell} = -D_i^P S_i \frac{\Delta p}{\ell} \tag{8}$$

The product of the solubility and the coefficients of diffusion is known as the permeability coefficient, re-written as:

$$J_i = \frac{q_i.\ell}{A^P.\Delta p.t} \tag{9}$$

Permeability is the amount of substance that crosses a film of thickness (ℓ) per unit area, time and concentration gradient. This coefficient is an intensive magnitude for a material and a permanent gas, being independent of the shape or thickness of the object manufactured, and, generally considered exclusively dependent on the temperature according to the Arrhenius Law. Nonetheless, it must be remembered that in the definition of permeability the compliance of Henry's Law and constancy of D in Fick's Law has been considered. Therefore, in certain polymeric systems/permanent, specifically in edible biopolymers where moisture effect causes large deviations from these laws. In general, the values of effective permeability are determined.

Transport through Edible Film and Coatings

Transport phenomena in biomaterials applied as packaging depends on many factors, which are studied from certain tests described below:

Factors Affecting the Transport

The mechanisms of mass transfer depends on many factors, including molecular structure of biopolymers, for example, materials of biopackaging based on proteins increase their cohesive forces due to the hydrogen bond interactions, which reduce the diffusion of polar compounds, since they are retained due to intermolecular forces established between the biopolymer macromolecules and polar additives incorporated. By contrast, in materials lacking such interactions, the diffusion of polar molecules is favored. For example, Fig. 3 shows how the *yerba mate* extract diffuses faster towards the medium compared to the propolis extract (Chang-Bravo et al. 2014).

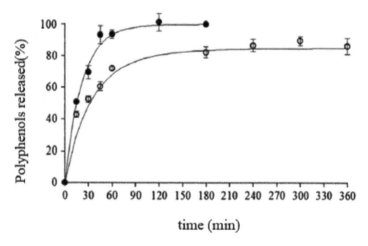

Fig. 3. Release kinetics of the extracts of propolis (○) and yerba mate (•) in aqueous medium. Symbols represent experimental data, while lines corresponding to predicted values by the first-order kinetic model.

Likewise, it is well known that most polymers are mainly constituted by an amorphous phase, which may contain a crystalline phase in a minor proportion. This crystalline phase can create physical barriers, thus impeding mass transport. In the same way, in the amorphous regions, microcavities can be created which distort the sorption and permeation of low molecular weight compounds in these materials. Similarly, as these behaviors are dependent on the characteristics of each biopolymer, therefore, the transport phenomena also depend on type of biopolymer used for the production of food packaging (Lagaron et al. 2004).

For example, for a particular material, the transport phenomena increases with decreasing the crystallinity and the cohesive forces, as well as when the free volume and chain flexibility increases. The addition of plasticizers has these effects while the use of cross-linking agents has the opposite effect.

Likewise, penetration properties of an additive within the chains of biopolymers change these processes. For example, chemical compatibility with the polymeric matrix increases the solubility of the permeant in the film, and thereby increases the coefficients S or K. In the same manner, water absorption in hydrophilic materials as polysaccharides is very high. On the contrary, the water absorption is smaller in hydrophobic materials from lipids, which also affects the transport properties. Therefore, the size and molecular form are factors which modify the diffusion of a substance. Small and flexible molecules have greater mobility through the free volume of the polymer molecules than the large and rigid molecules.

There are also important environmental factors affecting the transport properties in edible and biodegradable films such as moisture, temperature and type of food. Moisture can be absorbed into hydrophilic polymers, due to the hydrogen bond interactions that are established with the water adsorbed from the environment. This results in the adsorbed water being able to compete with the plasticizer of thermoplastic materials based on biopolymers, which could modify the transport properties. The same effect can occur when there is a high level of compatibility with food or with any of its major components. With respect to temperature, an increase of this parameter produces an acceleration of all kinetic processes, and thus increases the diffusion coefficient and mass transfer properties.

In the same sense, the condensation (low volatility) of a substance on the surface of a biodegradable film also increases its solubilization. For this reason, carbon dioxide is much more absorbed (than is oxygen) in the polymers (Lewis et al. 2003).

A representative scheme of the possible transport phenomena that can occur in a thermoplastic based on biopolymers can be seen in Fig. 4 (Belbekhouche et al. 2011).

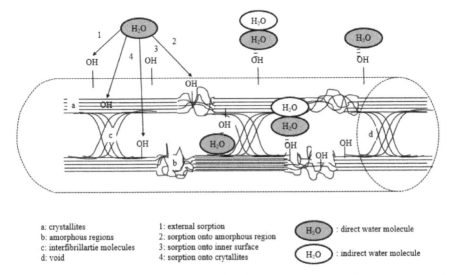

a: crystallites
b: amorphous regions
c: interfibrillartie molecules
d: void

1: external sorption
2: sorption onto amorphous region
3: sorption onto inner surface
4: sorption onto crytallites

H_2O : direct water molecule

H_2O : indirect water molecule

Fig. 4. Schematic diagram of direct and indirect moisture sorption onto external surface (1), amorphous regions (2), inner surface of voids (3) and crystallites (4).

Methods for Characterization of the Mass Transfer Processes

To characterize the barrier properties different methods have been designed to simplify the number of variables and to allow the repeatability of the tests. In general, the film is introduced into a known environment in which the initial conditions are established. However, these conditions are modified until equilibrium or a steady state of the system is reached.

Permeability of Gases and Vapours

To determine the permeability of a gas or vapour, a film of uniform thickness is placed in permeation cell where the film separates two rooms or chambers. All constants and variables during the test remain identical in both chambers, except the partial pressure of the gases or vapours to be tested, which is higher in one chamber than the other, which must be maintained until the end of the test. Figure 5 shows schematically a complete permeation test.

At baseline (Step 1: balance), the two chambers are pressurized with an inert gas at low pressure (or zero) in order to balance the pressures of different components in both chambers as well as in the interior of the film; therefore, the concentration of the permeant must be identical in any point of the film. Once stabilized (Step 2: transition state), a flow of high pressure gas is incorporated into one of the chambers marking time "0" of the experiment, and is maintained under these conditions until the end of the experiment.

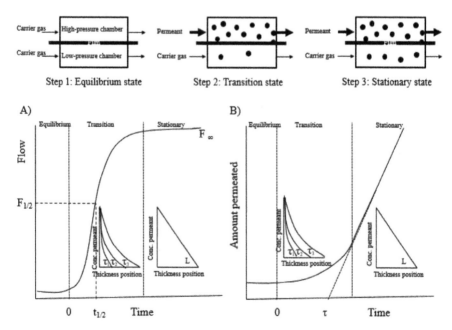

Fig. 5. Schematic of a full permeation test with a description of the transfer processes that occur.

During this step, the permeant gas crosses the films, reaching the other surface of the film. In time, a stationary state is reached (Step 3) in which the pressures along the film are stabilized (not equal) and the permeant gas flow is constant. Upon arriving at this point, the experiment is finalized. Finally, the passage of permeant gas from one chamber to another is determined by a detection system, where the gas is removed through a gas flow (f), or accumulation of gas in a chamber. In the first case, the concentration of permeant in the carrier gas (c) is determined. This value being related to the flow through the film (F) by:

$$F = c \cdot f \tag{10}$$

To calculate the permeability (P), simply multiply the gas flow at the stationary state (F_∞) by the film thickness (ℓ) and divide by the area of the exposed film (A) and the difference of partial pressures of permeant gas between the chambers (Δp):

$$P = \frac{F_\infty \cdot \ell}{A \cdot \Delta p} \tag{11}$$

The other magnitudes related to permeability, which is widely used, are the transmission rate (TR) and permeance ($\wp$):

$$TR = \frac{F_\infty}{A} ; \wp = \frac{F_\infty}{A \cdot \Delta p} \tag{12}$$

Permeance measures the amount of permeant that crosses a film, which is determined per unit area, time and pressure difference, while the transmission rate measures the amount of permeant that crosses a film per unit area and time, under specific conditions. These magnitudes should be used instead of permeability as long as some of the conditions imposed in the development of the definition of permeability have not been met, such as the homogeneity of the film (multilayer, coated films), the dependence on the concentration or pressure of the permeant at which the film is exposed and breach of the Henry or Fick Laws, among others. In these cases, in addition to the TR or $\wp$ data, the description of the film (thick and layers) and the concentration of the permeant must be included for testing.

Permeation tests can also provide information on the kinetics of the process, provided that the Fick and Henry laws have been met for these system films/permeant. Considering that the films are homogeneous and of constant thickness, and realizing a complete permeation test, that is, including the equilibrium stage and knowing time $t = 0$, the value of diffusion coefficient in these conditions can be calculated. Flow curve *versus* time (Fig. 5A) satisfies the following equation (Crank 1975):

$$\Phi = \frac{F}{F_\infty} = \left(\frac{4}{\sqrt{\pi}} \right) \left(\sqrt{\frac{\ell^2}{4Dt}} \right) \sum_{n}^{\infty} = 1,3,5\ldots\exp\left(\frac{-n^2 \cdot \ell^2}{4Dt} \right) \tag{13}$$

For Φ values below 0.95, the sum can be reduced to the first term, and therefore:

$$\Phi = \left(\frac{4}{\sqrt{\pi}}\right)\sqrt{X} + \exp(-X); \; X = \frac{\ell^2}{4Dt} \tag{14}$$

The resolution of the X values for each value of Φ, and representing these values *versus* l/t, allows scientists to obtain a straight line passing the origin, one whose slope obtained the D values. Another simpler way to obtain D is from the time at which Φ = 0.5, $t_{1/2}$ is obtained.

$$D = \frac{\ell^2}{7.205 \cdot t_{1/2}} \tag{15}$$

Gavara and Hernández (1993) developed a consistency test for this type of tests, which allow the validity of Eq. 15 for calculating D.

Once registered the final pressure in the low pressure chamber, a similar curve to Fig. 5B can be observed. At stationary state, the increase in mass permeated with time becomes constant. Therefore, the permeability can be calculated from slope of linear part of curve (m/t):

$$P = \frac{m}{t} \frac{\ell}{A\Delta p} \tag{16}$$

Obtaining expressions for validating transmission and permeance are obvious. It is also possible to calculate the diffusion coefficient from the cutoff point of prolongation of straight section with the time axis; this point is known as delay time or *lag time* (τ).

$$D = \frac{\ell^2}{6 \cdot \tau} \tag{17}$$

According to the ASTM E96 standard, determining water vapour permeation (WVP) is established for films of constant thickness, similarly as described above, although realizing a record of the weight gain of the WVP cell. Permeability is obtained according to Eq. 16 or alternatively, the water vapour transmission rate ($WVTR$):

$$WVTR = \frac{m}{\tau} \frac{1}{A} \tag{18}$$

For the development of the practical test, a disc of permeation is used to fix the film, which allows scientists to separate the inner chamber the from external environment, as shown in Fig. 6.

The disc is deposited inside a desiccant material (silica gel) or a saturated saline solution of constant and known water activity (a_w^i). The cell is introduced into a container which is maintained to constant moisture (a_w^e).

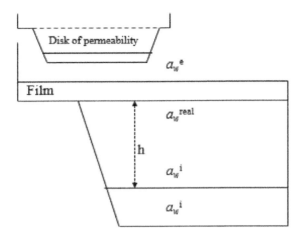

Fig. 6. Disc design of permeability described in ASTM E96 standard.

To ensure the correct maintenance of the external moisture, it is desirable to have a hygrometer at the container and set an air flow through a fan to ensure an identical atmosphere in the entire container. McHugh and Krochta (1994) observed that flows with lower speed at 150 m/min lead to a value of WVTR dependent on the air speed established. Inside the disk convection flows cannot be set; thus, water molecules diffuse from the interior interface of film to upper surface of absorbent. This inevitably leads to a concentration gradient of water in the air with distance from the adsorbent. This gradient can be neglected for many trials, but in the case of very hydrophilic edible films, high *WVTR* values are expected. This gradient can produce significant effects of said value. McHugh and Krochta (1994) suggest a correction of experimental value, considering the difference in water activity, which is established between the internal interface of the film (a_w^{real}) and the adsorbent (a_w^i). From the *WVTR* value obtained experimentally, the actual value of activity on the inner surface of the film is calculated as follows:

$$a_w^{actual} = 1 - (1 - a_w^i) \cdot \exp\left(\frac{WVTR \cdot R \cdot T \cdot h}{P_T \cdot D_{agua \to aire}}\right) \qquad (19)$$

In this equation P_T, R and T, are the total pressure, the gas constant and the test temperature, respectively; h is the distance between the film and the adsorbent surface (Fig. 6) and $D_{water\text{-}air}$ is the diffusivity of water in the air, which can be obtained by the equation of Fuller, Schettler and Giddings (Perry and Green 1984).

With the value of actual water activity, the corrected permeability or permeance can be obtained by the following equations:

$$P = \frac{WVTR \cdot \ell}{P_T \cdot (a_w^e - a_w^{actual})}; \quad \wp = \frac{WVTR}{P_T \cdot (a_w^e - a_w^{actual})} \qquad (20)$$

Release or Retention of Vapours and Solutes

Furthermore, for the transfer of substances between two mediums separated by the film, as well as for many other applications, the release or retention of vapours or solutes, especially water, flavorings, antimicrobials, antioxidants, among others, has great importance in the transport phenomena of films based on biopolymers. Therefore, the determination of the concentration of these compounds during the time (diffusion rate) is essential for determining transport phenomena processes in edible and biodegradable films. For example, for compounds in gaseous state (vapour), often the sorption isotherms are determined to evaluate the transport phenomena in edible films (Hernández-Muñoz et al. 1999). In the case of water, gravimetric tests are performed, and for other substances, chromatographic or spectroscopic methods are common (Cava et al. 2004; López-Carballo et al. 2005).

Just like in the permeation tests, the sorption essays allow obtaining information of kinetic processes, as long as the increase in additive concentration with time is found registered and the test begins from an equilibrium state. The evolution of the amount of penetrant, which is released or withheld from the film when the diffusivity is not depend on the concentration, which can be described by the following equation:

$$\Phi = \frac{M}{M_\infty} = 1 - \frac{8}{\pi} \sum_n^\infty 1,3,5\ldots \frac{1}{n^2}\exp\left(\frac{-n^2 \cdot \pi^2 \cdot D \cdot t}{\ell^2}\right) \tag{21}$$

The diffusion coefficient can be obtained from the time when $\Phi = 0.5$ is reached:

$$D = \frac{0.049 \cdot \ell^2}{1_{1/2}} \tag{22}$$

Considerations for Edible Films and Coatings

The above expressions are analytical solutions to the Fick Laws, in ideal contour conditions, which are very different of actual conditions. For example, in ideal contour conditions, films are assumed to be isotropic systems (when in fact they are anisotropic systems), which conform to the Henry's Law, as well as the independence of the diffusion coefficient over time, the penetrating concentration or the position in the film. These expressions have been used successfully in describing numerous mass transfer processes, especially in those cases in which participates inert agents as permeant gases in inert synthetic polymeric materials.

Edible films and coatings based on biopolymeric materials are far from the characteristics described above. In general, the biodegradable and edible films are totally or partially composed by hydrophilic polymers such as proteins or polysaccharides accompanied of polar plasticizer compounds. With these features, the edible films and coatings can absorb amounts of water directly proportional to ambient relative humidity (RH) until reach values exceed the dry weight of the original film, which can lead to the loss of integrity of the same or even its dissolution. In addition, this progressive increase in water gain from environmental moisture is not linear. On the contrary, these curves are described by a sigmoid function, similar to most foods as it shown in Fig. 7, which presents an exponential growth in the section of greater RH. Henry isotherm does not provide an adequate description of the behavior of these films. Instead, the isotherms with more parameters are used, or the D`Arcy and Watts isotherm with five parameters.

$$u_i = \frac{k_1 k_2 a_w}{1 + k_1 a_w} + k_3 a_w + \frac{k_4 k_5 a_w}{1 - k_4 a_w} \tag{23}$$

The mass of water held per unit weight of the film (u) is a function of water activity (a_w) through three terms. The first corresponds to the Langmuir isotherm and describes the initial part of the isotherm, while k_1 and k_2 are associated with the binding energy and the number of active sites of the film, as well as molecules that interact with macromolecules of the film and that are difficult of eliminate during the drying process.

The second term corresponds to the Henry isotherm, and describes the middle part of the curve, k_3 being related to the solubility coefficient, which describes the formation of a monolayer of water molecules. These water molecules coated the biopolymer chains and interact with each other through hydrogen bonds, thus reducing cohesive energy between polymer-polymer chains. The third term is known as term of deviation and describes the concave curvature at high humidities. The k_4 and k_5 parameters represent

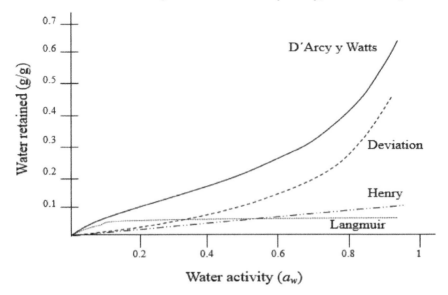

Fig. 7. Representation of a water sorption isotherm of a polysaccharides-based film by D'Arcy and Watts isotherms, and its three components.

the binding energy and the formation of multilayers of water molecules. At high humidities, new water molecules penetrate the matrix, which interact together to form self-associations or *clusters*, leading to a high swelling in the film.

Several authors using the grouping function (G/Φ_w) developed by Zimm and Lundberg (Hernández and Gavara 1994) to estimate the activity to which the self-association of water in the film begins:

$$\frac{G}{\Phi_w} = -(1-\Phi_w)\frac{\partial(a_w/\Phi_w)}{\partial a_w} - 1 \tag{24}$$

where Φ_w is the volume fraction of water in the film, which can be obtained from the isotherm, determining the density of the film at different humidities. When the *cluster* function exceeds the value of −1, the water self-association begins.

The polymer-water interactions in films also affect the diffusion kinetics of water in the matrix. The breaking of hydrogen bonds as a result of entrance of the first water molecules produces a reduction of the cohesive energy of the polymer and increases the flexibility and mobility of their chains. The result of these processes in the mass transport is an increase of diffusivity of small size molecules like water, and other permeant compounds such as oxygen, carbon dioxide or food aroma components (López-Carballo et al. 2005). With the formation of water *clusters*, the swelling of the matrix is produced, leading to the formation of water microregions, which increase the mobility of the substances and their diffusion coefficient. Therefore, the diffusion coefficient of water in majority of the edible films and coatings are dependent on the moisture to which they are exposed, increasing with water activity. In particular, the diffusion coefficient increases exponentially to high moisture contents as a cause of the self-association of water. This dependence of D on water activity also extends to other permanent gases.

Given the above dependencies, both water absorption and diffusion, it is clear that mass transfer through the edible films do not meet the requirements listed in the definition of the permeability coefficient. Therefore, the permeability value reflects the dependencies on the water activity of both parameters. However, the water vapour transmission rate (*WVTR*) is a very important parameter for many edible films and coating applications.

The *WVTR* values are obtained by permeation tests as described above. In these, the film separates two fluid mediums with different water activities. In the stationary state, the film interfaces with both mediums are balanced, while inside the film occurs a concentration gradient of water, whose profile is kept constant (Fig. 5). This variation in water concentration along the thickness produces a water film that increases with proximity to the interface of the chamber of higher concentration, so that the film becomes an anisotropic material. For this type of matrices it is irrelevant to know the permeability, since this parameter is not intensive, therefore, showing numerous dependencies. Therefore, it is advisable to determine the *WVTR* in moisture conditions closer to the actual application.

Hauser and McLaren (1948) developed a method for estimating permeability for any relative humidity gradient from the profile of permeability values determined experimentally, using gradients of water activity of $a_w^i = 0$ in the chamber of low concentration and different values a_w^e in the chamber of high concentration (essay according to ASTM E96 standard).

Figure 8 shows a typical curve of water permeability values *vs.* a_w; in the same an exponential dependence is observed (Hernandez-Munoz et al. 2003). The permeability values of Fig. 8 do not correspond to a particular material, but describe the general behavior of the hydrophilic films in arbitrary units of permeability. As already noted, the concentration gradient of water (or water activity), which is set along the thickness of the film, is responsible for water diffusion processes, which makes the molecules find in their passage through the films different water activities, and consequently different diffusion and sorption capacities (*D* and *S*) are produced. Thus, actually the measure of permeability corresponds to an average value (P_{mean}) that can be expressed as:

$$P_{mean}(a_w^e - a_w^i) = \int_{a_w^i}^{a_w^e} P \cdot da_w \qquad (25)$$

where the value is expressed as the integral of the permeability respect to water activity, which represent the area under the permeability curve in Fig. 8. The application of this is that the permeability can be determined in a range of required water activities, with only determine the area under the curve in the selected region. In Fig. 8, it can provide an example in which the dark area is the region where obtaining the permeability is required, then the calculation mechanism is as follows:

$$P_{mean}(a_w^e - a_w^i) = P_{mean}(a_w^e - 0) - P_{mean}(a_w^i - 0)$$

$$P_{mean} = \frac{P_{mean}(a_w^e - 0) - P_{mean}(a_w^i - 0)}{(a_w^e - a_w^i)} = \frac{2.3 \cdot (0.75 - 0) - 0.91 \cdot (0.5 - 0)}{(0.75 - 0.5)} = 5.1 \qquad (26)$$

It is worth noting that the values obtained in this manner are much higher than the experimentally determined values for the gradient of activities between 0 and 0.75. This occurs because along the film thickness are found regions highly plasticized by water in the area closest to the interface with the outside, while in the closest area to the interface inside the film is completely dry, thus slowing the diffusion of water molecules in the film. This retention is reflected in the measured average value. In contrast, when the film is exposed to humidities between 50 and 75%, the water molecules found less opposition to diffusion, resulting in a far superior average permeability.

In general, as discussed in previous sections, it is considered that the *D*, *S* and *P* coefficients are only dependent on temperature. Nonetheless, as is usual, other dependencies are presented, as those mentioned above,

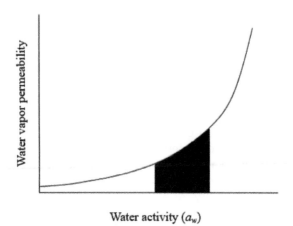

Fig. 8. Effect of water activity on the water vapour permeability
of hydrophilic edible films.

e.g. moisture conditions. Another parameter that has significant influence
is the film thickness. During the production process of films, it is inevitable
that changes occur in the polymer morphology *versus* position along the
thickness. In extrusion processes, the film surface cools faster than the
interior, producing differences in packing of macromolecules that can affect
crystallinity and the free volume in the amorphous regions; as a result, the
transport properties are modified. Likewise, during the coatings or casting
processes, drying the film-forming solution (*FFS*) occurs faster on the surface
in contact with the drying air than inside the film or in surface contact with
the support or the product to be coated. It is therefore evident that the
thickness is a factor to consider, thereby, performing the tests on films with
similar thickness to the intended application is convenient.

The permeance of a film ($\wp_T$) is related to the respective layers that
conform, by the following expression:

$$\frac{1}{\wp_T} = \frac{\ell_T}{P_T} = \frac{1}{\wp_r} + \frac{1}{\wp_s} = \frac{\ell_r}{P_r} + \frac{\ell_s}{P_s} \rightarrow \wp_r = \frac{1}{\dfrac{\ell_T}{P_T} - \dfrac{\ell_s}{P_s}} \tag{27}$$

The T, r and s subscripts denote the corresponding values to the coated
film, the coating and the film support, respectively. By means of two
independent tests, the permeance of the carrier film and the coated film
can be determined. These values allow to obtain the corresponding value of
the coating. In order to improve the quality of the determination of coating
properties, it is convenient to select a carrier film very permeable to the
gas under study. Reducing the value of subtracting from the denominator;
smaller is the effect of the substrate on the multilayer properties as well as
smaller is the error made in calculating the coating permeance.

To determine the permeability of other compounds, alone or in
combination, depending on the relative humidity and under actual

conditions in which there may be simultaneous concentration gradients of several substances (water, oxygen, carbon dioxide, etc.) along the thickness, mathematical models can be used, allowing researchers to interpolate and extrapolate the mass transfer properties under closest contour conditions to the real (i.e. in anisotropic materials), based on the experimental information of the transport parameters following the techniques described above.

Prediction and Simulation of the Mass Transfer Processes

Advantages and Limitations of Mathematical Simulation

Although edible films and coatings based on polymeric materials represent one of the simplest physical systems to study from the point of view of transport phenomena, the great variety and complexity of mass transfer mechanisms, usually prevent to know exactly the behavior of these systems in real situations with the simple realization of experimental tests, as mentioned in the previous section.

In this way, the interest in performing mathematical simulations of these transport phenomena lies mainly in obtaining exhaustive descriptions of all physicochemical processes through computerized calculations, fed by a minimum inventory of experimental data, with the consequent saving costs and human or material resources. In addition, a deeper analysis of the physicochemical properties of the systems under study allows researchers to reach a double objective:

(1) The prediction with reasonable accuracy of its long-term evolution when subjected to different external actions, and,
(2) The design of new systems whose long-term evolution can be adjusted to fulfill a certain aim.

Despite these advantages, the mathematical simulation has limited ability to represent the mass transfer processes that occur in real systems. On the one hand, there is the classic dilemma "accuracy-speed", since the number of variables and equations increases with the aim of reducing the number of hypotheses and obtaining more precise and accurate results, but the processing time of the results increases exponentially, thereby increasing the time required for the computational analysis. In addition, this increase in complexity also increases the need to have sufficient coefficients and fundamental parameters that satisfy all integrated equations in the model. Due to the scarcity of bibliographic references on this large number of parameters. This often causes the data to be experimentally determined. Therefore, at the time to give an answer to a mathematical simulation of any mass transfer process, the correct compromise between accuracy, speed of response and the minimum size of their inventory of experimental data should always be determined.

Modeling of Biodegradable Films and Coatings

The first step to simulate a real system involves the construction of a model whose geometric characteristics and physicochemical properties can be as close as possible to the actual case, but maintaining at the same time sufficient simplicity to be treated mathematically; reaching the commitment "accuracy-speed".

In the case of the system models constituted by edible films and coatings based on polymeric materials, the characteristics and properties that usually occur with more frequency, are:

- The geometry of the "infinite lamina", that is, a symmetric parallelepiped-shaped body, finite thickness and long and widths infinite.
- Isotropic and constant physicochemical properties throughout the system.

With these premises, the mathematical model for the transfer of a component i through an edible film or coating of thickness 2L, to or from the medium formed by the adjacent physical phases is shown schematically in Fig. 9. The most important variables of the system are the initial concentrations (i) in the film and in the medium (C_{i0}^P and C_{i0}^M), the diffusion coefficients in the film and in the medium (D_i^P and D_i^M) and the individual coefficient of mass transfer of component i through the medium (k_{ci}^M).

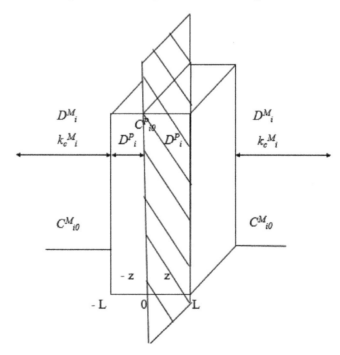

Fig. 9. Scheme of a symmetrical film with assigning transport variables.

Hypotheses and General Equation

In order to correctly propose the general mass balance to control the volume of the film studied, it is necessary to specify in advance all fundamental hypothesis of the model to be taken into consideration:

- The solid and fluid phases at rest, static (no net flow of matter in any direction) and isotropic (identity properties at any point).
- Fluid phases perfectly agitated and dynamics, therefore, isotropic.
- Interfaces in thermodynamic equilibrium (therefore, equality of chemical potential at the phase boundaries).
- Temperature and pressure, uniform and constant throughout the system.
- The total concentration of constant chemical species throughout the system (absence of generation or disappearance of material).
- Non-stationary state (evolution of the system over time to reach a stationary state) with absence of an induction period.
- Unidirectional matter transfer, and of molecular or diffusive type, always governed by the Fick's Law.

Taking into account all the characteristics and properties of the constructed model and the hypothesis made, and combining the mass conservation law with the Law of continuity of Fick, we can deduct the general equation of the microscopic mass balance applied to component i:

$$\frac{\partial c_i^P}{\partial t} = D_i^P \frac{\partial^2 c_i^P}{\partial z^2} \tag{28}$$

This mathematical expression is a partial differential equation (PDE) parabolic type. Its resolution allows knowing the concentration profile of component i. (CPi (z, y)) at each point inside the film and for each instant of time elapsed.

Solution of Equation of the Mass General Balance

Unfortunately, the PDE (Eq. 28) can only be solved analytically for simplified systems and in situations where it can be used in very specific and simple integration conditions. For more general circumstances, an alternative method should be used such as graphical or numerical methods.

In any case, both analytical and numerical resolution require the help of advanced computer systems to find and represent the solutions with sufficient accuracy, and always within a reasonable time interval. Finally, depending on the method chosen, the number and complexity of the calculations to be carried out will be different and, therefore, the minimum software and hardware requirements of the computer system will also need to be conditioned. With regard to software solutions available nowadays, the calculation and programming suites such as MATLAB (The Mathwork INC, Natic, MA, E.U.) and COMSOL Multiphysics (COMSOL Inc, Burlington,

MA, E.U.), are generally the most used applications and recommended by its power and versatility.

Analytical Solutions of the General Equation

In the literature, the mathematical calculation can be found in numerous analytical methods for solving differential equations in partial derivatives. Of these, the most widely used due to its simplicity and applicability are: separation of variables, change of variables and integral transform (usually of Fourier or Laplace). The applicability of each method for the model in study is always subject to the fulfillment of certain and determined mathematical requirements, this is the "severability" of the variables of the PDE, the existence of a bijective function between the variables to exchange or the existence of an inverse Fourier or Laplace transform, respectively. In the case where at least two analytical methods can be applied, selecting the simplest to perform is recommended, since the result should always be the same.

To apply the method selected to the above differential equation, all conditions of the integration model must be established so that the resulting integral solutions remains correctly defined. These conditions emanate from the characteristics and properties of the model in the study and/or the previously established fundamental hypotheses, which are usually of two types:

- Initial conditions: the distribution of component i is established in all phases of the system and for the initial time of the transport process.
- The contour, boundary or limit conditions: the evolution of component i is established over time for each boundary point or system limit.

In the case of edible polymeric films, which have been used as an example of a mathematical model in this section, they have been considered to be immersed in an infinite medium constituted by a stirred fluid phase. In such a way, that initially, the concentration of component i in the film and in the medium are C^P_{i0} and C^M_{i0}, respectively. With regard to the boundary condition, in this case, there are "symmetry" conditions in the origin of coordinates, where it holds that $\dfrac{\partial c^P_i}{\partial z} = 0$ and other "frontier" conditions in $z = L$, where the interface between the polymer film and the fluid medium, and whose expression is based on the level of agitation of this latter, among other factors.

Assuming that hypothesis described is met, before the fluid phase is infinite and is always found perfectly agitated, it can be inferred that the concentration of i in its bosom ($C^M_i(t)$) is uniform in space and constant in time, and therefore of known value. Thus, the second boundary condition at $z = L$ is $C^P_i = K \cdot C^M_i$, where K makes reference to the partition constant or Henry constant; this if the fluid phase is liquid or gaseous, as already was highlighted in section of chemical balance.

Finally, through the integration of the differential equation in partial derivatives with all initial and boundary conditions described above, and then by applying one of the previously established analytical methods, the general expression of macroscopic mass balance applied to component i is obtained (Crank 1975).

$$C_i^P = K \cdot C_i^M + \frac{2(c_{io}^P - K \cdot c_i^M)}{\pi} \sum_{v=1}^{\infty} \frac{(-1)^{v-1}}{v - \frac{1}{2}} \cdot e^{\dfrac{-\pi^2 \left(v - \frac{1}{2}\right)^2 \cdot D_i^P \cdot t}{L^2}} \cdot$$

$$\mathrm{Cos}\left[\frac{z.\pi}{L}\left(v - \frac{1}{2}\right)\right]. \tag{29}$$

Unfortunately, in most real systems involving solid bodies immersed in agitated fluid phases, the hypothesis of perfect mixing is inadmissible, since its properties can never be considered uniformly or consistently throughout the medium. In these cases, the problem can be addressed using a parameter called as local individual coefficient of mass transfer (k_c), which is a function of the fluid characteristics (density, viscosity, diffusivity of components (i), among others) and the nature of its flow (laminar and/or turbulent). Although this parameter is extremely useful to simplify a great number of calculations, its value is hardly localizable in the literature, so it almost always can be quantified through reliable estimation methods, or should ultimately be determined experimentally under conditions appropriate for the model.

Therefore, using the above coefficient (k_c) and proposing a matter balance around the system interface, situated at $z = L$, and applying the component i, the second boundary condition can be deduced (Crank 1975).

$$-D_i^P \cdot \frac{\partial c_i^P}{\partial z} = k_{ci}^M \cdot (C_i^P - K \cdot C_{i\infty}^M) \tag{30}$$

Which establishes that at $z = L$ and for any instant of time, it should always comply that the flux density of the matter of component i is transferred by diffusion from the inside the film to the surface of the interface, which is equal to the flux density that is transferred by convection from the interface toward bosom of the fluid phase.

In this way, following a procedure analogous to development in the previous section, the general expression of macroscopic mass balance can be achieved for edible films immersed in infinite fluid mediums and subjected to any level of agitation (Crank 1975).

$$C_i^P - K \cdot C_{i\infty}^M + 2\left(C_{i0}^P - K \cdot C_{i\infty}^M\right) \cdot \sum_{v=1}^{\infty} \frac{\sin \psi_v}{\psi_v + \sin \psi_v \cdot \cos \psi_v} \cdot$$

$$e^{\dfrac{\psi_v^2 \cdot D_i^P \cdot t}{L^2}} \cdot \cos\left(\psi_v \frac{z}{L}\right) \tag{31}$$

where $C^M_{i\infty}$ makes reference to the concentration of component i in the fluid medium and at an infinite distance from the interface, while ψ symbolizes the successive solutions of the equation:

$$\psi_v \cdot \tan \psi_v = \frac{k^M_{ci} \cdot L}{D^P_i} \tag{32}$$

For each value obtained by the index v in the previous sumatory. It is also very common to find real cases where the hypothesis of perfectly agitated fluid phase has been totally acceptable, but the hypothesis of infinite extension is inadequate. Usually, these situations arise when a film is immersed in a fluid medium whose volume is small enough to start a perfect agitation, but not sufficiently high compared to the volume of the film, and therefore an infinite amplitude on the effects of matter transfer can be considered.

In such circumstances, obtaining an analytical solution following a procedure analogous to development in previous cases it is still possible, with a simple adaptation of the second boundary condition such that is expressed as an identity between the flux density of matter of the component i, leaving the film and the flux density that is inserted into the fluid phase. Therefore, the expression resulting from those modifications is as follows (Crank 1975):

$$-D^P_i \cdot \frac{\partial c^P_i}{\partial z} = \frac{L_M}{K \cdot L_P} \cdot \frac{\partial c^P_i}{\partial t} \tag{33}$$

where L_M makes reference to the thickness of the layer of fluid medium which is located in the interface of the edible film.

Finally, if the PDE is integrated again, by means of some of the analytical methods already described, but taking into account the new condition of contour, it is deduced the general expression of the macroscopic mass balance applied to component i for edible films immersed in finite fluid mediums and perfectly agitated (Crank 1975).

$$C^P_i = K \cdot C^M_i - 2\left(C^P_{i0} - K \cdot C^M_i\right)\sum^{\infty}_{v=1} \frac{1 + \dfrac{L_M}{K \cdot L_P}}{\left(\dfrac{L_M}{K \cdot L_P}\right)^2 \psi^2_v + \dfrac{L_M}{K \cdot L_P} + 1}$$

$$\cdot e^{\dfrac{-\psi^2_v \cdot D^P_i \cdot t}{L^2_p}} \cdot \frac{\cos\left(\dfrac{\psi_v \cdot z}{L_P}\right)}{\cos \psi_v} \tag{34}$$

where ψ_v again is used to symbolize the successive solutions of the equation:

$$\text{Tan } \psi_v = \frac{L_M}{K.L_P} \cdot \psi_v \tag{35}$$

For each value acquired by the index v in the sum above.

Another of the most common features usually present in edible films and coatings, and that also greatly complicates the mathematical modeling, is the high hydrophilicity of most polymeric materials that constitute them. This characteristic is responsible for major water absorption observed in certain polymeric matrices when they are immersed in aqueous or gaseous fluids located at high relative humidities.

In these circumstances, the hydrophilic polymers tend to weaken intermolecular interactions, specifically hydrogen bond-type interactions, leading to a relaxation of their matrices; and therefore, a significant decrease in its glass transition temperature.

As the polymer matrix locally reaches this "rubbery" state, the transfer of small molecules by diffusion through its interstices ceases to be only governed by the phenomena of polymer relaxation.

These two different matter transfer mechanisms have been widely proposed in the literature with some theoretical and semi-empirical approximations, at the time of mathematically modeling the behavior of the hydrophilic edible films and coatings. The Long and Richman model (1960) is one of simplest ways to combine both phenomena, and is based on the hypothesis that the relaxation of the polymer matrix occurs only at the interface with the environment.

Therefore, the authors propose an adaptation of the second boundary condition for infinite fluid mediums and perfectly agitated, so that an exponential evolution of concentration of component i in the interface level is described, at $z = L$, until asymptotically reaching an equilibrium value, $C_{i\infty}^P$ such as shown in the following equation (Long and Richman 1960):

$$C_i^P = C_{i\infty}^P + \left(C_{i0}^P - C_{i\infty}^P\right) \cdot e^{\frac{-t}{\tau}} \tag{36}$$

where τ is a parameter that controls the rate of relaxation of the concentration of component i in the interface, and it is assumed constant.

Lastly, if the above differential equation in partial derivatives by any analytical method is integrated again and taking into account the last boundary condition, the general expression of macroscopic mass balance is obtained for hydrophilic edible films immersed in infinite fluid mediums and perfectly agitated (Long and Richman 1960):

$$C_i^P = K \cdot C_i^M + \frac{2\left(C_{i0}^P - K \cdot C_i^M\right)}{\pi^2} \sum_{v=1}^{\infty} \frac{1}{v - \frac{1}{2}} \cdot$$
$$e^{\frac{-\pi^2\left(v-\frac{1}{2}\right)^2 D_i^P \cdot t}{L^2}} + C_{i\infty}^P + \left(C_{i0}^P - C_{i\infty}^P\right) \cdot e^{\frac{-t}{\tau}} \tag{37}$$

where C_i^P represents the total concentration of component i in the edible film.

Numerical Equations of the General Equation

The mathematical calculation is defined as those numerical approximation algorithms that allow the simulation of complex mathematical processes which are associated with the resolution of continuous functions of the real world. In other words, the numerical analysis of a continuous function, as is differential equation in partial derivatives proposed in the previous section, always allow the mathematical resolution in all cases where, due to the complexity of the physical systems or their integration conditions, it is impossible to reach or computationally very expensive.

Nonetheless, given the complexity of most algorithms or the high number of interactions necessary to achieve convergence, the use of these methods always require the availability of advanced computer systems to find and represent solutions with accuracy and within a reasonable time. In this sense, the analytical methods compared to these algorithms only have a range of values that includes the solution instead of an exact value; therefore, it is very important to correctly define the magnitude of this range based on the precision required in order to control the number of interactions necessary to achieve convergence, and therefore, the total computation time.

In the case of the edible film used as an example of the mathematical model in the previous section, all analytical solutions developed could have achieved a shape much more quickly and easily by any method of computer-aided numerical analysis. In addition, such analysis would also be useful in many other more complex cases (and more realistic cases) where the system acquires different geometrics or anisotropic properties, or where it is impossible to admit some hypotheses. Examples of these cases are films and multilayer coatings. Films used in packaging construction, coatings on three-dimensional or irregular bodies (where mass transfer can be multidirectional), the materials with non-uniform or constant properties, the multicomponent transport phenomena or non-regents of second Fick Law, or including turbulence or other adventive flows of matter besides the molecular transport by diffusion, the systems in which mass transfer of a component is found linked to that of another compound or for simultaneous transport of energy or in short, any possible combination of all them.

Among the abundant numerical algorithms collected from the literature, the most notable by their reasonable simplicity and their proven efficacy are the finite difference method, the finite volumes method and finite elements method. In all them, the differential equations are solved in partial derivatives previously discrete in finite subspaces. Therefore, the problems of simple systems of algebraic or differential equations are easily resolved by conventional analytical methods. The fundamentals of the finite difference methods and the finite elements are show below:

Finite difference method: Basically, the technique used in the finite difference method consists in approximating the solution of the differential equations in partial derivatives by replacing the expressions of the derived by differential quotients of approximately equivalent value, for example, in the above case $C_i^P(z, t)$.

$$\partial C_i^P = \lim_{k \to 0} \frac{c_i^P(t+k) - c_i^P(t)}{k} \approx \frac{c_i^P(t+k) - c_i^P(t)}{k} \tag{38}$$

On the one hand, if the domain of function $C_i^P(z, t)$ in the space is divided by a network as $z_0, z_1, z_2 \ldots z_j$, and over time, through another network as $t_0, t_1, t_2 \ldots t_{n'}$ and considering that these divisions are uniform in the two subdomains, the h and k value, respectively, such as is shown schematized in Fig. 10. It was agreed that the point called $\left[C_i^P\right]_j^n$ always represents the approximate numerical value of $C_i^P(z_j, t_n)$. On the other hand, considering that the magnitude of the temporal steps (k) is small enough, researchers can apply the known Crank-Nicolson method (Leveque 2007), which is an implicit and second order numerical algorithm in time, although unconditionally stable, accurate and convergent, which is based on the first-order central differences for each time instant $t_{n+\frac{1}{2}}$, and the second-order central differences for each z position corresponding to a spatial derivative.

Thus, using the finite difference method developed by Crank and Nicolson to above differential equation in partial derivatives, the following recurrence equation is obtained (Leveque 2007):

$$\frac{\left[C_i^P\right]_j^{n+1} - \left[C_i^P\right]_j^n}{k} = \frac{D_i^P}{2.h^2} \cdot \left(\left[C_i^P\right]_{j+1}^{n+1} - 2.\left[C_i^P\right]_j^{n+1} + \left[C_i^P\right]_{j-1}^{n+1} + \left[C_i^P\right]_{j+1}^{n}\right.$$
$$\left. - 2.\left[C_i^P\right]_j^{n} + \left[C_i^P\right]_{j-1}^{n}\right) \tag{39}$$

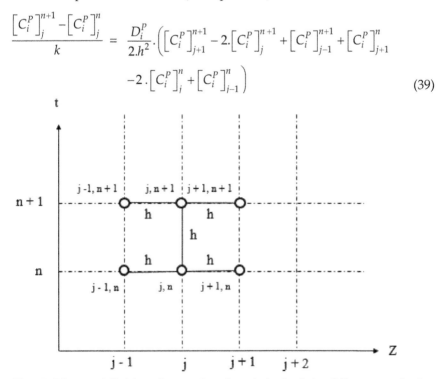

Fig. 10. Scheme of division of space-time domain in the finite difference method.

Whose application to all modes of known value of space network, gives rise to a system of algebraic equations of simple resolution allowing advancement in the algorithm to another step further in time, and thus achieving a new interaction.

Finite element methods. Like the previous technique, the finite element method allows researchers to numerically solve any system of differential equations in partial derivatives, previously transformed into an algebraic equations system or ordinary differentials, easily treatable through traditional analytical methods or by other numerical methods of easier implementation. In this technique, the product of the discretization of domain of the PDE also consists of a multitude of mesh nodes and small subdomains of finite size, called "elements", but in this case are completely interconnected by very simple models and easy to analyze. In this way, applying the finite or subdomain element analysis, this allows researchers to obtain an approximate value of the real solution for each node, and will be as accurate when denser has been the mesh generated in the domain of the function that is intended to simulate. The usual procedure for resolving a PDE by this numerical method consists essentially in (Beers 2007):

- Expression of the PDE in its weak form (integral form), separation of the boundary conditions of Dirichlet (related to the concentration of property, C_i^p) and of Neumann (related to the flow property, $\dfrac{\partial c_i^p}{\partial z}$).

- Discretization of finite elements domain, creating for this a mesh of nodes interconnected in a polygonal shape (usually tetrahedral or hexahedral in 3D, and quadrangles or triangles non-overlapping in 2D).

- Approximation of the functions as interpolation of nodal values with shape functions, usually n-linear or n-quadratic.

- Selection of the test functions for verifying the weak form (e.g. such as is defined in the Galerkin method, based on the expression of the function test of the solution as a linear combination of the product of the value of each node by the corresponding global function).

- Obtaining the corresponding linear system (with a non-regular matrix obtained from the assembly of the matrices associated with each element).

- Imposition of the boundary conditions.

- Resolution of the matrix of the linear system by conventional methods of linear algebra in a finite dimensional space.

The main advantages of the method of the differences or finite volumes reside in the possibility of using irregular mesh and/or poorly structured (that is, finite elements with variable size and shapes), with the possibility of imposing the contour conditions of systemic way (without casuistry), and its growing convergence to divisions of the successively smaller finite elements.

Analysis of a Real Case: Analytical Solution versus Numerical Solution

As discussed above, the numerical analysis of a differential equation in partial derivatives always has many advantages compared to its resolution by analytical methods, achieving from reducing the computation time in the simplest systems for approximating accurate solutions in most complex systems.

A simple way to evaluate the behavior of both methods could be based on the comparison of the results obtained during its application in resolving a real case of matter transfer in an edible film. To this end, we propose to study again the system previously showed in Fig. 10, consisting in an edible film or coating of 10 μm of thick, initially saturated of component i in a concentration of 10%, which is immersed within a perfectly agitated and infinite fluid phase that is found free of said compound.

In addition, it can also be admitted that all the fundamental hypothesis established in the corresponding section, except only one: the diffusion coefficient of component i is not uniform in space, nor constant in time, but can vary depending on water activity (a_w) of the film, according to the following equation:

$$D_i^P = 10^{-19} \cdot e^{\ln(10^4)} \cdot a_w \tag{40}$$

Typically, this situation occurs in the study of edible films, which are composed of hydrophilic polymers since they have a tendency to adsorb water from the ambient humidity, thus altering their properties, according to the prediction of phenomena of polymeric relaxation, as explained in the section "analytical solutions of the general equation". For example, in this case study, it can be assumed that the fluid medium is found permanently saturated with moisture, thereby allowing the humidification of the edible film driving gradually from a state of complete dehydration to a certain equilibrium moisture estimated of 1 kg water/kg dry polymer, and with a speed of transport determined by the diffusion coefficient of water through the material (D_w^P).

Therefore, taking into account all these characteristics and properties, it can apply and resolve the general equation of the microscopic mass balance developed in the section 4.3 (hypothesis and general equation), in a simultaneous way for water and the component i, if the following conditions previously established of integration:

$$C_i^P (z, 0) = 10 \qquad\qquad C_w^P (z, 0) = 0$$

$$C_i^P (10, t) = 0 \qquad\qquad C_i^P (10, t) = k \cdot C_w^M = 1 \tag{41}$$

$$\frac{\partial C_i^P}{\partial z} (0, t) = 0 \qquad\qquad \frac{\partial C_w^P}{\partial z} (0, t) = 0$$

Unfortunately, the system equation cannot be resolved analytically in any case of the proposed way, because the solutions corresponding to both PDE require the fulfillment of the hypothesis formulated where the properties should be isotropic and constant throughout the system, contrary to what occurs during the diffusion of component i. As can be appreciated in Eq. 40, the diffusivity value of the substance through the film should be between 10^{-19} and 10^{-15} m^2/s, depending on water activity in its interior, which in turn may increase from 0 to 1 kg water/kg dry polymer during the process. In these circumstances, the only possible application of the analytical solution known as the PDE resides in the establishment of an average value of D_i^P located between the limits of the above range. As expected, the goodness of this approach depends on the development of water transfer processes within the film, among other factors.

However, if the system of PDE is introduced together with the integration conditions in a computer programmed with numerical algorithms, the machine will be able to converge quickly to the solutions ($t_c < 10$ s), achieving an excellent level of precision. In this way, numerical methods allow the simultaneous resolution of two microscopic mass balances applied to two different substances such as water and the component i, whose mass transfer is interrelated according to the terms of Eq. 40 ($D_i^P(a_w)$).

In order to compare the results achieved by both methodologies, the graph shown in Fig. 11 was developed. In this sense, curves in the semi-logarithmic scale are shown for PDE analytical solutions applied to the component i to the interval extreme values of diffusivities marked by $D_i^P(a_w)$. Therefore, any solution obtained from an average coefficient of said ends must have the same shape of the curves represented and positioned in some intermediate point in the space between them. On the other hand, curves corresponding to PDE numerical solutions applied to component i have also been represented in the same scale for different values of the diffusion coefficient of water through the film, from 10^{-14} to 10^{-20}.

Graphically in Fig. 11, the superior ability of numerical approximation methods to find precise answers to the complex problems of real world solutions can be seen. In the example of this section, curves obtained by these methods have a considerable evolution over time with differences from the analytical curves due to the continuous variations of D_i^P depending on the a_w between said extreme values (10^{-19} and 10^{-15} m^2/s). Logically, if the basic parameters governing water transfer through the film such as the diffusion coefficient (D_w^P), among others, are known, it is possible to obtain accurate, analytical or numerical solutions of the PDE applied to water, thus providing the profiles of water activity in the film and its evolution over time ($a_w(z, t)$), which was introduced in $D_i^P(a_w)$, thus allowing researchers to know in detail the variation of this coefficient and avoiding committing the error associated with choosing a simple average value.

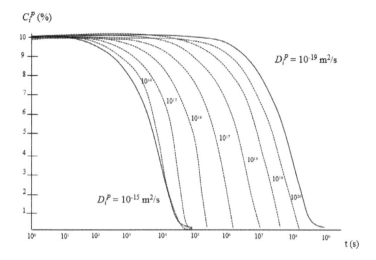

Fig. 11. Example of releasing a compound from a hydrophilic film exposed to a humid environment, whose diffusion coefficient is dependent on water activity. Curves in the black lines are analytical solutions of extreme cases. Segmented curves are exact solutions obtained according to the diffusion coefficient of water in the film.

Conclusions and Outlooks

Standard methods for the characterization of mass transfer properties were developed for the analysis of the synthetic polymer films, but have been applied to all kinds of films and coatings but nevertheless do not always lead to consistent results. Therefore, it is necessary to review and adapt the method for each test with edible films and coatings. New methods must take into account the particular characteristics of these films, which in summary are highly hydrophilic and anisotropic, to which must be added the reduced thickness of these materials.

The difficulty arises from the changes in behavior with the boundary conditions, therefore, the use of predictive models is required to describe the actual behavior of the data obtained in ideal conditions. These models are particularly important in the numerical resolution methods.

Acknowledgements

The authors would like to thank the Consejo Nacional de Investigaciones Científicas y Técnicas (CONICET) (Postdoctoral fellowship internal PDTS-Resolution 2417), Universidad Nacional de Mar del Plata (UNMdP) for the financial support, and to Dr. Mirian Carmona-Rodríguez.

References

ASTM. 1999. Standard methods for water vapor transmission of materials (E96-00) Annual Book of ASTM Standards. American Society for Testing and Materials, Philadelphia, PA.

Beers, K. 2007. Numerical Methods for Chemical Engineering. Cambridge University Press, Cambridge, pp. 181–200 and 313–316.

Belbekhouche, S., J. Bras, G. Siqueira, C. Chappey, L. Lebrun, B. Khelifi, et al. 2011. Water sorption behavior and gas barrier properties of cellulose whiskers and microfibrils films. Carbohyd. Polym. 83(4): 1740–1748.

Cava, D., J.M. Lagaron, A. López-Rubio, R. Catala and R. Gavara. 2004. On the applicability of FT-IR spectroscopy to test aroma transport properties in polymer films. Polym. Test. 23(5): 551–557.

Chang-Bravo, L., A. López-Córdoba and M. Martino. 2014. Biopolymeric matrices made of carrageenan and corn starch for the antioxidant extracts delivery of Cuban red propolis and yerba mate. React. Funct. Polym. 85: 11–19.

Crank, J. 1975. The Mathematics of Diffusion. Clarendon Press, Londres.

Gavara, R. and R.J. Hernández. 1993. Consistency test for continuous flow permeability experimental data. J. Plast. Film Sheet. 9(2): 126–138.

Hauser, P.M. and A.D. McLaren. 1948. Permeation through and sorption of water vapor by high polymers. Ind. Eng. Chem. 40(1): 112–117.

Hernández, R.J. and R. Gavara. 1994. Sorption and transport of water in nylon-6 films. J. Polym. Sci. Pol. Phys. 32(14): 2367–2374.

Hernández, R.J. and R. Gavara. 1999. Plastics Packaging: Methods for Studying Mass Transfer Interactions: A Literature Review. Pira International, Leatherhead, UK.

Hernández-Muñoz, P., R. Gavara and R.J. Hernández. 1999. Evaluation of solubility and diffusion coefficients in polymer film-vapor systems by sorption experiments. J. Membrane Sci. 154(2): 195–204.

Hernández-Muñoz, P., A. Kanavouras, P.K. Ng and R. Gavara. 2003. Development and characterization of biodegradable films made from wheat gluten protein fractions. J. Agr. Food Chem. 51(26): 7647–7654.

Lagaron, J.M., R. Catalá and R. Gavara. 2004. Structural characteristics defining high barrier properties in polymeric materials. Mater. Sci. Tech-Lond 20(1): 1–7.

LeVeque, R.J. 2007. Finite Difference Methods for Ordinary and Partial Differential Equations. Society for Industrial and Applied Mathematics, Philadelphia.

Lewis, E.L.V., R.A. Duckett, I.M. Ward, J.P.A. Fairclough and A.J. Ryan. 2003. The barrier properties of polyethylene terephthalate to mixtures of oxygen, carbon dioxide and nitrogen. Polymer 44(5): 1631–1640.

Long, F.A. and D. Richman. 1960. Concentration gradients for diffusion of vapors in glassy polymers and their relation to time dependent diffusion phenomena. J. Am. Chem. Soc. 82(3): 513–519.

López-Carballo, G., D. Cava, J.M. Lagarón, R. Catalá and R. Gavara. 2005. Characterization of the interaction between two food aroma components, α-pinene and ethyl butyrate, and ethylene-vinyl alcohol copolymer (EVOH) packaging films as a function of environmental humidity. J. Agr. Food Chem. 53(18): 7212–7216.

McHugh, T.H. and J.M. Krochta. 1994. Permeability properties of edible films. pp. 139–188. *In*: Krochta, J.M., E.A. Baldwin and M. Nisperos-Carriedo (eds.). Edible Coatings and Films to Improve Food Quality. Technomics Pub., Lancaster, P.A.

Peelman, N., P. Ragaert, B. De Meulenaer, D. Adons, R. Peeters, L. Cardon, et al. 2013. Application of bioplastics for food packaging. Trends Food Sci. Tech. 32(2): 128–141.

Perry, R.H. and D.W. Green. 1984. Perry´s Chemical Engineers Handbook, 6a. ed. McGraw-Hill, New York.

Salame, M. 1989. The use of barrier polymers in food and beverage packaging. pp. 132–145. *In*: Finlayso, K.M. (ed.). High Barrier Plastic Films for Packaging. Technomic Publishing, Lancaster, PA.

Skurtys, O., C. Acevedo, F. Pedreschi, J. Enrione, F. Osorio and J.M. Aguilera. 2010. Food Hydrocolloid Edible Films and Coatings. Nova Science Publishers, Hauppauge, New York, USA.

Formulation, Properties and Performance of Edible Films and Coatings from Marine Sources in Vegetable and Fruits

Armida Rodríguez-Félix[1], Tomás J. Madera-Santana[1,]* and Yolanda Freile-Pelegrín[2]

[1] Centro de Investigación en Alimentación y Desarrollo (CIAD), A.C. CTAOV. Km. 0.6 Carr La Victoria 83304 Hermosillo, Sonora México
[2] Departamento de Recursos del Mar. CINVESTAV-IPN, Unidad Mérida Carr Antg Progreso 97310 Mérida, Yucatán México

Introduction

One of the most important processes for maintaining the quality of fresh produce for storage, transportation, and end-use is packaging (Greener-Donhowe and Fennema 1994). The main functions of packaging are protection, containment, to provide information, convenience, preservation and safe delivery of fresh produce until consumption. However, during the distribution process, the quality of fresh produce can deteriorate due to biological and chemical factors as well as physical factors. The food packaging contributes to extending the shelf life and maintaining the quality and safety of the food products. Packaging technology uses several packaging materials to analyze their performance with respect to fresh produce and to reduce the volume or weight of materials, costs and resources.

Edible films and coatings have attracted attention due to their potential to offer solutions regarding health, food quality, convenience, safety, etc. However, these materials could be used to solve technical challenges to a greater extent than they currently are. For fresh fruits and vegetables, between

*Corresponding author: tmadera.santana@gmail.com

25 and 80% of harvested produce is lost due to spoilage. It has a huge impact on developing and emergent economies. Therefore, edible films and coatings are considered a viable alternative method to extend postharvest shelf-life of fresh produce. Logically, to know how the edible film or coating can affect harvested fresh produce, it is necessary to have background knowledge of postharvest fruit physiology and necessary storage techniques (Baldwin 1994).

In this chapter, a general overview of the quality features of fruits and vegetables, as well as edible films and coatings, is described. Particular attention is given to edible films and coatings from marine sources, and in this sense, two polysaccharides, agar and chitosan, are described in detail in terms of sources, physicochemical properties and their application in postharvest handling of fruits and vegetables. Finally, a conclusion and a general perspective on future applications are given.

Quality Features of Fruits and Vegetables

Physiological Changes (Respiration and Ripening–Maturity)

Fresh horticultural crops are diverse in morphology, composition and physiology. Because fresh produce is living tissue, its metabolic activity continues after harvest, affecting its quality and causing deteriorative changes that comprise its postharvest life (Kader 2002; Kays 1997). When fresh commodities are harvested, the supply of mineral nutrients and water from soil is eliminated, as well as the flow of carbon and energy from photosynthesis. Therefore, once produce is harvested, the stored nutrients represent the reserves that will be used to maintain metabolic activity or sustain essential chemical and physiological activities (Kays 1997).

Respiratory metabolism is one of the most important characteristics of fresh produce due to its influence on postharvest life. Respiration is a series of oxidation-reduction reactions that use stored organic materials of produce (carbohydrates, proteins and fats) as substrates, and oxygen from the environment; those substrates are oxidized to form various intermediate compounds and carbon dioxide (CO_2), whereas oxygen (O_2) is reduced to form water. The intermediate compounds produced by respiration, also called carbon skeletons, are precursors required for the formation of plant products such as amino acids, nucleotides, pigments, lipids and flavor compounds. In addition, carbon dioxide and water are produced during the respiration process. An important amount of energy produced in respiration is retained by the cell as high energy bonds in compounds that are used in other metabolic essential processes. However, the remaining energy produced in respiration is lost as heat (Kays 1997; Kader 2002; Saltveit 2014). This heat, known as vital heat, varies with the commodity and increases as the temperature increases, up to about 38 to 40 °C, and is part of the refrigeration load that must be considered during cold storage of horticultural products (Hardenburg et al. 1986).

There is a close relationship between the respiration rate and storage life. It is related to the intermediate compounds formed during respiration, which affect the rate of metabolic processes related to produce quality such as flavor, sugar content, aroma, and others. At higher respiration rates of the commodity, there is a greater rate of deterioration or perishability, and shorter storage life can be observed in the produce. Asparagus, broccoli, mushroom, pea, spinach and sweet corn are vegetables classified with an extremely high respiration rate (higher than 60 mg CO_2 kg^{-1} h^{-1} at 5°C) that are considered very perishable with a potential storage life of less than two weeks at near optimal temperature and relative humidity; whereas commodities such as apple (some cultivars), lemon, winter squash, sweet potato, dry onion, and potato (mature), classified as low respiration rate (5 to 10 mg CO_2 kg^{-1} h^{-1} at 5 °C) produce, are considered to have low perishability with a potential storage life of 8 to 16 weeks (Kader 2002; Saltveit 2014).

Table 1. Classification of fruits according to their respiratory behaviour during ripening.

Climacteric Fruits		Non-Climacteric Fruits	
Apple	Papaya	Blackberry	Pomegranate
Apricot	Passion fruit	Cactus pear	Raspberry
Avocado	Peach	Carambola	Strawberry
Banana	Pear	Cherry	Tangerine
Blueberry	Persimmon	Cranberry	Mandarin
Breadfruit Cherimoya	Plum	Cucumber	Watermelon
Feijoa	Quince	Grape	
Fig	Sapote	Grapefruit	
Guava	Soursop	Lemon	
Jackfruit	Tomato	Lime	
Kiwifruit		Lychee	
Mango		Olive	
Muskmelon		Orange	
Nectarine		Pepper	
		Pineapple	

Source: Kader (2002).

The respiration is affected by different environmental factors, but the most important are temperature, atmospheric composition and physical stress. However, temperature is the factor that has the greatest influence on the storage life of commodities, due to its effect on respiration and metabolic reactions; there is an exponential rise (roughly doubled or tripled) in respiration at increased temperatures from 0 to 30 °C (32 to 86 °F), considered the physiological range of most crops. The changes in the respiration rate with temperature follows van't Hoff's rule, which states that the rate of most chemical or biochemical reactions increases two or three times with every 10 °C increment in temperature. The respiration rate is reduced when the atmospheric composition is modified and O_2 levels are lowered to values around 2 to 3% (Hardenburg et al. 1986; Saltveit 2014).

The respiration rate changes depending on the stage of development of the horticultural product; it declines as the plant organ matures. Therefore, a horticultural product harvested during active growth such as vegetables or immature fruits have a high respiration rate, whereas mature fruits, dormant buds and storage organs have a low respiration rate. After harvesting, a decline in the respiration rate is observed. This is slow in non-climacteric fruits and storage organs and rapid in immature fruit; it is related to loss of substrates used in respiration (Saltveit 2014).

The respiratory behavior of fleshy fruits shows differences during ripening, and based on this, fleshy fruits have been classified into two classes: climacteric and non-climacteric (Table 1). Climacteric fruits, such as tomato, banana, avocado, apple, and others, show an increase in the respiration rate and in ethylene production at the beginning of ripening. This type of fruit can be harvested at physiological maturity and the ripening will continue off the plant or begins after treatment with exogenous ethylene. On the contrary, in non-climacteric fruits, there is not an increase in the respiration rate during ripening, and ethylene does not play a critical role in the ripening process. These fruits have a low respiration rate and ethylene production during ripening. In these fruits, the ripening process does not continue after the fruits are removed from the plant or harvested; therefore, these fruits should be harvested when fully ripe to ensure good flavor quality (Kader 2002; Seymour et al. 2013; Saltveit 2014).

Water Release (Dehydration)

Water is the principal component of fruits and vegetables varying from 85 to 90%, and reaching up to 95% by weight in lettuce. Water from the cells vaporizes into the intercellular spaces and maintains a saturated atmosphere inside almost all fruits and vegetables. Water has a central role in the postharvest biology of produce; therefore, the movement of water to establish equilibrium between the product and its environment has great importance during postharvest handling of horticultural products. Another factor to be taken into account is the temperature management. Water is lost from produce as moisture (vapour) by evaporation, known as transpiration, and this process begins once the product is harvested, and is cumulative through the produce distribution chain. This process involves the transport of moisture through the skin of the commodity, the evaporation of this moisture from the commodity surface and the convective mass transport of the moisture to the surroundings. However, a minor weight loss occurs due to loss of carbon during the respiration process (Mitchell et al. 1972; Hardenburg et al. 1986; Kays 1997; Holcroft 2015).

Water loss is a major cause of produce deterioration, because of its impact on quality, which causes loss in salable weight (a decrease in produce value), loss in appearance (due to wilting and shriveling), loss in texture such as softening, flaccidity, loss of crispness and juiciness and in nutritional quality. The maximum percentage of water that can be lost from produce before it becomes unsalable varies widely depending on the species, and can be as low as 3% in spinach (*Spinacia oleracea,* L.) leaves to 41% in snap beans (*Phaseolus vulgaris,* L.). Water vapour is lost from horticultural products through the lenticels, stems or scars, any injured area or directly through the cuticle (Mitchell et al. 1972; Hardenburg et al. 1986; Kays 1997; Kader 2002; Holcroft 2015). Transpiration in leaves occurs mainly through stomata. However, transpiration through the cuticle is the primary mode of water loss in most fleshy fruits that have no or few stomata or non-functional stomata (Martin and Rose 2014).

The cuticle, a mostly lipidic protective hydrophobic layer, which covers the epidermal surfaces of all above-ground organs, is one of the most important plant barriers (its main function has been considered to minimize water loss). Other functions include limiting the loss of substances from internal tissues and providing mechanical support to maintain plant organ integrity. However, recent research considers that cuticles may regulate fruit quality and affect other features related to postharvest quality because it may play a major role in the protection against biotic and abiotic factors, as well as fruit appearance and textural properties. Cuticles are mainly composed of cutin, a lipidic polyester, which is impregnated and covered by hydrophobic compounds known as intracuticular and epicuticular waxes (Heredia-Guerrero et al. 2014; Martin and Rose 2014; Lara et al. 2014, 2015). Other constituents of some cuticles are cutan, a non-saponificable polymer, and a significant cell wall polysaccharide fraction (cellulose and pectin), and flavonoids that impart particular mechanical characteristics to cuticles (Martin and Rose 2014; Lara et al. 2015). It is important that the structural integrity of the cuticle be maintained for its function preservation. The chemical composition, thickness, structure and architecture of fruit and vegetable cuticles vary greatly among species and during development (Kader 2002; Martin and Rose 2014; Lara et al. 2015).

The net flux of water into or out of harvested produce is defined by the magnitude of the force's driving movement, which is the difference between water vapour pressure between the interior of the produce and the exterior or environment, and the composite of resistance to movement. The flux is equal to the water vapour pressure deficit divided by the cumulative resistance to movement (Kays 1997):

$$\text{Flux} = \frac{\text{Water vapor} - \text{Water vapor outside}}{\text{Resistance to water movement}}$$

The transpiration rate of fruits and vegetables is affected by internal or produce factors and external or environmental factors. The internal water vapour pressure is greatly influenced by the amount of moisture within the tissue, and in fruits and vegetables is considered to be at least 99%. The commodity evaporative surface is determined by commodity type and cultivar. The commodity evaporative surface has a great influence on the rate of transpiration or water loss. Some characteristics of the commodity surface such as natural waxes and the presence of trichrome increase the surface resistance to water loss. Additionally, the chemical composition and structure of the commodity affects the "retention" (how tightly water is held) water by the tissue and water loss. Transpiration coefficients (K) vary greatly among commodities from 42 up to 7400 mg/kg/sec/Mpa for apples and lettuce, respectively. Surface to volume ratio is a commodity factor of critical importance that affects transpiration. Horticultural products have a wide range in the surface to volume ratio from 50 to 100 cm^2/cm^3 in leafy vegetables, 0.5 to .5 cm^2/cm^3 in fruits like apple, pear, peach, citrus fruit, banana, and 0.2-0.5 cm^2/cm^3 in sugar beet, yam and pumpkin. Environmental or external factors that can have a direct effect on the force's driving movement are temperature, relative humidity (HR), air movement and atmospheric pressure (Hardenburg et al. 1986; Kays 1997; Paull 1999; Kader 2002; Holcroft 2015).

Relative Humidity and Temperature of Storage Conditions

During produce storage, it is very important to minimize water loss or transpiration by maintaining a uniform desired temperature and relative humidity (RH). Most fresh produce requires 85 to 95% RH, while dried commodities, such as onion and ginger, need a low RH. The amount of water vapour in the air surrounding a commodity greatly affects the movement of water into or out of the produce. This amount of water can be expressed as percent of relative humidity (% RH), absolute humidity, vapour pressure and dew point. The percent of RH is a ratio of the quantity of water vapour present and the maximum amount possible at that temperature and pressure. The amount of water vapour in the air is greatly affected by temperature and pressure. The difference in temperature during postharvest handling or storage has a major effect on the amount of water vapour in the air without changing the percent of RH. Water vapour pressure (*WVP*) is an equivalent measure of the water vapour in the air; it is a measure of the pressure exerted by the water vapour in the atmosphere. The water vapour pressure deficit (*WVPD*) is the difference between the water vapour inside the fruit or vegetable and that of the surrounding air (Hardenburg et al. 1986; Kays 1997; Holcroft 2015). The *WVPD* is greatly affected by temperature and humidity; it can increase exponentially as temperature increases or increase linearly as humidity decreases. The rate of water loss can be simplified and expressed as:

$$\text{Water loss} \left(\frac{\%}{\text{day}} \right) = K \text{ value } x \text{ WVPD}$$

Water loss of fruits and vegetables is reduced by a decrease of the transpiration coefficient (*K*) or the water vapour pressure deficit (*WVPD*).

The transpiration coefficient can be reduced by means of the application of waxes or edible coating, and/or packaging with plastics, or reducing mechanical damage. The *WVPD* is reduced by environment manipulation such as reducing temperatures, increasing RH and reducing air movement (Paull 1999; Kader 2002; Holcroft 2015).

Temperature is the most important factor that affects the deterioration rate of commodities. Therefore, good temperature management is critical for quality preservation between harvest and consumption, beginning with rapid removal of field heat and maintaining a low temperature during the distribution chain to retard produce metabolic changes. Products stored every 10 °C above their optimum temperature increase their deterioration rate by two to three fold. Cold storage is recommended for most commodities because it delays: a) ripening, softening and colour changes, b) water loss with its consequent shriveling and wilting, c) decay by bacteria fungi and yeast, d) respiratory heat production and undesirable metabolic changes and e) undesirable growth. Low-temperature storage protects texture, nutrition, aroma and flavor of commodities. Temperature also affects the ethylene effect, due to low-temperature storage which decreases produce sensitivity to ethylene, and also influences the effect of low oxygen and high carbon dioxide. Cold storage should be done at the temperature required by the commodity because if the products are stored at undesirable temperatures can cause physiological disorders such as (1) freezing injury if the produce is stored below their freezing temperatures causing total loss of the commodity; (2) chilling injury if the produce is stored above its freezing point but below 5 to 15 °C, depending on the product; this type of disorder occurs mainly on tropical and subtropical produce; or (3) heat injury if produce is handled at very high temperatures (Hardenburg et al. 1986; Kays 1997; Paull 1999; Kader 2002; Thompson et al. 2002).

Gas Transfer Rate

Clearly, the respiration rate of fresh produce is expressed in terms of O_2 consumption rate and/or CO_2 production rate. Experimental methods to determine the respiration rate are: the permeable system, the flowing or flushed system, and the most common, the closed or static system. In the last system, a gas-tight container with a known volume is filled with the product and the container has ambient air, which is the initial atmosphere, and is closed (Fishman et al. 1996; Maneerat et al. 1997; Jacxsens et al. 1999). The changes in O_2 and CO_2 concentration over a certain period of time are measured to calculate the respiration rate. The absolute difference in gas concentration between the outlet and the inlet, when the system reaches steady state, is used to calculate the respiration rate. However, the difference of concentrations is enough to guarantee a noticeable modification in the container's atmosphere; this modification has to be minimal in order to avoid affectation in the respiration rate. A procedure to determine the period of time based on the accuracy of the gas measuring equipment was proposed by Talasila (1992). However, the gas concentration is associated with the respiration rate measured at the beginning and the end of the experimental period.

As we can observe, the estimation of the gas flow rate is often difficult to measure in the flow through the system. Moreover, the gas flow rates must be carefully chosen to measure accurately the differences in gas concentration between outlet and inlet. Gases move by bulk flow, and this movement is produced by a pressure gradient and all of the gases present that move together rather than independently as with diffusion. In postharvest conditions, the gas movement is followed by bulk flow, and it has low pressure (hypobaric) storage. This means that the atmospheric pressure is reduced around the fresh produce; the pressure differential produces an outward flow of gases from the product to equilibrate the internal and external pressures (Kays 1997).

Vegetables and fruits have different metabolic activities, and consequently, different respiration rates and gas exchange. Generally, non-climacteric produce has higher respiration rate in the early stage of development that steadily declines during maturation (Lopez-Galvez et al. 1997). In contrast, climacteric products do not follow this behavior, and they exhibit a peak of respiration and ethylene production (C_2H_4) associated with senescence or ripening. Nevertheless, it is not related to changes in the respiratory response under modified atmosphere (MA) and controlled atmosphere (CA) during the climacteric period.

A particular observation should be taken into account when fresh produce is packaged in modified atmosphere packaging (MAP) due to alterations of the respiration rate over time. Fresh produce in MAP is usually for short-term storage (distribution and retailing), thus, the influence of storage time due to senescence may be considered negligible. The climacteric changes are important in the long term and are not considered relevant to MAP, according to Fishman et al. (1996). However, the gas transfer during the respiration process under a MAP is quite complex, because it involves several factors (gas composition, temperature), mechanisms (enzymatic reactions), etc., where many of them are very difficult or even impossible to model. The usual alternative strategy is to develop empirical models as a function of controllable variables (temperature, gas concentrations, volume, etc.) (Fonseca et al. 2002). These authors mentioned that particular variables that influence the O_2 uptake and CO_2 production rate should be identified and quantified for each fruit or vegetable product.

Overview of Edible Films and Coatings for Foods

Definition

Recently, edible films and coatings have received considerable attention because they offer extra advantages such as edibility, biocompatibility, appearance, barrier properties to gasses and moisture, mechanical properties, microbial protection, non-sensory perceptions, non-toxicity, non-polluting and low cost (Debeaufort et al. 1998; Han 2000; Krochta 2002). Moreover, increased interest in edible films and coatings have been motivated by the environmental impact produced by the use of non-biodegradable plastic-based packaging. Therefore, the biopolymers such as starches, cellulose derivatives, proteins (from animal and plant-based), gums, lipids, chitin/chitosan, and phycocolloids (agar, carrageenans, alginate) offer suitable advantages to get the thin film to form and coatings to adhere and cover surfaces of fruits and vegetables, as well as of processed food in order to extend their shelf life.

An edible film or coating is a primary packaging made primarily from edible components from biopolymers. A thin layer of edible material can be directly coated on a fruit or vegetable and it can be used as a food wrap without changing the original ingredients or the processing method. The main difference between an edible coating and edible film is the form of application. An edible coating is applied in liquid form on a food product, but an edible film is obtained as solid sheets and then applied to food products (Falguera et al. 2011; Kang et al. 2013). Nevertheless, edible films and coatings have the ability to improve food quality in different aspects, which will be described in detail in the following sections of this chapter.

Performance and Advantages

The most important characteristics of edible films and coatings are probably the edibility and inherent biodegradability (Guilbert et al. 1996; Han 2014), and that they can be consumed with the packaged products. Therefore, the package (edible film or coating) does not need to be disposed of even if the package is not consumed, so without a doubt, this kind of packaging material contributes to the reduction of environmental pollution. Moreover, the edible films and coatings can enhance organoleptic properties of packaged foods (fresh produce, meat, etc.), such as flavorings, colourings, sweeter, etc. Also, they can be applied for individual packaging of small pieces of food (fruits, vegetables, beans, nuts, meat pieces, etc.), and used inside heterogeneous foods at the interfaces between different layers of components. The films can be tailored to avoid deteriorative inter-component (moisture) and solute migration of different foods (pizzas, pies or candies) (Bourtoom 2008).

The use of edible films or coatings has attracted attention because they can act as carriers for antimicrobial and antioxidant agents, and also, when applied on the surface of any fresh produce, can control the diffusion rate of the preservative agent to the interior. Recently, the application of multilayer edible films and coatings has been an alternative to improve the performance of these materials in fresh produce.

The performance of edible films and coatings is strongly dependent on their mechanical and barrier properties. To determine these properties, standard methods have been developed to determine material performance, such as permeability parameters, where these values could measure the water vapour and gas transfer through the film. The water vapour transmission rate (WVTR) is used to calculate water vapour permeability (WVP), this process consists of solution and diffusion of the solute vapour through the membrane (film). To determine the permeability parameters several methods have been reported (Gennadios et al. 1993). The "cup method" is the most popular method to determine WVP in films, and it is based on the gravimetric technique (Kester and Fennema 1989; Martin-Polo and Voilley 1990); however, it is possible to find standardized methods to determine WVP in films and thin sheets of packaging (ASTM E96, 2015). For fruits and vegetables, the gas permeability of edible films and coatings cannot use procedures described by the ASTM D1434 and by ISO 2556. However, some instruments such as Oxtran (Mocon, Modern Control Inc.) can be used to determine the oxygen permeability. This instrument has a colourimetric sensor or a manometer gauge used by Lyssy L100 (Lyssy, Zurich Switzerland) (Debeaufort et al. 1998). Some reports on measuring gas permeability in edible films have used the gas chromatographic method (Karel et al. 1963; Hagenmaier and Shaw 1992; Debeaufort and Voilley 1995).

On the other hand, an edible film or coating can have good barrier properties but poor mechanical properties, where they do not maintain the film integrity in packaging, handling and carrying processes. The methods used for mechanical characterization in edible films and coatings are those for plastic packaging described by the standard test method ASTM D882 (ASTM 882, 2012). Usual mechanical parameters measured in edible films and coatings are a tensile strength, elongation, deformability, elastic modulus, tear strength, etc. (Debeaufort et al. 1995). However, mechanical properties are affected by the conditions of testing, such as temperature and relative humidity, irradiation, etc. (Debeaufort and Voilley 1997).

Edible films and coatings from carbohydrates have shown an improvement in the quality and stability of meat during storage and the commercialization process (Debeaufort et al. 1998). Polysaccharide films and coatings can be applied on fresh and frozen meat, poultry and fish to prevent superficial dehydration. Mixtures of alginates and starch to coat beef steaks, pork chops and skinned chicken drum sticks have shown excellent performance in the shelf life of these food products. In fresh produce, the edible films and coatings are developed to control the ripening by reducing oxygen diffusion in the produce and increasing CO_2 and ethylene release. Moreover, spoilage can be retarded or prevented by the incorporation of antimicrobial agents within the film; it means that the produce can be preserved by the coating during and after the manufacture process (Debeaufort et al. 1998).

Additives and Functionalities

Edible films and coatings are generally classified according to the structural material: proteins, polysaccharides, lipids or composite. The main advantage of such structures as potential food packaging materials is their biodegradability. Starch, cellulose and its derivatives, pectin, chitosan, alginate, carrageenan, pullulan and gellan gum are the main polysaccharide materials tested as edible packaging materials (Han and Gennadios 2005). Plasticizers (i.e. glycerol, sorbitol, monoglycerides, polyethylene glycol, glucose) are usually used to increase flexibility and elasticity of biopolymers. Plasticizers are non-volatile and low-molecular weight compounds, which are added to polymers in order to reduce brittleness, impact flow, and flexibility, and enhance toughness and strength of films. As a specific definition for coatings, the plasticizers modify the impact resistance of the coating and reduce flaking and cracking by improving coating flexibility and toughness. As a disadvantage, plasticizers usually increase the oxygen permeability, moisture, aroma, and oils because, the intermolecular attractions along the polymer chains are reduced (it means an increase of free volume) (Sothomvit and Krochta 2005; Rojas-Graü et al. 2007; Madera-Santana et al. 2011).

The plasticizer characteristics are its small size, high polarity, more polar groups per molecule and more distance between polar groups within a molecule (Tavassoli-Kafrani et al. 2016). Some common plasticizers used in edible films and coatings are monosaccharides, disaccharides or polysaccharides (most of them are present in the syrup of fructose-glucose, corn, sucrose or honey). Furthermore, traditionally, polyols are used (glycerol, sorbitol, glyceryl derivatives, polyethylene glycol (PEG), etc.) and lipids and derivatives (fatty acids, phospholipids and surfactants) (Sothomvit and Krochta 2005; Hamzah et al. 2013; Rodríguez-Núñez et al. 2014). In fact, water could be used in some biopolymers as a plasticizer because the polymer chains are disrupted in their hydrogen bonds or form hydrogen bridges (Xiao et al. 2012; Rodríguez-Núñez et al. 2014).

The antimicrobial agents are incorporated into edible films or coatings in order to decrease pathogen growth and to protect foodstuff against spoilage flora. These additives are classified as chemical and natural antimicrobial agents. The chemicals are sodium benzoate, propionic, sorbic and benzoic acid. Among the most widely studied antimicrobials are the organic acids (acetic, lactic, propionic and malic), enzymes (lysozyme, lactoperoxidase), bacteriocins (nisin, lacticin), peptides and natural antimicrobials (spices, essential oils, propolis, etc.) (Sánchez-González et al. 2011a). However, consumers have had a particular concern about the health effect of these agents. Therefore, the demand to consume natural and health-safe antimicrobial agents, they are classified as GRAS (generally recognized as safe), has recently increased. Some examples of these agents are organic acids (malic, citric, acetic and lactic acid), biopolymers such as chitosan, plant-derived secondary metabolites as phytoalexins and essential oils (garlic, clove, oregano, thyme, rosemary, lemongrass, cassia, etc.).

Nowadays, nanotechnology has revolutionized several industries, where food packaging has not been an exception. However, the application of nanomaterials in edible coatings is not at all accepted by international agencies, and the reports on metal nanomaterials have recently increased in research papers. One of the common metal nanomaterials used as an antimicrobial agent is silver nanoparticles because the silver ions are able to release from the film or coating support and attach to the cell membrane and then start to penetrate into the bacteria. This antimicrobial agent has high efficiency to interrupt bacterial process, such as cell division and respiration, producing cell death (Jiang et al. 2013).

Edible films and coatings can carry several active agents, such as antioxidants, nutraceuticals, flavors and colourants, besides the additives previously mentioned. They can enhance and improve the food quality and safety, without any interference with the physicochemical properties of the films (Howard and Gonzalez 2001; Han 2003; Sánchez-González et al. 2011b). Moreover, probiotics are functional ingredients incorporated in edible films and coatings, as well as minerals and vitamins (Robles-Sánchez et al. 2013; Rhim and Wang 2014), in order to increase or improve the functionality of films. Flavor, colourants or pigments are also usually incorporated into edible films and coatings to improve sensory quality in vegetables or fresh-cut food products (Skurtys et al. 2010; Tavassoli-Kafrani et al. 2016). The combination of antioxidants and antimicrobial additives in edible films and coatings develops functions of active film packaging. These films and coating systems are able to protect food products from microbial spoilage, fungi attack, oxidation, among other external agents, producing a significant improvement in quality and safety in foodstuff (Lee et al. 2012; Kang et al. 2013). Recently, an increase in the number of patents and research articles on active packaging topics can be observed in databases. A combination of several antimicrobials and other additives such as bioactive compounds in modified atmosphere packaging has been investigated because these films inhibit the development of spoilage and pathogenic bacteria by controlling the release of the bioactive compound (Serrano et al. 2008).

Agar and Chitosan as Edible Films and Coatings from Marine Sources

Agar: Sources and Physicochemical Properties

Macroalgae, or seaweeds, are one of the most important marine resources and represent a crucial component of the marine ecosystem as primary producers in oceanic food webs. Many species of seaweeds occur along the seacoast, mostly in the intertidal zone. Seaweeds comprise a complex and heterogeneous group of organisms characterized by their photosynthetic nature and their simple reproductive structures. They are classified into three divisions according to, mainly but not exclusively, their pigment composition: red, brown and green. Botanists refer to these broad groups as Rhodophyta, Heterokontophyta and Chlorophyta, respectively (Armisén et al. 2000).

The principal commercial seaweed extracts are the three hydrocolloids (also called phycocolloids): alginates extracted from brown seaweeds, and agar and carrageenans extracted from red seaweeds. The processed food industry is still the primary market for the phycocolloids where they serve as texturing agents and stabilizers, although agar and its derivative products agarose and bacteriological agar have long enjoyed attractive markets as microbiological and electrophoresis media, respectively (Bixler and Porse 2011).

Related to agar, the main sources of this phycocolloid are two different agarophytes *Gelidium* and *Gracilaria*, which are the dominant industrial seaweeds for the agar extraction. *Gelidium* was the original species used, until shortages during World War II that led to the discovery of *Gracilaria* as a suitable alternative to *Gelidium* (Imeson 2009). However, *Gracilaria* is only suitable as an alternative to *Gelidium* if it is treated with alkali prior to the extraction, which causes a chemical change that leads to increased gel strength (McHugh 2003). Other species that contribute to the agar market are *Pterocladia* and *Gelidiella*, although in a much smaller amount (Armisén et al. 2000). About 9,600 tons of agar (valued at US$173 million) were produced worldwide in 2009, and *Gracilaria* has become the preferred seaweed for the production of food grade agar. This is due to the success of its cultivation in Indonesia and Chile, and to a far lesser degree, in Malaysia, Thailand and China. The cultures of *Gracilaria* increase its availability and therefore, more competitive prices (Bixler and Porse 2011). On the other hand, *Gelidium* is harvested from wild beds in Spain, Portugal, Morocco, Japan, Korea, Mexico, France, USA, China, Chile and South Africa (Hernández-Carmona et al. 2013).

Agar is defined as a mixture of polysaccharides that occur in the cell matrix of the marine red algae (Rhodophyta). Its biological function consists of offering a flexible structure to the algae, helping them stand the varying stresses of currents and wave motion (Craigie 1990). The traditional concept of agar has been comprised in neutral agarose and negatively charged agaropectin. The agarose is a polymer that contains agarobiose (1,3-linked β-D-galactose and 1,4-linked 3,6-anhydro-α-L-galactose (Fig. 1).

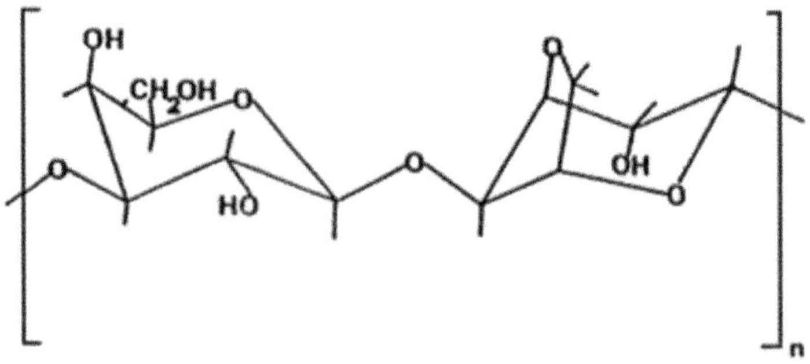

Fig. 1. Chemical structure of an agarobiose unit (Armisén et al. 2000).

Substitution with sulfates, methyl ethers and/or pyruvate ketals can occur at various sites in the polysaccharide chain. The substitution pattern of these groups depends on both the algae species as well as the extraction method used, which can promote desulfation, causing an increase in agar quality since agar gelation is produced through a helical conformation of agar polysaccharides (Murano 1995; Hernández-Carmona et al. 2013). These helices can be aggregated by hydrogen bonds and, therefore, thermo-reversible gels are formed (Lahaye and Rochas 1991).

Gel strength of agar depends on the species. Desulfation in agar molecules in species of *Gelidium*, *Gelidiella* and *Pterocladia* occurs through a natural internal transformation through an enzymatic process that can be considered a maturing of the polysaccharide in the algae. On the other hand, agar isolated from *Gracilaria* species are typically more sulfated with the pattern of sulfation dominated by the esterification of C-6 of the linked galactose L-unit. This L-galactose 6-sulfated residue is synthesized in *Gracilaria* as a biological precursor of the 3,6-anhydro-L-galactose, and is enzymatically converted to the anhydrous form by sulphohydrolases (Murano 1995). However, this enzymatic activity seems to be lower than that which occurs in the other genera, therefore, a higher number of unfinished 6-sulfated molecules producing non-gelling or weak gel-forming polysaccharides. It becomes necessary to produce desulfation industrially by means of a chemical method before extracting the agar from the algae. Sulfate units at points in the polysaccharide chains comprising the gel cause kinks in the helical structure responsible for gel formation. This results in agars of lower gel strength (Lahaye and Rochas 1991). It can be improved by alkali treatment of agar molecules eliminating the unstable axial sulfate groups at C6 of L-galactopyranosyl units and converts it into the anhydride units, giving rise to a more stable 3,6-anhydro-galactose (Armisén et al. 2000).

Agar is insoluble in cold water, but it swells considerably, absorbing as much as twenty times its own weight of water. It dissolves readily in boiling water and sets to a firm and transparent gel at concentrations as low as 0.50%. Regarding its gelling power, agar is outstanding among other hydrocolloids. Not needing any other substances to produce strong gels, agar has an enormous potential in applications such as a foodstuff ingredient. A 1.5% solution of agar forms a gel on cooling to about 34 to 42 °C that does not melt below 85 °C. This hysteresis interval is another interesting property of agar that finds many uses in food applications since it lends a suitable balance between easy melting and good gel stability at relatively high temperatures. The gelling temperature is affected by the methoxyl groups, which vary depending on the algal species used (Guiseley 1970). In particular, the extent of methoxylation of agar polymers from *Gracilaria* is significantly higher than *Gelidium* and *Pterocladia*, and this difference is reflected in their gelling temperatures, which are in the range of 40-42 °C in *Gracilaria* species, compared to 34-36 °C for Gelidiaceae (Hernández-Carmona et al. 2013). Thus, physical properties of the resulting agar (gel strength, gelling and melting temperatures) are linked to their chemical structures, and therefore, to the biological species.

Agar-based Edible Films and Coatings

Biopolymers from several natural sources are renewable, environmentally friendly, abundant, inexpensive, biocompatible and biodegradable (Tang et al. 2012). They are also considered as attractive alternatives for non-biodegradable petroleum-based plastic packaging materials (Freile-Pelegrín et al. 2007; Madera-Santana et al. 2011). In this sense, agar is a biopolymer that has shown particular characteristics. For example, due to its ability to form very hard gels at very low concentrations, agar has been used extensively as a gelling agent in the food industry and in other applications such as microbiology and molecular biology techniques. Other uses of agar include dental moulds, casting of archaeological pieces and sculpture moulds (Freile-Pelegrín et al. 2007). Due to its combination of renewability and biodegradability, its enormous gelling power, and the simplicity of the extraction process, agar has been singled out as a promising candidate for future use in plastic materials. Therefore, the use of agar in combination with natural or synthetic polymers to produce materials with degradable properties is highly promising (Madera-Santana et al. 2009, 2010). Packaging materials based on biopolymers have beneficial properties (such as shelf-life of food, securing food safety and improve food quality) when bioactive compounds are incorporated. Naturally derived polymers, such as polysaccharides (starch, cellulose, pectin, alginate, agar, chitosan, among others), are attractive due to a good film forming, moderate oxygen and moisture permeability, and particular colloidal nature. Agar is a promising polysaccharide for developing biodegradable, bioactive, antimicrobial packaging films (Madera-Santana et al. 2011; Gimenez et al. 2013). This polysaccharide is extracted from marine red algae; it can be used to produce films. These films have been characterized by mechanical, thermal and transport properties (Madera-Santana et al. 2011, 2014; García-Baldenegro et al. 2014). In order to improve the properties of agar films, the addition of additives (plasticizers) has been reported (Madera-Santana et al. 2014). Furthermore, agar has also been blended with other biopolymers such as starch, protein, carrageenan, gelatin and with synthetic polymers such as polyvinyl alcohol (PVOH) (Letendre et al. 2002; Phan et al. 2005; Rhim et al. 2013; Gimenez et al. 2013; Madera-Santana et al. 2014). Agar films are able to carry semiochemical or natural extracts, and antimicrobials (silver or zinc oxide nanoparticles) (Garcia-Baldenegro et al. 2014; Kanmani and Rhim 2014a, 2014b).

Chitosan: Sources and Physicochemical Properties

In recent years, a remarkable development in edible films and coatings has been developed using biopolymers. These materials can be prepared in thin films from edible materials such as cellulose, chitosan, alginate and starch that act as a barrier to the external elements (such as moisture, oil and vapour), thereby protecting the products and extending their shelf life (Ziani et al. 2008).

Chitosan is usually isolated from the exoskeletons of crustacean, mollusks, insects and certain fungi. It is a natural polymer derived through the deacetylation of chitin, which is the second most abundant biopolymer after cellulose and has a unique nature for a cationic polysaccharide. Chemically it is a copolymer consisting of chitosan β-(1–4)-2-acetamido-D-glucose and β-(1–4)-2-amino-D-glucose units (Fig. 2) with the latter usually exceeding 60%.

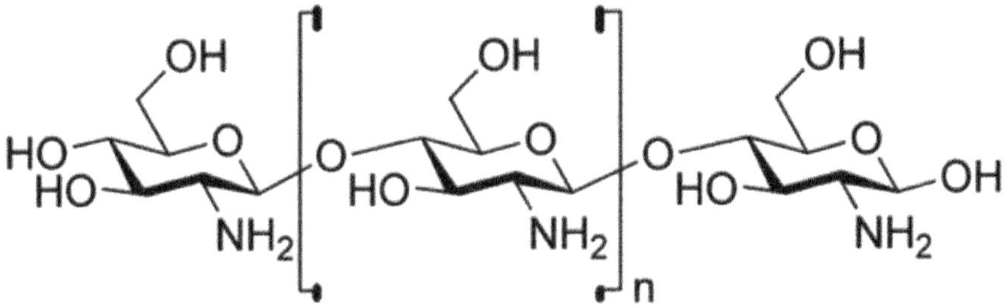

Fig. 2. Chemical structure of chitosan (Younes and Rinaudo 2015).

Chitosan has been demonstrated to be nontoxic, biodegradable and biocompatible, and it is insoluble in water, but soluble in acidic solvents such as dilute hydrochloric, formic and acetic acids. In acidic solutions, the amine groups on the chitosan molecule are protonated to NH^{3+} and thus acquire a positive charge (Ye et al. 2008). Because of the film forming properties of chitosan, it has been reported to be a potential material for producing food packaging, particularly as edible films or coatings (Rivero et al. 2009). Chitosan films are generally clear and transparent, and they do not possess any pores. In addition, Suyatma et al. (2005) reported that the production of chitosan from shrimp shells, which is a by-product from the seafood industry, is economically feasible.

The properties of chitosan films depend on the morphology of the chitosan, which is affected by its molecular weight, degree of N-acetylation, solvent evaporation and the free amine regenerating mechanism. Additionally, chitosan films have intrinsic antimicrobial activity and inhibit the growth of a wide variety of bacteria (*Escherichia coli*, *Salmonella typhimurium*, *Staphylococcus aureus* and *Listeria monocytogenes*) and fungi (*Botrytis cinerea*, *Fusarium oxysporum* and *Piricularia oryzae*) (Rabea et al. 2003; Holappa et al. 2006). Due to this property, chitosan has been successfully used to maintain the quality of post-harvest fruits and vegetables. One of the reasons for the antimicrobial character of chitosan is its positively charged amino group which interacts with the negatively charged microbial cell membrane, leading to the leakage of proteinaceous and other intracellular constituents of the microorganism (Pranoto et al. 2005).

Although chitosan has a good film-forming ability, the chitosan film is brittle. The brittleness of the films is primarily determined by the strength of polymer-polymer interactions, which can be controlled through the polymer chemistry and the addition of a plasticizer. Plasticizers are generally small molecules, such as polyols including sorbitol, glycerol and polyethylene glycol (PEG), which disperse and intercalate between the polymer chains, thereby disrupting hydrogen bonding and spreading the chains apart to increase the flexibility, water vapour transmission rate and gas permeability of the films. Glycerol (GLY) and sorbitol (SOR) are considered to be good plasticizers for use in edible films (Rodríguez-Núñez et al. 2014). The films plasticized with GLY are susceptible to water, therefore some limitations of use are environments with high relative humidity. Furthermore, SOR can crystallize when the films are stored under low and intermediate relative humidity conditions (Ray et al. 2009).

Chitosan-based Edible Films and Coatings

Chitosan is a polysaccharide derived from chitin by deacetylation in alkaline media. Chitosan can be described in terms of degree of deacetylation and average molecular weight, as well as in its antimicrobial properties in conjunction with its cationicity and film-forming properties. Chitosan has potential as a food preservative of natural origin, and it has been reported on the basis of *in vitro* trials as well as through direct application on real complex matrix foods (Zheng et al. 2003; Younes et al. 2014). Chitosan is also an excellent film forming material, and these films have a selective permeability to gases (CO_2 and O_2), resistance to fat diffusion and good mechanical properties. Nevertheless, chitosan films are highly permeable to water vapour, which means high interaction with water molecules, an important drawback for an effective control of moisture transfer, because it is a desirable property for most foods (vegetables and meats). Hence, some strategies have been developed to improve the physical properties of chitosan-based films. The addition of fatty acids waxes, neutral lipids and minerals such as clays, multilayer systems have been added to increase hydrophobicity. Reasonably, the addition of these additives compromises the mechanical properties, stability or film forming and organoleptic characteristics.

In contrast to agar films, chitosan films show antifungal and antimicrobial activities due to the polyprotonated nature of the amino groups (NH_2). Antimicrobial activities of chitosan in solution, powder and edible films and coating against spoilage bacteria, foodborne pathogens and pathogenic viruses and fungi in different food categories have been reported by Friedman et al. (2010). These authors suggested that the low molecular weight chitosan at a pH below 6 shows the optimal condition for a suitable antioxidant, antimicrobial and preservative effect in solid and liquid foods.

The antimicrobial activity of chitosan is not proportional to its degree of deacetylation (DD) values, as was reported by Park et al. (2004).

Chitosan has been blended with pectins, proteins (soy proteins, milk proteins, gelatin and collagen) and polysaccharides (starch, cellulose, alginates, carrageenan, etc.). Chitosan-soy protein films are not completely miscible, therefore the films are brittle, and segregation due to phase separation among blend components can be observed on the film surface (rough surface morphology). The compatibility of the components and improvement of properties (mechanical and transport) can be reached with compatibilizers or cross-linking agents. Di Pierro et al. (2006) reported the preparation of chitosan-whey protein films at pH 6 with different protein concentrations and transglutaminase as a cross-linking agent. Among polysaccharides, starch is the most used to produce films with chitosan. The main difference between these biopolymers is the glucoside linkage in α (1,4) for starch and β (1,4) for chitosan, and the hydroxyl group of the second carbon is replaced by the amine group, which appears acetylated in the case of the natural polymer chitin (Elsabee and Abdou 2013). The properties of chitosan-starch films plasticized with polyols (glycerol and i-erythritol) using different molecular weights of chitosan (low, medium and high) and starch with amylose (28 and 70%) have been reported by Fernandez et al. (2004). Another application form of chitosan is a coating of starch film, as was reported by Bangyekan et al. (2006), where they coated cassava starch films containing glycerol as a plasticizer with chitosan solution. They found higher gloss values (132-145 units) in starch films coated with chitosan. The characteristic of these films was a smooth surface and an enhanced regularity in the surface. The gloss values are an aesthetic factor related to the general appearance and customer acceptance. The mechanical, dynamic-mechanical, water vapour permeability (WVP) and colour properties of films with tapioca starch-chitosan and glycerol have been reported by several authors (Xu et al. 2005; Chillo et al. 2008; Liu et al. 2009). They found that the tensile strength and elastic modulus is positively influenced by chitosan content and an opposite behavior was observed in elongation at the break parameter. The elongation at break parameter usually indicates the flexibility and extensibility (stretchability) of a film subjected to a tensile test. However, to explain at a molecular level it is important to take into account the hydrogen bonds between NH_3^+ of protonated chitosan and OH^- of the starch molecule. When the amount of starch is high, a significant reduction in the elongation at break is observed, which is also explained by the continuous and homogeneous phase of the starch/chitosan system. Geraldine et al. (2008) characterized coatings based on agar-chitosan in minimally processed garlic cloves. The authors reported that this active coating guarantees the quality of minimally processed garlic for an extended period. This coating did not modify the colour, moisture loss or respiration, extending the shelf life of the garlic clove.

The mechanical properties of chitosan edible films and coatings are important because they can guarantee the performance of these materials to handle foods (vegetables and meats), pharmaceutical products or biomaterials (medical textiles, sutures, wound healing or dressing). To the best of our knowledge, the Food and Drug Administration (FDA) has concluded that shrimp-derived chitosan is a substance Generally Recognized as Safe (GRAS), through scientific procedures, for multiple technical effects in several food categories (GRAS Notice No. GRN 000443). Chitosan has been approved by the FDA for medical uses such bandages and drug encapsulation (Raafat and Sahl 2009).

Edible Agar and Chitosan Films in Postharvest Fruits and Vegetables

Performance and Applications of Agar Films

For several years, agar has had a wide use in a variety of industries including food industry products (processed cheese, ice cream, bread and soft candy), dental moulds, casting of archaeological pieces and sculpture moulds (Armisén et al. 2000). Contrary to chitosan, blending agar with other polymers or the incorporation of other additives (antioxidants, antimicrobials, bioactive, etc.) is still very rare. In this section, we will describe some results from the literature. Letendre et al. (2002) reported the microbial resistance of films based on milk proteins (calcium caseinate and whey protein isolate) blended with pectin and agar. They found that the films showed an improvement in puncture strength and better water vapour permeability (WVP) when they are produced by solutions subjected to autoclaving and gamma irradiation. Agar has been blended with rice starch and cassava starch to produce edible films (Phan et al. 2005). These authors found that the moisture barrier properties of the agar-starch films are comparable to other films from polysaccharides and their derivatives. These films showed a high hygroscopic characteristic, suggesting a possible application as a high-moisture gelatinous coating that could retard the water loss in some foods during short-term storage. Agar films plasticized with glycerol are transparent, clear and homogenous, flexible and easy to handle.

Lopez et al. (2014) developed a bioactive film composed of agar, incorporating green tea extract and probiotic strains (*Lactobacillus paracasei* L26 and *Bifidobacterium lactis* B94) that were applied on hake fillets in order to evaluate the effect of the films during 15 days of storage. They reported that the application of green tea films on hake delayed the growth of microorganisms in fish and therefore, reduced the spoilage indexes. Furthermore, agar films with green tea and probiotic films could extend the shelf-life of hake at least for a week and, at the same time; it could be a way to incorporate beneficial probiotic bacteria to the fish.

The agar is extracted from marine algae; it is a living organism that can show variation between seasons as Madera-Santana et al. (2011) reported. These authors compare properties of biopolymeric films formulated with agar extracted from *Hydropuntia cornea* collected during two different seasons (rainy and dry), and polyvinyl alcohol (PVOH) as a synthetic matrix. The films were formulated at different agar contents (0%, 25%, 50%, 75% and 100%) and prepared by the solution cast method because water solubility of both polymers facilitates blending and processing. They determined the optical, mechanical, thermal, morphological and chemical properties in the PVOH–agar films in order to evaluate their suitability to be used as a biodegradable material. From the results, they reported that the season to obtain agar from *H. cornea*, where this agar blends with PVOH, has a significant effect on the mechanical properties of biodegradable film produced. Moreover, they stated that formulations of 50-75% agar from *H. cornea* collected in the rainy season with PVOH showed suitable properties for possible applications in the biodegradable package industry. Recently, the addition of glycerol on the physicochemical and morphological properties of PVOH–agar films prepared by solution cast method was reported by Madera-Santana et al. (2014). The addition of glycerol in PVOH-agar films altered the optical properties, resulting in a decrease in opacity values and in the colour difference (ΔE) of the films, because this plasticizer reduces the intermolecular forces of the components and introduces transparency in the PVOH-agar films. The authors performed structural analyses and demonstrated that the plasticization effect and the interaction of glycerol with PVOH and agar molecules produced arrangements among these components. The mechanical (tensile and tear) tests performed on PVOH-agar films provided information about the performance of these films under both tests, as well as the effect of glycerol on this property. The authors concluded that the formulation with 75 wt. % agar and 15% GLY has suitable properties for possible applications in the biodegradable package industry.

On the other hand, environmental interest has emerged, leading to research on biodegradable polymeric materials (chitosan, agar, starch, cellulose derivatives, alginates, carrageenan, pectin, etc.) as alternatives to non-biodegradable synthetic products. These biopolymers have been used to produce films for food coating, tissue engineering and regenerative medicine, and food packaging (Rhim et al. 2006) and as supports for the retention and controlled release of active compounds (Mayachiew and Devahastin, 2010). Among the active agents that can be incorporated into an agar matrix, plant extracts have been found to be promising for replacing synthetic agents. Natural products from plant extracts can be used for controlling herbivorous insects in crops and are considered an interesting alternative due to their lower negative impact on the environment. In particular, the semiochemical compounds can act as a messenger and they are able to identify insect life situations, such as feeding, mating and egg-laying (ovipositing). García-Baldenegro et al. (2014) reported that agar films can be impregnated with natural semiochemicals, such as pecan extracts, where they are potential agents for the selective control of pest insects. Semiochemicals from plant extracts are liquid at room temperature and can easily be transformed from a liquid to a gaseous state at room or slightly higher temperature without undergoing decomposition. The authors pointed out that the impregnation process modifies the structural and morphological characteristics of the agar films, and the phenolic compounds of the pecan extract interact with the agar functional groups (ester carbonyl groups). However, volatile compounds in the pecan extract are released in good standing from the films. The measurements were performed using desorption kinetic of pecan extracts in agar films and it demonstrated that followed a Fickian diffusion. These results have also shown that pecan extract on agar films are completely released within 70 h. It is the first report of a biodegradable polymer film for active agricultural films and other wares used in short durations, which would be safe for the environment upon disposal.

Performance and Applications of Chitosan Films

Edible coatings of chitosan have been applied to diverse fruit including non-climacteric (Table 2), climacteric (Table 3) as well as fresh-cut fruits and vegetables (Table 4) with different beneficial effects such as maintaining conservation, overall quality and reducing microbial growth. Several studies have inoculated fruit with different microorganisms and evaluated chitosan disease control. Chitosan application has been made mainly by immersion using pure chitosan, modified chitosan, chitosan derivatives, or chitosan commercial formulations at concentrations from 0.2% to 3%. However, in some studies chitosan application has been done by spraying. Generally, immersion times vary from 10 s to 5 min. Chitosan edible coatings have been applied as a single coating alone or combined with essential oils, bioactive compounds (phenolic compounds), antioxidant compounds (citric acid, ascorbic acid and vitamin E), zeolite, pullulan, putrescine, grapefruit seed extract, calcium compounds (calcium chloride, calcium gluconate, calcium lactate), chlorine dioxide, preservatives (potassium sorbate), oils (olive oil), sodium caseinate, oleoresins, 1-methyl cyclopropene (1-MCP); or as bilayer or multilayer composite coatings combined with carboxy methyl cellulose (CMC), gelatin, beeswax, hydroxyl-propyl methyl cellulose (HPMC), sodium alginate, pectin, methyl cellulose, starch (cassava and yam) and Arabic gum.

Recent research has reported the effect of chitosan nanocoating alone (Mustafa et al. 2014) or with essential oils (Mohammadi et al. 2015) or as nanomultilayer coating (Medeiros et al. 2012) on the quality of whole or fresh-cut fruits and vegetables during storage.

Table 2. Edible coatings of chitosan applied to diverse non climacteric fruits.

Product	Coating	Application of Edible Coating	Packaging	Storage Conditions	Benefits	Reference
Strawberries (*Fragaria x ananassa*).	1. Control: a) uncoated, b) Sodium benzoate (0.05% w/w) + potassium sorbate (0.05% w/w) for microbiological analysis. 2. Pectin-Edible Active Coating (EAC): Pectin (3.5% w/w) + (A). 3. Pullulan-EAC: Pullulan (6.5% w/w) + (A). 4. Chitosan-EAC: Chitosan (1.5% w/w) + (A). (A) Sodium benzoate (0.05% w/w) + potassium sorbate (0.05% w/w) + glycerol (0.5% w/w).	Immersion for 5 min, and then dried at room temperature on polypropylene trays for 30 min.	Packed in polyethylene terephthalate clamshell containers (178 × 127 × 89 mm and capacity of 454 g) with venting holes.	15 days at 4 °C and 90% RH.	The chitosan-EAC was the most effective in reducing the microbial growth (total aerobic counts, moulds and yeasts). All EACs reduced weight loss and fruit softening. They also delayed alteration of colour and maintained sensory quality during storage.	Treviño-Garza et al. 2015
Cucumber (*Cucumis sativus*).	1. Control: Distilled water. 2. Chitosan nanoparticles (CSN) at 1.5 g/L. 3. *Cinnamomum zeylanicum* essential oil (CEO) loaded CSN. (CEO-CSN) at weight ratios of CS to CEO of 1:0.25 at 1.5 g/L.	Immersion for 2 min and drained on a nylon filter.	5 cucumbers were packed in polypropylene trays.	21 days at 10 ± 1 °C and 90–95% RH.	CEO-CSNs was more effective than CSNs coating to reduce respiration rate, decay, and weight loss, and total aerobic bacteria, yeasts and mould growth, and extend the microbial and postharvest shelf life of cucumbers.	Mohammadi et al. 2015

(*Contd.*)

Table 2. (*Contd.*)

Product	Coating	Application of Edible Coating	Packaging	Storage Conditions	Benefits	Reference
Litchi (*Litchi chinensis* Sonn.) cv. Purbi at a fully mature stage (90–100% of the pericarp showing red colour).	1. Control: Distilled water. 2. Salicylic acid (SA): 0.5 mM and 1.0 mM with Tween-20 (2 g/L). 3. Chitosan: 2% with Tween-20 (2 g/L). 4. SA + CH: SA (0.5 mM and 1.0 mM) as 2 + CH (2%) as 3.	Immersion for 5 min at 25 °C and air-dried		6 days at 4 ± 1 °C	The combination of salicylic acid (1.0 mM) with chitosan (2%) was the best treatment to reduce pericarp browning, weight loss, decay loss and to preserve the highest values of anthocyanins, phenolics, flavonoids, ascorbic acid and antioxidant capacity. This treatment maintained colour and preserved the quality of litchi fruit during cold storage.	Kumari et al. 2015
Pomegranate (*Punica granatum* L.) cv. Malase Torshe Saveh at commercial ripening stage (180 days after fruit set).	1. Control: Distilled water. 2. Resin wax (Britex Ti). 3. Carnauba wax (Xedasol M14). 4. Chitosan (1 and 2% w/v) in acetic acid (1% v/v).	2) and 3): Manually applied using a brush at ambient temperature. 1) and 4): Immersion for 2 min. All treatments were		120 days at 4.5 ± 0.5 °C with 90 ± 5% RH, and 3 additional days at 20 °C.	Chitosan was less effective than resin wax and carnauba wax in reducing fruit water loss and respiration rate. Carnauba wax was the most effective coating on reducing respiration rate and weight loss,	Meighani et al. 2015

(*Contd.*)

		dried at room temperature.			and maintaining bioactive compounds.	
Carambola (*Averrhoa carambola* L.) Sweet variety at a mature stage (colour index 3).	1. Control: uncoated. 2. Chitosan (CH): 0.3 and 0.5% (w/v) in acetic acid (0.5% v/v) + glycerol monostearate (0.75%) + Tween-80 (0.1%), pH 5.6. 3. Gum Arabic (GA): 1 and 3% (w/v) + glycerol monostearate (0.75%), pH 5.6. 4. Alginate (AL): 1.5 and 2% (w/v) with calcium chloride (2% w/v).	Immersion for 2 min and air-dried at room temperature for 60 min.		12 days at 26 ± 1 °C, 65–70% RH.	All coatings reduced decay. CH (0.3%) had the highest weight loss reduction. CH (0.3%) and GA (1%) were the best treatments in reducing cell wall degrading enzyme activities and maintained highest sensory attribute values (colour, texture, flavor) reaching the best overall acceptability.	Gol et al. 2015
Mandarins (*Citrus reticulata* Blanco) cv. Rishon and Michal.	1. Control: uncoated. 2. Polyethylene based commercial wax (PE-CW). 3. Carboxy methyl cellulose (CMC): 1.5% (w/v). 4. Chitosan: 1.5% (w/v). 5. Composite polysaccharide coatings by the Layer-by-layer (LBL) approach coatings: CMC (1.5% w/v) as an internal layer and Medium	1 to 4. Manually applied with paint brushes. Then fruits were dried in a tunnel at 37 °C. 5. CMC coating was applied first and fruits were dried for 2-3 min at room temperature, and	25–30 mandarins were packed in cardboard boxes.	10 days at 20 °C and 80–85% (RH) (shelf-life conditions).	Although chitosan coatings increased gloss, they caused accumulation of ethanol in the fruit juice impairing fruit flavor. Composite polysaccharide coatings (LBL), CMC and chitosan, showed the best performance, improved the	Arnon et al. 2015

(Contd.)

Table 2. (*Contd.*)

Product	Coating	Application of Edible Coating	Packaging	Storage Conditions	Benefits	Reference
	molecular weight chitosan (200–800 cP) at 0.5, 1.0 and 1.5% (w/v) as an external layer.	then chitosan was applied and dried as (4).			physiological quality of mandarins. The higher the chitosan concentration the higher the increase in gloss and firmness improvement.	
Table grapes (*Vitis labrusca* L.) cv. Isabella at commercially mature stage.	1. Control: Acetic acid (1 mL/100 mL). 2. Chitosan from *Cunninghamella elegans*: 3.75, 7.75, 15 mg/mL.	Immersion for 1 min in inoculum of *Botrytis cinerea* and *Penicillium expansum* (about 10⁶ spores/mL), followed by immersion in CHI, and then dried on a nylon filter.	Packed in commercially sterile polyethylene containers with a lid.	A. 12 days at 25 °C and 85% RH. B. 24 days at 12 °C and 85% RH.	Coatings of chitosan from *C. elegans* inhibited the *B. cinerea* and *P. expansum* growth on table grapes stored at room or cold temperature. Chitosan coating maintained the physical, physicochemical and sensory attributes during storage of grapes.	Vasconcelos de Oliveira 2014a
Table grapes (*Vitis labrusca* L.) cv. Isabella at commercially mature stage.	1. Control: Acetic acid (1 mL/100 mL). 2. Chitosan from *Mucor circinelloides*: 3.75, 7.75, 15 mg/mL.	Immersion for 1 min in inoculum of *Rhizopus stolonifera* and *Aspergillus niger* (about 10⁶ spores/mL), and	Packed in commercially sterile polyethylene containers with a lid.	A. 12 days at 25 °C and 85% RH. B. 24 days at 12 °C and 85% RH.	Chitosan coating at 7.5 and 15 mg/mL maintained the physical, physicochemical and sensory attributes of grapes, and delayed	Vasconcelos de Oliveira 2014b

(Contd.)

Product	Formulation	Application	Storage	Results	Reference
Raspberries (*R. idaeus* L.) cv. Autumn Bliss.	1. Control: Acetic acid (5%). 2. Chitosan: 0.5, 1.0 or 2.0% (w/v) in 30 ml/L of acetic acid (5%) per gram of chitosan, pH 5.5.	4.2 kg were immersed into 10 L of solution for 5 min and then allowed to dry for 40 min on absorbent paper at room temperature. then immersion in CHI, then dried on a nylon filter.	Packed in plastic containers. 15 days at 0 ± 1 °C and 90 ± 5% RH.	Chitosan coatings (1 or 2%) maintained titratable acidity and decreased respiration and ethylene production, reduced weight loss, and decay. Chitosan at 2% was most effective in reducing darkening and maintaining the anthocyanin content. However, the higher the chitosan concentration the higher the decrease of ascorbic acid contents. the fungal growth during low-temperature storage of grapes.	Tezotto-Uliana et al. 2014
Red bell peppers (*Capsicum annum* L.) cv. Vergasa.	1. Control: Uncoated. 2. Chitosan (CH): 2% (w/v). 3. Gelatin (GL): 1% (w/v). 4. CH (2%, w/v) + GL (1%, w/v).	Manually with paint brushes, and dried in a drying tunnel for 2 min at 38 °C.	15-20 fruit were packed in 3–5 kg cartons. a) Regular storage (14/5): 14 days at 7 °C and 95% RH, then 5 days at 20 °C and 75% RH. b) Prolonged cold storage	The composite CH + GL coating decrease (two fold) microbial decay, enhanced fruit texture and prolonged the period of cold storage up to 21 days and fruit shelf-life up to 14 days, without	Poverenov et al. 2014a

(Contd.)

Table 2. *(Contd.)*

Product	Coating	Application of Edible Coating	Packaging	Storage Conditions	Benefits	Reference
				(21/5): 21 days at 7 °C and 95% RH, then 5 days at 20 °C and 75% RH. c) Long shelf storage (14): 14 days at 20 °C and 75% RH with no prior cold storage.	affecting the respiration or nutritional content of the fruit.	
Blackberry (*Rubus* spp.) cv. Tupy with 100% of the surface in black.	1. Control: Uncoated. 2. Chitosan: 1.5% in ascorbic acid (0.8%) + SG. 3. Cassava starch: 2.5% + SG. 4. Kefir grains in water 20% + SG. SG: Sorbitol/glycerol 1.0% (w/v).	Immersion for 2 min at room temperature and then dryed over nylon screens.	250 g were packed in transparent plastic boxes (190 cm×120 cm×63 cm) without perforations, and sealed with PET (polyethylene terephtalate) cover.	18 days at 0 and 10 °C.	Chitosan (1.5%) in combination with storage at 0 °C was the best treatment to decrease weight loss, to maintain fruit firmness, and to reduce rot incidence without affecting physical and chemical parameters.	Oliveira et al. 2014
Luffa [*Luffa cylindrica* (L.) Roe.] Roem.	1. Control: Acetic acid (1% v/v). 2. Chitosan: 0.5% and 1.0%	Immersion for 1 min, and then dried for 2 h at 25 °C and 50% RH.	Packed in plastic boxes with polyethylene film bags	25 ± 1 °C and 90–95% RH.	Chitosan reduced weight loss and respiration rate, maintained chlorophyll,	Han et al. 2014

(Contd.)

Commodity	Treatment	Application	Packaging	Storage	Results	Reference
Tianxiangrou] or sponge gourd at a commercial young immature stage.	(w/v) of chitosan acetate, in acetic acid (1%), pH 6.0.		(0.03 mm in thickness).		ascorbic acid, and total phenolic content, as well as firmness and visual appearance. It also delayed the increase of PPO activity. Besides 1.0% w/v chitosan inhibited POD and PAL activities and decreased cellulose content. Chitosan maintains postharvest quality and prolonged shelf life of sponge gourd.	
Strawberry (*Fragaria x ananassa* Duchesne) cv. Selva with red colour on more than 75% of the surface.	1. Control: Uncoated. 2. Chitosan nanoparticles: 50–110 nm at 2500 ppm loaded with copper (0 and 25 ppm).	Immersion for 2 min, and allowed to dry for 1 h at 20 °C.	200 g were packed in polypropylene container	20 days at 4 ± 1 °C and 70% RH.	Nanochitosan coatings reduced respiration rate, weight loss and softening, as well as PPO and POD activities, while maintained the visual quality, inhibited fungal decay, and did not affect sensory quality.	Eshghi et al. 2014
Mandarins (*Citrus reticulata* Blanco) cv. Or and Mor. Oranges (*Citrus sinensis* [L].	1. Control: Uncoated. 2. Commercial polyethylene (PE) wax: a) 'Ziv-dar' wax for	Applied manually with paint brushes, and then dried	30 fruits were packed in cardboard boxes	Mandarins and oranges: 4 weeks at 5 °C and ~90–95%	CMC/chitosan bilayer edible coating was as effective as commercial PE wax in increasing	Arnon et al. 2014

(Contd.)

Table 2. (*Contd.*)

Product	Coating	Application of Edible Coating	Packaging	Storage Conditions	Benefits	Reference
Osbeck) cv. Navel Grapefruit (*Citrus paradisi* Macf.) cv. Star Ruby.	grapefruit and oranges. b) 'Tag' wax for mandarins. 3. CMC/chitosan bilayer coating: a) First CMC 1.5% (w/v). b) Chitosan 1% (w/v) in acetic acid 0.7% (v/v).	by passing fruit through a hot-air tunnel at 37 °C.		RH plus 5 days at 20 °C and ~80–85% RH. Grapefruit: 4 weeks at 10 °C and ~90–95% RH plus 5 days at 20 °C and ~80–85% RH.	fruit gloss and appearance. However, it did not reduce postharvest weight loss. Additionally, this coating did not affect flavour in oranges and grapefruit, but slightly impaired flavour in mandarins.	
Dragon fruit (*Hylocereus polyrhizus* (Weber) Britton & Rose) at purplish pink fruit colour mature stage.	1. Control: Purified water. 2. Conventional Chitosan: 1%. 3. Double Layer Coating: Submicron chitosan: 200, 600 and 1000 nm droplet size + conventional chitosan 1.0%.	1. and 2.: Immersion for 5 min, and then dried at ambient temperature. 3. Immersion in submicron chitosan for 15 min, and dried at room temperature overnight. Then chitosan immersion as 2.	Packed in cardboard boxes.	28 days at 10 ± 2 °C and 80 ± 5% RH.	Double layer coatings with submicron chitosan 600 nm droplet size +1.0% conventional chitosan reduced weight and firmness loss, as well as maintained the highest retention of total phenols and flavonoids, lycopene and total antioxidant activity. This coating reduced decay incidence during storage.	Ali et al. 2014

(*Contd.*)

Commodity	Formulation	Treatment	Packaging	Storage	Results	Reference
Strawberries (*Fragaria x ananassas* Duchesne ex Rozier) cv. Earliglow	1. Control: Acetic acid (0.5% v/v) + Tween-80 (0.05% w/v), pH 5.6. 2. Chitosan: 0.5, 1.0 and 1.5% w/v in acetic acid (0.5% v/v) + Tween-80 (0.05% w/v), pH 5.6.	Immersion for 5 min at 20 °C and then dried for 2 h at 20 °C.	Fifteen strawberries were packed into 1 L polystyrene containers with snap-on lids.	12 days at 5 °C or 9 days at 10 °C.	Chitosan coatings maintained high antioxidant levels (phenolics, anthocyanins, and flavonoids), high antioxidant enzyme activity [catalase (CAT), glutathione-peroxidase (GSH-POD), guaiacol peroxidase (G-POD), dehydroascorbate reductase (DHAR) and monodehydroascorbate reductase (MDAR)], and high total antioxidant activity ORAC, DPPH, and HOSC, while reduced the increase of oxidative enzyme [guaiacol peroxidase (G-POD) and PPO) activity, weight loss and fruit decay during storage at 5 °C or 10 °C.	Wang and Gao 2013
Strawberries (*Fragaria x ananassa*) cv. Camarosa at a ripening stage of	1. Control: uncoated. 2. Monolayer chitosan film: Chitosan (CH) 0.8% (w/w) in acetic acid (1%, v/v) + glycerol (0.2%, v/v) + Tween 80 (0.2%, w/w) + glycerol (0.2%, w/w).	2) and 5): Immersion for 30 s and draining for 2 to 4 h. 3. Wax: Immersion for	250 g of fruit were packed as one layer in perforated (6 perforations) plastic boxes.	7 days at 20 °C and 35-40% RH.	Three-layer coatings (3 and 4) were the best treatments in reducing the decay and weight loss of the fruits. These	Velickova et al. 2013

(Contd.)

Table 2. (*Contd.*)

Product	Coating	Application of Edible Coating	Packaging	Storage Conditions	Benefits	Reference
>75% red surface colour.	3. Wax coated multilayer film: Beeswax (BW) (0.5% w/w) + CH (as 2) + BW (0.5% w/w). 4. Crosslinked chitosan, wax coated, multilayer film: BW (as 3) + sodium tripolyphosphate, TPP (0.3: 1 TPP: Chitosan ratio) crosslinked chitosan (as 2) + BW (as 3). 5. Composite film: CH (as 2) + BW (10% w/v).	30 s and then dried for 5 min, then CH as (2), followed by wax. 4. Wax as 3), then CH as 2), followed by crosslinking by immersion for 30 s in TPP and 5 min for crosslinking, then wax as 3).			coatings maintained the highest values on colour and firmness values. However, their sensorial quality was not acceptable. Monolayer chitosan and composite film coated gave the highest sensorial quality with the best visual appearance and taste and highest acceptability values.	
Table grapes cluster (*Vitis vinifera*) cv. Shahroudi with a 19.5% total soluble solids (TSS) ripening stage.	1. Control: Acetic acid (0.5% v/v), pH 5.2. 2. Chitosan: (Low molecular weight chitosan 0.5 and 1% w/v) in acetic acid (0.5% v/v), pH 5.2.	Immersion for 1 min and then drying for 2 h at room temperature.	One cluster was individually packed in polypropylene bags (20 × 20 cm²) with an oxygen transmission rate of 29.2 pmol/s m Pa.	60 days at 0 °C and 90 ± 5% RH, plus 5 days at 24 °C and 70 ± 5% RH.	Chitosan reduced weight loss, decay, browning, shattering, and cracking in grape berries; and also maintained higher TSS values, titratable acidity (TA), and TSS/TA ratio levels than control fruit, while delayed the total phenolic, catechin, and antioxidant capacity changes.	Shiri et al. 2013a

(*Contd.*)

Table grapes cluster (*Vitis vinifera*) cv. Shahroudi at ripe stage.	1. Control. 2. Putrescine (PUT): 1.0 and 2.0 mM +Tween 20 (0.01%). 3. Chitosan (CH): Low molecular weight chitosan 1% (w/v) in acetic acid (0.5% v/v), pH 5.2. 4. Putrescine (PUT) + Chitosan (CH): PUT as 2) + CH as 3).	1. and 2. Immersion for 3 min, and drained in baskets for 2 h at room temperature. 3. Immersion for 1 min. 4. First treated as 2. followed by 3.	One cluster was individually packed in polypropylene ± 5% RH. bags (20 × 20 cm^2) with an oxygen transmission rate of 29.2 pmol/s m Pa.	60 days at 0 ± 1 °C and 90 ± 5% RH, plus 5 days at 24 °C and 70	Grapes treated with PUT (1 mM) + CH (1%) had less weight loss, decay incidence, browning, and berry shattering and cracking, while grapes treated with PUT (1 and 2 mM) maintained the highest values of total phenolic content, catechin, total quercetin and antioxidant activity.	Shiri et al. 2013b
Strawberry (*Fragaria* × *ananassa* Duch) cv. Camarosa at 2/3 red on the surface ripening stage.	1. Control: Distilled water. 2. Chitosan acetic: Chitosan (CH) 1% (w/v) in acetic acid (1%, v/v). 3. Chitosan chloride: CH 1% (w/v) in hydrochloric acid (1%, v/v). 4. Chitosan formate: CH 1% (w/v) in formic acid (1%, v/v). 5. Chitosan glutamate: CH 1% (w/v) in glutamic acid (1%, v/v). Solutions 2 to 5 were at pH 5.6 + Triton X-100 (0.05%, w/v).	Immersion for 10 s in 5 L of solution, and then dried in air for 1 h.	Fruits were individually packed in small plastic boxes and these were placed in covered plastic boxes.	7 days at 0 ± 1 °C and 95–98% RH, plus 3 days shelf-life at 20 ± 1 °C and 95–98% RH.	Commercial chitosan controlled gray mould and *Rhizopus* rot during storage of strawberries for four days at 20 ± 1 °C in a similar way than practical grade chitosan. Commercial chitosan followed by benzothiadiazole, calcium, and organic acids were the best treatments to reduce *Rhizopus* rot, gray mould and blue mould during storage of strawberries for 7 days at 0 ± 1 °C plus 3 days at 20 °C.	Romanazzi et al. 2013

(Contd.)

Table 2. (*Contd.*)

Product	Coating	Application of Edible Coating	Packaging	Storage Conditions	Benefits	Reference
	6. Commercial chitosan: Chitosan 1% (w/v) in distilled water. 7. Oligosaccharides: 1% (v/v). 8. Benzothiadiazole: 0.2% (w/v). 9. Calcium and organic acids: 1% (v/v). 10. Soybean lecithin: 1% (v/v). 11. *Abies sibirica* extract: 1% (v/v). 12. *Urtica dioica* extract: 1% (w/v).					
Strawberry (*Fragaria* × *ananassa* Duch.) cv. Camarosa.	1. Control: Distilled water. 2. Carboxymethyl cellulose (CMC): 1% (w/v) in water–ethanol (3:1). 3. Hydroxypropyl-methyl cellulose (HPMC): 1% in water. 4. CMC (1%) as (2) + Chitosan (CH) 1% (w/v) in acetic acid 0.5% (v/v), pH 5.6.	Immersion for 2 min, then drained off and the fruit was dried at 26 ± 2 °C for 2 h.	Packed in plastic boxes.	11 ± 1 °C and 70–75% RH.	The combination of chitosan with CMC and HPMC were the best coatings to reduce weight loss, decay percentage, and TSS, pH, TA, and ascorbic acid content changes. These coatings maintained the highest content of total phenolics and total	Gol et al. 2013

(*Contd.*)

Fruit	Treatments	Application	Packaging	Storage	Results	Reference
	5. HPMC (1%) as (3) + CH (1%) as (4). Glycerol monostearate (0.75%) was added to all solutions.				anthocyanins, and the lowest activities of cell wall degrading enzymes such as polygalacturonase (PG), cellulase, pectin methyl esterase (PME) and β-Galactosydase (β-Gal) improving postharvest quality and shelf-life.	
Table grape cv. Muscat Hamburg	1. Control 2. Chitosan (CH): 1 (w/v) in acetic acid (1% v/v). 3. Glucose: 1% (w/v) in deionized water. 4. Chitosan + Glucose: CH as (2) and glucose (1% v/v), pH 6.0.	Immersion for 1 min and then clusters were dried over shelves under fans.	5 kg of fruit were packed in plastic boxes and sealed with PE films. (0.03 mm thickness).	60 days at 0 °C and 95% RH plus three days at 20 °C.	Chitosan–glucose complex was the best treatment to reduce decay, weight loss and respiration rate, and to retard the decrease of total soluble solids, ascorbic acid and titratable acidity, while increasing POD and SOD activities. This treatment also had the best fruit texture and highest sensory quality.	Gao et al. 2013
Pomegranate (*Punica granatum* L.) cv. Rabbab-e-Neyriz.	1. Control: Distilled water with acetic acid (1%). 2. Chitosan: 1% and 2% (w/v) in acetic acid (1% v/v).	Immersion and then dried for 12 h at room temperature.	15 fruits were packed in baskets.	45 days at 2 ± 0.5 °C; or 5 ± 0.5 °C and 90 ± 5% RH, then transferred for 3 days at 20 °C.	The combination of chitosan coating and storage at 2 °C was the most effective to delay anthocyanin degradation and to prevent colour deterioration of pomegranates.	Varasteh et al. 2012

(Contd.)

Table 2. (*Contd.*)

Product	Coating	Application of Edible Coating	Packaging	Storage Conditions	Benefits	Reference
Strawberries (*Fragaria × ananassa*) cv. Camarosa.	1. Control: Acetic acid 0.5% (v/v). 2. Chitosan (CH): 1% (w/w) in acetic acid (0.5% v/w). 3. Chitosan-essential oil (CH-LO): CH as (2) + lemon essential oil (3% w/w). 4. Microfluidized chitosan–lemon essential oil coating as (3).	Immersion for 1 min and then dried by natural convection for 1 h at 20 °C. A batch was inoculated with *Botrytis cinerea* (10^5 spores/mL) before coating to evaluate decay.	10 strawberries were packed in a PET tray.	14 days at 4 ± 1 °C and 90% RH.	Chitosan with lemon essential oil coating reduced respiration rate and increased the chitosan antifungal activity during cold storage of strawberries inoculated with *Botrytis cinerea*.	Perdones et al. 2012
Clementine mandarins cv. Oronules.	1. Control: Water. 2. Chitosan: at different solid content (0.6%, 1.2% and 1.8%) + Tween 80 (0.1%, v/v). 3. Polyethylene-shellac commercial wax: Citrashine 10%.	1. and 2. Immersion for 20 s, then drained and dried for 2 min in a tunnel at 55 °C. 3. Spraying with nozzles and dried as (2).		a. 28 days at 5 °C and 90–95% followed by seven days at 20 °C. b. Nine days at 20 °C and 75% RH.	Chitosan coating maintained firmness and the highest solid content of mandarins after cold storage. Chitosan and commercial wax did not affect the sensory quality even though they modified the internal atmosphere of mandarins, this effect was greater at higher chitosan concentration.	Contreras-Oliva et al. 2012

(*Contd.*)

Sweet pepper (*Capsicum annuum* L.) at 80% maturity (green stage).	1. Control: Distilled water. 2. Chitosan (CH): 1.0% in acetic acid (0.5%) + glycerol (0.75%). 3. Cinnamon oil (CO): 0.25% + Tween 80 (0.2%). 4. CH coating + CO: CH as (2) and CO as (3).	Immersion for 5 min then dried for 50 min over a plastic sieve with a low-speed air, followed by drying over a tissue paper.	Peppers were wrapped with a polypropylene film (600 mm × 800 mm).	35 days at 8 °C and 95% RH.	Chitosan-oil coatings reduced decay and maintained good sensory acceptability, while increased the activities of defense related enzymes such as SOD, POD and CAT. This coating also reduced the electrolyte leakage and MDA content, maintaining quality and extending pepper shelf life.	Xing et al. 2011
Strawberry (*Fragaria* x *ananassa* Duchesne).	1. Control: Uncoated. 2. Modified chitosan (MC) (N-acylated chitosan) 2% in acetic acid 3%, pH 3.5. 3. MC + Limonene oil (LIM): MC as (2) + LIM 0.02% (w/v) + Tween®80 (0.4%), pH 3.5. 4. MC + Peppermint (PM): MC as (2) + PM (0.02%, w/v) + Tween®80 (0.4%), pH 3.5.	Each side of each strawberry was sprayed twice. Then dried for 15 min on sterile aluminum sheets.	Fruit was covered with a sterile aluminum sheet.	14 days at 4 °C.	Modified chitosan containing LIM and Tween®80 coating was the best treatment to reduce decay during storage of strawberries.	Vu et al. 2011

(Contd.)

Table 2. (*Contd.*)

Product	Coating	Application of Edible Coating	Packaging	Storage Conditions	Benefits	Reference
Table grapes (*Vitis vinifera*) cv. Muscatel.	1. Control: Uncoated. 2. Hydroxypropyl methyl cellulose (HPMC): 1% (w/v) in distilled water. 3. Chitosan (CH): CH (1%, w/v) in acetic acid (1%, v/v). 4. HPMC + bergamot essential oil (BEO): HPMC (1%, w/v) + BEO (2%, w/v). 5. CH + BEO: CH (1%, w/v) + BEO (2%, w/v). 6. HPMC + BEO: HPMC (1%, w/v) in distilled water + BEO (2%, w/v).	Clusters immersion for 1 min, then they were hanged up and dried for 2-3 h at room temperature under natural convection.	Clusters were packed in perforated PET trays.	24 days at 1-2 °C and 85-90% RH.	Even though HPMC and chitosan coatings with and without bergamot essential oil reduced weight loss and maintained fruit firmness, chitosan with bergamot oil was the best coating which had the highest antimicrobial activity and the greatest reduction of respiration rates, and a decrease of weight loss.	Sánchez-González et al. 2011
Litchi (*Litchi chinensis* Somn.) cv. Heli at 80–90% red colour.	1. Control: Acetic acid (2%). 2. Chitosan: 1% in acetic acid (2%) + ascorbic acid (0.05–0.5%).	Immersion for 1-2 min, and then dried for 1 h.	1. Static stage (SS): 100 fruit were packed in a modified polyethylene bag 2. Dynamic stage (DS): 300 fruit were packed in a plastic box covered with a plastic bag.	1. SS: Room temperature and 70–80% RH. 2. DS: 25–35 °C and 70–80% RH during car transportation 1300 km.	Chitosan coating reduced respiration rate, sarcocarp temperature, PPO activity and weight loss, extending the postharvest life of litchi 5 days longer than the uncoated fruit.	Lin et al. 2011

(*Contd.*)

Commodity	Formulation	Application	Packaging	Storage	Results	Reference
Sweet cherries (*Prunus avium* L.).	1. Control: distilled water. 2. Chitosan acetate (CA): 0.1, 0.3, 0.5, 1.0 and 2.0% (w/v) in distilled water.	Immersion for 3 min, then air-dried for 30 min at 25 °C.	Cherries were packed into plastic trays, over-wrapped with 30 × 30 cm PVDC film.	40 days at 4 °C.	CA coating at 0.3 and 0.5% were the best treatment to decrease weight loss, to maintain the highest sensory quality values, and to delay the spoilage rate of fruits, while maintaining the titratable acidity, total soluble solid, reducing sugars and ascorbic acid content of sweet cherries. Also increased the peroxidase and catalase activities and extended postharvest life of the sweet cherries more than 40 days at 4 °C.	Dang et al. 2010
Cherries cv. 0-900 Ziraat.	1. Control: Uncoated. 2. Chitosan (deacetylation degree more than 90%, low viscosity): 3% (w/v) in acetic acid (1%, v/v) + glycerol (0.75%, v/v). 2. Whey protein isolate (WPI): 12.5% (w/w protein) + glycerol 2:1 (w/w) in solution. 3. Shellac mixed with ethyl alcohol (12.5:87.5 v/v).	Immersion for 1 min, drained, and immersion again for 1 min, and then dried at ambient temperature.	250 g of fruit was packed into 20 × 30 cm polyethylene (PE) trays and heat sealed. The O_2 and CO_2 transmission rates of PE were 4050 and 14 000 cc/m²/day, respectively.	11 days at 20 °C and 55 ± 5% (RH).	All coatings maintained quality, firmness, total soluble solids (TSS), ascorbic acid (AA), and reduced weight loss, but shellac coating showed better performance.	Aday and Caner 2010

(Contd.)

Table 2. (*Contd.*)

Product	Coating	Application of Edible Coating	Packaging	Storage Conditions	Benefits	Reference
Japanese loquat (*Eriobotrya japonica* T.).	1. Control: Water. 2. Commercial chitosan: 0.6% soluble solids (v/v). 3. sucrose ester fatty acids: 1% soluble solids (v/v).	Immersion for 1 min, and then dried for 1 h at ambient conditions.		16 days at 20 °C and 90% HR.	Chitosan was the best coating to decrease weight loss, respiration rate, and ethylene production. Both coatings maintained flesh firmness, sensory quality, and appearance.	Márquez et al. 2009
Table grapes (*Vitis vinifera*) cv. Redglobe with 19% soluble solids and 0.4% tartaric acid.	1. Control. 2. Grapefruit seed extract (GSE, Citricidal™: 0.05% (v/v) in distilled water + Tween-80 (1%, v/v). 3. Chitosan (CH): 1% (w/v) in acetic acid (0.5%, v/v), pH 5.6 + Tween-80 (0.05%, v/v). 4. CH (1%) + GSE (0.1%). 5. Thiabendazole (TBZ): 0.1%.	Immersion for 1 min at 25 °C, then air-dried for about 1 h. A second group was first inoculated by spraying a suspension of 1.0×10⁵ spores/mL of *Botrytis cinerea*, and then dried at 25 °C for 2 h prior to treatment, followed by coating.	Clusters were packed in ventilated polyethylene bags (VPE), and then in commercial corrugated fibreboard boxes that were wrapped with polyethylene stretch film (20 μm).	4 weeks at 0–1 °C.	Chitosan and GSE alone or combined, reduced fungal rot, weight, and firmness loss, maintained rachis and berry appearance, and flavor acceptability, while reducing shatter and cracking. However, the combination of chitosan + GSE had the best results. It seems that GSE and chitosan have a synergistic effect in reducing decay and maintaining the keeping quality of 'Redglobe' grapes.	Xu et al. 2007

(*Contd.*)

Strawberry fruit (*Fragaria* × *ananassa*) cv. Camarosa.	1. Control: Uncoated. 2. Starch: 2% (w/v), pH 5.6 + sorbitol (2.0 g$_{solute}$/L$_{solution}$). 3. Chitosan: 1% (w/v) in hydrochloric acid (2.5 mL 10N HCl), pH 5.6 + Tween 80. 4. Carrageenan: 0.3%. (w/v) in distilled water, pH 5.6 + glycerol (0.75%, w/v) + Tween 80 (0.01 to 0.1%, w/v). All formulation was without and with 1% di-hydrated calcium chloride.	Sprayed in every 5 kg perforated tray, and then allowed to drip.		0-5 °C and 85-90% RH.	Carrageenan with calcium chloride coating was most effective to decrease firmness loss, whereas chitosan and carrageenan coatings added with calcium chloride showed the best reduction in weight loss. Even the addition of calcium to the coatings decreased the microbial growth rate the best result was shown for those fruit coated with chitosan and calcium chloride.	Ribeiro et al. 2007
Strawberries (*Fragaria* × *ananassa*) cv. Camarosa.	1. Control: Acetic acid (1%, v/v), pH 5.0. 2. Chitosan (1% w/v) in acetic acid (1%, v/v) + Tween 80 (0.1%, v/v) + oleic acid (0, 1, 2 and 4%, v/v), pH 5.0.	Immersion for 15 s and then dried for 1 h at 20 °C by natural convection.	10 fruit per box were packed in 750 mL perforated PET boxes.	14 days at 4 ± 1 °C	Chitosan coatings did not affect physicochemical quality, and maintained fruit colour. The addition of oleic acid in chitosan coating enhanced its effect to decrease the respiration rates and fungal infection. However, it affected strawberry sensory	Vargas et al. 2006

(Contd.)

Table 2. (*Contd.*)

Product	Coating	Application of Edible Coating	Packaging	Storage Conditions	Benefits	Reference
					quality, mainly aroma and flavour, especially at high ratios of oleic acid:chitosan in the film.	
Strawberries (*Fragaria* x *ananassa* Duch.) cv. Camarosa at 2/3 red colour ripening stage.	1. Control: Water. 2. Calcium gluconate (CG): 1%. 3. Chitosan (CH): 1.5% in acetic acid (0.5%). 4. CG (1%) as (2) + CH (1.5%) as (3).	Immersion for 35 min and air dried for 2 h at room temperature and 50% RH.		4 days at 20 °C and 70% RH.	Chitosan coating completely inhibited fungal decay compared to control that showed more than 80% disease incidence. Besides, reduced fruit weight loss, maintained firmness and delayed changes in fruit external colour. The addition of CG did not improve chitosan effect on fruit quality.	Hernandez-Muñoz et al. 2006
Strawberries (*Fragaria* x *ananassa* Duch) at 3/4 red colour ripening stage.	1. Control. 2. Chitosan (CH): 2% with sorbitol (S) (0.5%) in acetic acid (0.5%). 3. CH (2%) with S (0.5%) + potassium sorbate (PS) (0.3%) in acetic acid (0.5%). 3. HPMC (1%) with S	First fruits were inoculated by immersion for 30 s in a spore suspension of *Rhizopus* sp. or *Cladosporium* sp (1.1×10^4 spores/	Packed in recycled polyethylene terephthalate (PET) clamshell containers with venting holes ($120 \times 184 \times 90$ mm, 0.45 kg).	23 days at 6 ± 2 °C and 65% ± 5% RH for *Rhizopus* sp. and at 4 ± 2 °C and 50% ± 5% RH for	Chitosan coatings reduced the infection by *Rhizopus* sp. and *Cladosporium* sp. on inoculated fresh strawberries, with more reduction in	Park et al. 2005

(*Contd.*)

	(0.5%) + PS (0.3%) in distilled water.	mL) and dried for 1 h followed by coating application by immersion for 30 s and dried for about 1 h.		*Cladosporium* sp.	*Cladosporium* sp. than in *Rhizopus* sp. Chitosan and HPMC coatings reduced total aerobic count, coliforms, and weight loss of strawberries during storage. The addition of potassium sorbate did not improve chitosan' santifungal effect.	
Strawberries (*Fragaria* × *ananassa* Duch) cv. Diamante.	1. Control: Distilled water. 2. Chitosan: 1% in acetic acid (0.6%) and glycerol (0.5%), pH 5.2. 3. Chitosan: 1% in 0.6% lactic acid (0.6%) and glycerol (0.5%), pH 5.2. 4. Chitosan: 1% in lactic acid (0.6%) and glycerol (0.5%), pH 5.2 + vitamin E (dl-α-tocopheryl acetate) (0.2%).	Fruit with calyx removed was immersed for 10 s, drained for 30 min on stainless-steel screens, and dried for about 1 h under a fan. Fruit was immersed again, following the same procedure.	Fruit was packed in clamshell boxes.	1 week at 2 °C and 88% to 89% RH.	Chitosan coatings increased the appearance of strawberries, but coatings containing vitamin E decreased the acceptable appearance of strawberries. Chitosan coatings did not impart astringency nor affected the sensory acceptability of the fruit, which was similar to that of fresh strawberries, whereas coatings containing vitamin E decreased fruit appearance.	Han et al. 2005

(Contd.)

Table 2. (*Contd.*)

Product	Coating	Application of Edible Coating	Packaging	Storage Conditions	Benefits	Reference
Litchi (*Litchi chinensis* Sonn) cv. Kwai Mi at mature stage.	1. Control: Untreated. 2. Citric acid solution (CAS): at pH 0.8, 1.0 or 1.3. 3. Tartaric acid solution (TAS): at pH 0.8, 1.0 or 1.3. 4. Chitosan: 1% (w/w) + citric acid at pH 0.8, 1.0 and 1.3. 5. Chitosan: 1% (w/w) + tartaric acid at pH 0.8, 1.0 and 1.3.	Immersion for 5 s to 10 min, depending on the acidity of the solution.	Nine pieces of fruit were packed into small polyethylene baskets (PB). PB was placed into perforated plastic boxes (60 cm×40 cm).	Two weeks at 10 ± 2 °C.	Anthocyanins and total phenol contents are better maintained at high acidity treatment. Browning of treated fruit depended of the type of acid used, but mainly depended on the pH and level of dehydration of the pericarp after soaking. Chitosan coating reduced weight loss and allowed better acidification of the pericarp.	Joas et al. 2005
Litchi (*Litchi chinensis* Sonn) cv. Kwai Mi at mature stage.	1. Control: Uncoated. 2. Chitosan: 1% (w/w) + citric at pH 0.8, 1.0 and 1.3. 3. Chitosan: 1% (w/w) + tartaric acid at pH 0.8, 1.0 and 1.3.	First fruit was or not dipped in water for 15 s and drained-off, followed by coating application by immersion for 10 s (pH 0.8), 30 s (pH 1.0), and 4 min (pH 1.3), and air-dried.		Two weeks at 10 ± 2 °C.	Prehydrated fruit permitted better acid impregnation and reduced browning in fruit. This effect is enhanced at high pH (1.3). Chitosan plus citric acid resulted in less weight loss compared to chitosan plus tartaric acid.	Caro and Joas 2005

(*Contd.*)

Table 3. Edible coatings of chitosan applied to diverse climacteric fruits.

Product	Coating	Application of Edible Coating	Packaging	Storage Conditions	Benefits	Reference
Apples (*Malus domestica*) cv. Ligol.	1. Control. 2. Chitosan (CH): 1% (w/w). 3. CH + Quince juice (QJ): CH (as 1) in QJ (as A). 4. CH + Cranberry Juice (CrJ): CH (as 1) in CrJ (as B). 5. Whey protein (WP) + CH: CH (as 1) + WP (1% w/w). 6. WP + CH + QJ: WP (as 4) + CH (as 1) + QJ (as A). 7. WP + CH + CrJ: CH (as 1) + WP (as 4) + CrJ (as B). (A): QJ at 1:16, 2:15, 3:14, pH 2.0-2.3. B): CrJ at 1:19, 2:18, 3:17, pH 2.0-2.3.	Apples were inoculated with 20 µL of *Penicillium expansum* spore suspension (containing 10⁶ to 10⁷ spores/mL) wrapped with films.	Apples were placed into sterile containers (1000 mL).	10 days at 25 °C.	Wrapping apples with edible films made from chitosan, and whey protein and chitosan with the addition of cranberry and quince juice reduced the growth of *Penicillium expansum*.	Simonaitiene et al. 2015
Cherry tomato (*Lycopersicon esculentum* Mill).	1. Control. 2. Chitosan from *Mucor circinelloides* (CHI): 3.75 mg/mL	First wounded fruits were inoculated by immersion	Packed in commercially sterile polyethylene	12 days at 25 °C or 24 days at 12 °C.	Chitosan coating plus CAR inhibited the growth of *A. flavus* in artificially	de Souza et al. 2015

(Contd.)

Table 3. (*Contd.*)

Product	Coating	Application of Edible Coating	Packaging	Storage Conditions	Benefits	Reference
	+ Carvacrol (CAR) 2.5 µL/mL. 3. CHI (3.75 mg/mL) + CAR (1.25 µL/mL).	in a spore suspension (~10^6 spores/mL) of *Aspergillus flavus*, and dried for 30 min at 25 °C; followed by immersion for 1 min in the coating solution and air-dried on a nylon filter.	containers (with a lid).		contaminated fruits, as well as the native fungal microbiota of cherry tomato stored at 25 °C or 12 °C, with a major effect at the lower temperature. Besides, these coatings preserved the fruit physicochemical quality.	
Guava (*Psidium guajava* L.).	1. Control: uncoated 2. Cassava starch + Chitosan (CS-CH): Cassava starch [2% (w/v) + glycerol (0.64%, w/v)] + Chitosan [2.0%, w/v) in acetic acid (1.5%), + glycerol (1.28%, w/v)]. 3 and 4. Cassava starch + chitosan + Essential oil	Immersion for 3 min and then dried over nylon trays at room temperature.	Guavas were placed over aluminum trays and introduced into biochemical oxygen demand chambers.	10 days at 25 °C and 86–89% RH.	The addition of EOM to the CH-CS coating reduced the total aerobic mesophilic bacteria and mould and yeast counts during storage. The CS-CH-EOM coating improved the appearance of guavas due to it delayed the ripening process, reduced browning and inhibited colour development in	de Aquino et al. 2015

(*Contd.*)

	mixture (EOM) of *Lippia gracilis* Schauer: (CS-CH-EOM): I. Cassava starch (as 2) + Chitosan (as 2) + EOM I (1%), and III. Cassava starch (as 2) + Chitosan (as 2) + EOM III (3%).			guavas. EOM at a low concentration (1.0%) is recommended to be added to coatings to minimize its sensory impact.	
Mangoes (*Mangifera indica* L.) cv. Kent.	1. Control: Uncoated. 2. Chitosan (CH): (1 and 1.5%) in distilled water with lactic acid (0.7 mL) + glycerol (25% p/p of chitosan), pH 5.5. 3. Chitosan + Lactoperoxidase (LPO). 4. Chitosan + Lactoperoxidase + Iodine (LPOI).	Immersion for 1 min.	Eight days at 18 °C and 60% RH.	Chitosan coating alone reduced weight loss, and delayed the decline in firmness and respiration rate. Chitosan (1.5%) best maintained green colour. LPOS did not influence colour change. Chitosan (1%) with LPOS was effective to reduce microbial contamination and enabled delayed fruit ripening without affecting quality.	Cissé et al. 2015

(Contd.)

Table 3. (*Contd.*)

Product	Coating	Application of Edible Coating	Packaging	Storage Conditions	Benefits	Reference
Apricots (*Prunus armeniaca*) cv. Sufeda.	1. Control: Uncoated 2. Zinc fortified alginate based coatings: Alginate (1 and 2%, w/v) + citric acid (CA) (1%, w/v) + glycerol (G) (1.5%, w/v) + sunflower oil (SO) (0.025%, w/v) + N-acetyl L-cysteine (1%, w/v) + calcium chloride (2%, w/v) + zinc sulfate (ZS) (30 and 50 ppm) or zinc chloride (ZC) (30 and 50 ppm). 3. Zinc fortified chitosan (CH) based coatings: CH (1 and 2%, w/v) in acetic acid (1%, w/v) + ascorbic acid (2%, w/v) +	Immersion and then were allowed to dry for 15-20 min.		60 days at 4 ± 1 °C and 85 ± 5% RH.	Chitosan maintains keeping quality and prolongs the storage life of apricots. Zinc-fortified chitosan coating (2%) with either 50 ppm zinc sulfate or 50 ppm zinc chloride were the best treatments to reduce weight loss of apricots during storage. The zinc fortified chitosan coatings were better than alginate coatings to maintain total soluble solids, pH and acidity.	Younas et al. 2014

(*Contd.*)

	CA, G and SO (as 1) + ZS (30 and 50 ppm) or ZC (30 and 50 ppm).					
Blueberries (*Vaccinium* spp.).	1. Control: Distilled water. 2. Chitosan (CH): 2% (w/v) in acetic acid 1% (w/v). 3. CH (2%) + Blueberry leaf extracts (BLE): 4, 8, and 12% (w/v) of BLE. 4. CH (2%) + BLE 12% (w/v) + MAP (3 kPa O_2 + 12 kPa CO_2).	Immersion for 15 s, and then drained and dried at room temperature for 30 min.	250 g were packed in clamshell containers with vents.	35 days at 2 ± 1 °C and 95 ± 2% RH, plus 3 days at room temperature.	The addition of BLE into chitosan coatings improved its effect to decrease weight loss and decay, to maintain total phenolic content and radical scavenging activity of fruit during storage, this effect is highest when combined with MAP.	Yang et al. 2014
Banana cv. Cavendish.	Chitosan of 70% and 80% deacetylation degree at 1, 1.5 and 2% (w/w) in acetic acid (0.6% w/w) with triethanolamine (TEA) at (0 and 0.15%, w/w)	Immersion for 1 minute.		7 days at 30 ± 2 °C.	Chitosan at increasing concentration and deacetylation degree reduced weight loss and vitamin C loss, and maintained sensory quality. The highest benefit was obtained with 2% (w/w) chitosan with DD of 80% without TEA.	Suseno et al. 2014

(Contd.)

Table 3. (*Contd.*)

Product	Coating	Application of Edible Coating	Packaging	Storage Conditions	Benefits	Reference
Tomatoes at mature green stage.	1. Control: a) Water, and b) Emulsifier. 2. Chitosan–surfactant nanostructure assembly: Chitosan (1%) + Brij 56 and Span 20 (ratio of 1:1 w/w) to obtain the micelle sizes of 400, 600 and 800 nm.	Immersion for 1 min.		20 days at 15 ± 2 °C and 70–80% RH.	Chitosan–surfactant nanostructure coating delayed firmness and colour changes, and ripening for five days. It also decreased the decline of titratable acidity, and chlorophyll content, and the increase in soluble solid content, while decreased respiration rate and increased the phenolic content. However, it increased weight loss.	Mustafa et al. 2014
Pear (*Pyrus pyrifolia* Nakai), cv. Huang guan.	1. Control: Deionized water. 2. Calcium chloride: 2% (w/v). 3. Pullulan: 1% (w/v).	Immersion for 15 min at 20 °C and then dried for 24 h at 20 °C.	140 pears were packed into cartons with a nanoplastic film.	210 days at 0 ± 1 °C	Chitosan and pullulan were the most effective treatments to reduce the decline of total phenolic and flavonoid contents, chlorogenic acid,	Kou et al. 2014

(*Contd.*)

4. Chitosan: 2% (w/v) in acetic acid (0.5%, v/v), pH 5.6.			arbutin, catechin and caffeic acid, and to increase super oxide dismutase (SOD) and CAT activities, and total antioxidant activity, while decreased POD activity.	
Tomato (*Lycopersicon sculentum* cv. FA-180 HAZERA) at breaker stage of ripening.	1. Control 2. Chitosan (1.5% w/v) in lactic acid (1% v/v). 3. Chitosan as (2) + Tween 80 (0.1% w/v). 4. Chitosan (as 2) + zeolite (3% w/w based on chitosan).	Double immersion for 2 min, and then dried by forced convection for 2 h at 25 °C.	The combination of zeolite or Tween 80 with chitosan delayed the ripening of tomatoes. However, these treatments did not reduce weight losses.	37 days at 10 °C. Garcia et al. 2014
Avocado (*Persea americana* Mill.) cv. Hass.	1. Control: Sterile distilled water. 2. Prochloraz: 0.05%. 3. Thyme oil (TO): 1%. 4. Chitosan, (CH): 1% (w/v) in acetic acid (0.5%, v/v) + TO (0 and 1%) [3:1 v/v].	Immersion for 5 min and then dried for about 3 h at 20 °C, followed by inoculation with *Colletotrichum gloeosporioides* suspension (10⁵ spores/mL).	Packed in standard corrugated cardboard cartons. A major reduction in anthracnose disease incidence was obtained with the CH + thyme oil and AL + thyme oil coatings. CH + thyme oil coating increased total phenolic content	Five days at 20 °C. Bill et al. 2014

(Contd.)

Table 3. (*Contd.*)

Product	Coating	Application of Edible Coating	Packaging	Storage Conditions	Benefits	Reference
	5. Aloe vera (AV): 2% (w/v) + TO (0 and 1%) [3:1 v/v]. As anthracnose preventive treatment.				and the activities of POD, PAL, β-1,3-glucanase, chitinase, CAT and SOD, while reduced softening and flesh colour change.	
Plum (*Prunus domestica* L.) cv. Stanley and Giant.	1. Control: Acetic acid (0.5%), pH 5.6 + Tween-80 (0.5%). 2. Chitosan: 1% in acetic acid (0.5%), pH 5.6 + Tween-80 (0.5%).	Immersion for 1 min and then dried for 2 h at 25 °C.	2 kg of plums were packed in polypropylene baskets.	40 days at 0-1 °C and 90 ± 5% RH.	Chitosan coating reduced weight loss, respiration rate, ripening and decay rate, while it maintained higher titratable acidity, pH and firmness values.	Bal 2013
Tomato cv. Saladette at a Breaker, pink and red ripening stage.	1. Control: water. 2. Chitosan (1%) + oleic acid (1%). 3. Chitosan (1%) + oleic acid (1%) + lime essential oil (0.1%). All formulations contained glycerol at 0.3%.	Immersion and dried for 2 h. Inoculation with *Rhizopus stolonifer* (10^5 spores/ml) was done before (preventive) and after (curative) fruit		Four days at 12 °C and 25 °C.	Chitosan (1%) + beeswax (0.1%) + lime essential oil (0.1%) completely inhibited the development of *R. stolonifer* and completely.	Ramos-Garcia et al. 2012

(*Contd.*)

Mangoes (*Mangifera indica*) cv. Tommy Atkins at a semi-ripe stage.	1. Control: Distilled water, pH 3.0 or 7.0. 2. Nanomultilayer coating (5 layers): a) Pectin (PEC): 0.2% (w/v) in distilled water, pH 7.0. b) Chitosan (CH): 0.2% (w/v) in lactic acid 1.0% (v/v), pH 3.0.	a) Immersion for 15 min in PEC, rinsed with water (pH 7.0) and dried under a nitrogen flow. b) Then immersion for 15 min in CH rinsed with water (pH 3.0) and dried as before. a) and b) were repeated until completing five layers. Scars were first inoculated with a four-strain cocktail of *Salmonella* (serovars Montevideo,	Packaged in plastic boxes.	4 °C and 93% RH.	were coated. Inoculation with *E. coli* DH5α (10^5 cfu/ml). controlled *E. coli* DH5α during storage at 12° and 25 °C. Nanomultilayer coating reduced weight loss, pulp browning, shriveling, and microbial spoilage, as well as maintained titratable acidity and appearance.	Medeiros et al. 2012
Red round tomatoes.	1. Control. 2. 3 Acids (3A): Lactic acid (2%) + Acetic acid (2%) + Levulinic acid (2%).			14 days at 10 °C.	3A + chitosan coatings reduced *Salmonella* growth after two days of storage and completely inhibited its growth on tomato stem scar after the	Jin and Gurtler 2012

(*Contd.*)

Table 3. (*Contd.*)

Product	Coating	Application of Edible Coating	Packaging	Storage Conditions	Benefits	Reference
	3. 1 Acid (1A) + Chitosan (CH): Acetic acid (2%) + CH (low molecular weight, 75 to 85% deacetylation) 2% in Acetic acid (2%). 4. 3A + CH: CH (2%) in acetic acid (2%) + lactic acid (2%) + levulinic acid (2%). 5. 3A + CH + Allyl isothiocyanate (AIT): Formulation 4 + AIT (10 ppm).	Newport, *Saintpaul,* and *Typhimurium)* and dried for 2 h at 22 °C. Then 1 mL of the coating was dropped over the scar and dried for 24 h at 22 °C.			remaining time of storage. This coating has potential as an alternative antimicrobial for decontamination of tomatoes.	
Guava (*Psidium guajava* L.) cv. Pearl at 80% maturity (green stage).	1. Control: Acetic acid. 1% + Tween 80 0.1%, pH 6.0. 2. Chitosan: 0.5, 1.0 and 2.0% in acetic acid (1%) + Tween 80 (0.1%), pH 6.0.	Immersion for 1 min and then dried for 30 min at room temperature.	Five guavas were packaged in unsealed plastic bags (0.04 mm thick).	12 days at 11 °C and 90–95% RH.	Chitosan coating (2.0%) decreased firmness and weight loss, delayed changes in chlorophyll, malondialdehyde (MDA), and soluble solids content (SSC); also retarded the loss of titratable	Hong et al. 2012

(*Contd.*)

Commodity	Treatments	Method	Storage conditions	Results	Reference
Banana (AAA group) cv. Pisang Berangan Papaya (*Carica papaya* L.) cv. ksotika II.	1. Control. 2. Gum Arabic (5%) in water + chitosan (1.0%) in acetic acid (0.5%, v/v), pH 5.6. 3. Gum Arabic (10%) + chitosan (1%). 4. Gum Arabic (15%) + chitosan (1%). 5. Gum Arabic (20%) + chitosan (1%). Solutions 3, 4 and 5 were prepared as 2.	Bananas and papayas were immersed in spore suspension of *Colletotrichum*. Fruit was packed into cardboard boxes. *musae* and *Colletotrichum gloeosporioides* (1×10^5 spores/ml), respectively. After 1 h fruit were immersed in coatings and dried.	28 days at 13 ± 1°C and 80-90% RH, for banana. 28 days at 12 ± 1°C and 80-90% RH, for papaya.	Gum Arabic (10%) plus chitosan (1%) was the best treatment to reduce anthracnose in artificially inoculated fruits. This coating also retarded the ripening processes reducing moisture loss, and changes in pH, colour, respiration and ethylene evolution. acidity (TA) and vitamin C losses during storage. This treatment increases the activities of POD, SOD and CAT, and inhibited superoxide free radical production.	Ali et al. 2012
Guava (*Psidium guajava* L.) cv. Pedro Sato.	1. Control: Acetic acid (0.4%, w/w). 2. Cassava starch (CS): 2.5% in acetic acid (0.4%, w/w) + glycerol (2%).	Immersion for 3 min and then dried for 3 h at 25 ± 2 °C.	12 days at 22 ± 2 °C and 62 ± 6% RH.	Cassava starch plus chitosan (1.5%) coating was the best treatment to maintain peel green colour, firmness and visual appearance, while	Soares et al. 2011

(*Contd.*)

Table 3. (*Contd.*)

Product	Coating	Application of Edible Coating	Packaging	Storage Conditions	Benefits	Reference
	3. CS + Chitosan (CH): CS as (2) + CH (1.0 and 1.5%) in acetic acid (0.4% w/w).				reducing weight loss and the growth of filamentous fungi and yeast and extending guava shelf life.	
Papaya (*Carica papaya* L.) cv. Eksotika II at mature-green ripening stage, colour index 2 (green with trace of yellow).	1. Control: Acetic acid (0.5% v/v), pH 5.6 + Tween-80 (0.1% v/v). 2. Chitosan (CH) (0.5, 1.0, 1.5, and 2.0% w/v), in acetic acid (0.5%, v/v), pH 5.6 + Tween-80 (0.1%, v/v).	Immersion for 1 min and air dried at ambient temperature.	Packed in commercial corrugated boxes.	Five weeks at 12 ± 1 ° C and 85–90% (RH).	Chitosan coatings (1.0%–2.0%) reduced weight loss and respiration rate, maintained firmness, and delayed changes in the peel colour and in total soluble solids (TSS). Chitosan coating (1.5%) showed the highest sensory quality values, followed by CH (1%).	Ali et al. 2011
Papaya (*Carica papaya* L.).	1. Control. 2. Aloe gel (AG): 50% + ascorbic acid (0.2% w/v) + citric acid (0.46% w/v), pH 4 + commercial gelling agent (1%).	Immersion for 15 min at room temperature, drained and then dried with forced convection air.	Fruits were packed in bamboo baskets covered with newspaper.	15 days at 30 ± 3 °C and 42–55% RH.	AG + PLE coated fruit had the lowest weight loss, the lowest decrease in pH, titratable acidity and total soluble solids, lowest	Marpudi et al. 2011

(*Contd.*)

	3. AG + Papaya leaf extract (PLE) (1:1). 4. Chitosan: 2.5% in hydrochloric acid (0.25 N), pH 5.6.			decay and highest sensory quality and marketability values, followed by aloe gel and chitosan coating.	
Bananas (AAA group) cv. Pisang Berangan.	1. Control: Purified water. 2. Gum Arabic (GA): at 5% in purified water, pH 5.6 + Chitosan (1%) in acetic acid (0.5%, v/v), pH 5.6. 3. GA (10%) + CH (1%). 4. GA (15%) + CH (1%). 5. GA (20%) + CH (1%). GA and CH in 3), 4) and 5) were prepared as 2).	Immersion for 2-3 min and then fruits were air-dried.	Fruit was packed in cardboard corrugated boxes.	28 days at 13 ± 1 °C and 80 ± 3% RH, followed by five days at 25 °C and 60% RH.	The composite coating GA (10%) plus chitosan (1%) fruit showed the lowest weight loss and soluble solids concentration, and the highest fruit firmness, total carbohydrates, and reduction in sugars values. This coating retarded colour development and decreased respiration rate and ethylene production during storage, and maintained the overall quality of banana fruits.	Maqbool et al. 2011
Highbush blueberry cv. Duke and Elliott.	1. Semperfresh™ (SF): 1% + GT. 2. Acid-soluble chitosan (ACH, 89.8%	Immersion for 30 s, and then drained over a stainless steel	120–130 g fruit were packaged in vented or non-vented clam-shell containers.	1 week at 2 ± 1 °C and 88% RH, followed	ACH, WCH, and WCH+ SA coatings reduced the decay	Duan et al. 2011

(Contd.)

Table 3. (*Contd.*)

Product	Coating	Application of Edible Coating	Packaging	Storage Conditions	Benefits	Reference
	deacetylation degree): 2% (w/v) in acetic acid (1%) +GT. 3. Water-soluble chitosan (WCH): 3% in deionized water + GT. 4. WCH (1.5%, w/v) + sodium alginate (SA) (1%, w/v) + GT. 5. Calcium caseinate (CC): 2% in deionized water + GT. GT: Glycerol (50%, w/w dry weight) + Tween 20 (0.15%, w/v).	screen for 30 min, and then the same procedure was repeated.		by 15 days at 20 ± 3 °C and 30% RH.	of 'Duke' or 'Elliott' fruit at 20 °C. SF coating reduced weight loss of 'Duke' during storage at 20 °C for 6 days. CC coating of 'Elliott' fruit maintained higher TA, lower pH, and greater firmness fruit than control fruits, also delayed fruit ripening. Packaging in non-vented containers reduced fruit weight loss and maintained firmness.	
Bananas (AAA group) cv. Pisang Berangan.	1. Control. 2. Arabic gum (AG): 5, 10, 15 and 20% (w/v) in purified water, pH 5.6.	Inoculation by immersion for 2-3 min in a spore suspension of	Fruit was packed in cardboard corrugated boxes.	28 days at 13 ± 1 °C and 80 ± 3% RH followed by 5 days at 25	Chitosan (1.5%) and the composite coating AG (10%) plus CH (1.0%) were the best treatments to	Maqbool et al. 2010

(*Contd.*)

Commodity	Formulation	Treatment	Storage conditions	Results	Reference
	3. Chitosan (CH): 0.5, 0.75, 1.0, and 1.5% (w/v) in acetic acid (0.5%, v/v), pH 5.6. 4. AG + CH: AG (5, 10, 15, and 20%, w/v) as (2) + CH 1% as (3).	*Colletotrichum musae* (1 × 10^5 spores/ml) and then dried at room temperature. Coating by immersion for 2-3 min in formulations, and then dried at ambient temperature.	°C and 60% RH, in order to simulate marketing conditions.	decrease the disease incidence by C. *musae* in artificially inoculated bananas. AG (10%) plus CH (1.0%) coating also had the lowest weight loss, and maintained fruit firmness, preserving the quality of banana for up to 33 days.	Muy Rangel et al. 2009
Mango (*Mangifera indica* L.) cv. Ataulfo at a physiologically mature stage.	1. Control: Untreated. 2. Commercial chitosan (CH): 1.2% soluble solids. 3. 1-methyl cyclopropene (1-MCP): 400 nL/L. 4. 1-MCP: 800 nL/L. 5. CH (1.2%) + 1-MCP: 400 nL/L 6. CH (1.2%) + 1-MCP 800 nL/L.	1. and 2. The fruits were sprayed with the solution. 3. and 4. Fruits were placed in hermetically sealed recipients and 1-MCP was applied as a gas for 24 h at 12 °C. 5) and 6): CH was applied first as 1) and then MCP as 3).	Seven days at 12 °C followed by 15 days at 20 °C.	All treatments reduced fruit respiration rate compared to control fruit. 1-MCP combined with chitosan delayed respiration climacteric peaks by three days. 1-MCP treatment reduced the activities of PG and cellulose, while treatment with 400 nL/L of 1-MCP showed the highest values of firmness during both storage conditions.	

(Contd.)

Table 3. (*Contd.*)

Product	Coating	Application of Edible Coating	Packaging	Storage Conditions	Benefits	Reference
Banana (*Musa paradisiaca,* L.) type Cavendish.	1. Control: Untreated. 2. Commercial (FreshSeal®) chitosan (CH): 3% (w/v). 3. 1-methyl cyclo-propene (1-MCP): 300 µg/L. 4. CH + 1-MCP: 300 µg/L.	1. and 2. Sprayed all over fruit surface 3. Fruit was placed in an airtight stainless steel container and gassed with 1-MCP for 12 h at 22 °C. 4. CH was applied first as 1) and then MCP as 3).		8 days at 22 °C, 85% RH.	1-MCP alone or combined with CH maintained fruit firmness, and delayed the incidence of sugar spots, while neither affected sensory quality. CH + 1-MCP prolonged the shelf life of bananas for up to four more days.	Baez-Sañudo et al. 2009
Pears (*Pyrus pyrifolia* Nakai) cv. Huanghua.	1. Control: Distilled water. 2. Shellac (14.3% w/v) + NH₃ (0.8%, w/v) + food grade poly-dimethylsiloxane antifoam (0.01%, w/v). 3. Semperfresh™: 1% (w/v).	Immersion for 15 s, then fruit was dried over stainless steel shelves under fans.		60 days at 4 °C and 95% RH.	All coatings maintained texture values and reduced respiration rate compared to the control fruit. However, shellac coating was better to decrease weight loss and to maintain the soluble solid content and titratable	Zhou et al. 2008

(*Contd.*)

	Coating formulation	Application	Packaging	Storage	Results	Reference
	4. N,O-Carboxymethyl chitosan (CMC): 2% (w/v) + glycerol (0.5%, w/v) + Tween 80 (1.2%, w/v) + dl-α-tocopheryl acetate (0.4%, w/v). All solutions were diluted in water.				acidity of pears than Semperfresh™ and CMC coatings.	
Hardy kiwifruit (*Actinidia argute* (Siebold & Zucc.) Planch. ex Miq) cv. Ananasnaya.	1. Control: Deionized water. 2. Calcium caseinate: 1% in deionized water. 3. Chitosan (CH): 3% (w/v) in acetic acid (1% v/v) + glycerol (10% w/w with CH) + stearic acid (25% w/w with CH). 4. PrimaFresh® 50-V: 1:6 with deionized water. 5. Semperfresh™: 1% in deionized water.	Immersion for 30 s, then dried for 30 min over a stainless steel screen under fans, followed by a second immersion and dried again under similar conditions.	Eight fruits (about 100 g) were packed in a plastic clamshell container (high vent, HV), standard berry containers with many air vents, or low-vent (LV) containers with individual wells with two small open air vents.	6 or 10 weeks at 2 °C and 88% RH or 22 ± 1 °C and 45% RH in the dark.	Even coatings did not reduce weight loss of kiwifruit. They imparted an attractive shiny appearance with good acceptability values. Coatings did not affect fruit ripening.	Fisk et al. 2008

(Contd.)

Table 3. (*Contd.*)

Product	Coating	Application of Edible Coating	Packaging	Storage Conditions	Benefits	Reference
Mango (*Mangifera indica* L.) cv. Tainong at mature stage.	1. Control: Acetic acid (1%), pH 6.0 + Tween-80 (0.5%). 2. Chitosan (CH): 2% in acetic acid (1%), pH 6.0 + Tween-80 (0.5%). 3. TP-Chitosan: CH (2%) as (2) + Tea polyphenols (TP) 1%.	Immersion for 1 min, and then dried under a fan at 25 °C.		35 days at 15 ± 1 °C and 85–90% RH.	Chitosan coating added with tea polyphenols was the best treatment to reduce respiration rate, weight loss and decay (stem end rot caused by *Dothoriella* spp) incidence. This coating delayed softening.	Wang et al. 2007
					maintained soluble solids content and reduced titratable acidity, starch content and chlorophyll content decrease, while enhancing sensory quality. TP–chitosan maintained keeping quality and prolonged storage life of mango.	

(*Contd.*)

| Mango fruits (*Mangifera indica*) cv. Alphonso at mature green ripening stage. | 1. Control: No cover. 2. Chitosan (CH) film: ~100 gauge. CH-acidic solution was spread on a leveled surface and dried 27 ± 1 °C and 65 ± 1% RH or using infrared heaters. 3. Low density polyethylene film (LDPE): ~100 gauge. | | Four fruits were packed in each carton box and top covered with films 2) and 3) or no cover. | 27 ± 1 °C and 65% RH. | Mango fruit packed into chitosan-covered boxes had an extended shelf-life of up to 18 days, without any microbial growth and off flavor, compared to control fruit with a shelf life of 9 days. | Srinivasa et al. 2002 |

Table 4. Edible coatings of chitosan applied in freshcut fruits and vegetables.

Product	Treatment	Application of Edible Coating	Packaging	Storage Conditions	Benefits	Reference
Sliced Bamboo shoots (*Phyllostachys praecox f. prevernalis*).	1. Control: Distilled water. 2. Chlorine dioxide (ClO_2): 28 mg L^{-1}. 3. Chitosan: 1.5%. 4. ClO_2 + chitosan (1.5%).	1. Immersion for 15 min then drained and air dried. 2. Immersion for 3 min at 20 °C then drained and air dried. 3. First treated as 2), and then treated as 3).	Shoots slices were packed in plastic trays, covered with unsealed individual polyethylene film.	6 days at 4 °C and 60–70% RH.	ClO_2 + chitosan coating showed better efficiency than individual ClO_2 or chitosan coatings in reducing the respiration rate and firmness, delaying browning and lignification and reducing total aerobic bacteria and yeasts and moulds counts.	Yang et al. 2015
Sliced 'Tommy Atkins' mangoes, semi-ripe stage.	1. Control: Uncoated. 2. Sodium alginate (Alg) (0.2% w/v), pH 7.0, 3. Chitosan (Ch) (0.2% w/v), pH 3.0.	Layer-by-layer technique: five nanolayers (Alg-Ch-Alg-Ch-Alg).	Zipped bags (15×15 cm).	14 days at 8 °C and 93% RH.	Alginate-chitosan-based nanomultilayer coating extended the shelf life of fresh-cut mangoes up to eight days at 8 °C, compared to uncoated fresh-cut mangoes (<2 days).	Souza et al. 2015

(Contd.)

Sliced (5-6 mm thick) pear, cv. Rocha, at a commercial maturity stage.	1. Control: Ascorbic acid (0.5%). 2. Chitosan (75-85% deacetylated): 0.7, 1, 1.5 and 2% (w/v), pH 6.0.	Immersion for 1 min, and drained for 3 min at 4° C.	Three slices in Petri dishes sealed with parafilm.	10 days at 4° C.	Chitosan coatings reduced browning, and maintained firmness. Chitosan (2%) reduced microbial growth.	Sánchez et al. 2015
Sliced apple, cv. Gala at commercial maturity stage.	1. Control: Uncoated. 2. Chitosan coatings with 110-nm chitosan tripolyphosphate (TPP) nanoparticles. 3. 300-nm chitosan (TPP) nanoparticles. 4. Chitosan: 0.2% (w/v), pH 4.16.	2) and 3) treatments were sprayed. 4) Immersion for 2 min and then drained.	200 g of slices were packed in polyethylene terephthalate trays (160 µm, 750 mL).	10 days at 5 °C.	Coatings with chitosan nanoparticles of 110 nm showed higher antimicrobial activity against moulds and yeasts, and mesophilic and psychrotrophic bacteria than the other treatments.	Pilon et al. 2015
Cubes (2 cm side) of quince, cv. Nishabur, at ripe stage.	1. Control: Distilled water + A. 2. Aloe vera gel (30% w/w) + A. 3. Chitosan (0.5 and 1.5% w/w) + A. 4. Carboxy methyl Cellulose (0.5 and 1.5% w/w) + A. A: Calcium chloride (2% w/v) and citric acid (4% w/v).	Immersion and air-dried for 30 min at room temperature.	Trays were over-wrapped with plastic films (0.02 mm thick polyethylene films).	Nine days at 4 °C and 85% RH.	*Aloe vera* coating delayed browning.	Noshad et al. 2015

(Contd.)

Table 4. (*Contd.*)

Product	Treatment	Application of Edible Coating	Packaging	Storage Conditions	Benefits	Reference
Commercial baby carrots (*Daucus carota* L.)	1. Control: Uncoated. 2. Chitosan (LMw): 1% with glycerol (15% wt).	Immersion or spraying for 30 s, followed by drying process into a convection oven at 20 °C.	120 g of produce was packed in polystyrene-based laminated trays (PS/EVOH/LDPE) and heat-sealed using a low O_2 permeable (3 cm³/m²/24 h at STP) laminate LDPE/EVOH/LDPE film (Cryovac).	15 days at 4° C under MAP; initial internal gas composition 10% O_2 and 10% CO_2.	Chitosan-based coatings delayed microbial spoilage, prevented surface whitening, provides better colour retention and preserved product texture.	Leceta et al. 2015
Cubes (2 cm side) of Honeydew melon (*Cucumis melo* L.), cv. Inodorus.	1. Control: Distilled water. 2. Chitosan: 2%. 3. Calcium chloride ($CaCl_2$): 1%. 4. $CaCl_2$ (1%) + Chitosan (2%, w/w).	Immersion for 10 min, followed by drying process for 1 h in air.	Ziploc bags (6 cm x 8 cm, and 0.05 mm thickness).	13 days at 7 ± 2° C.	$CaCl_2$ (1%) + chitosan (2%) was the most effective coating to reduce weight loss, to maintain firmness and colour and to reduce mesophilic and pyscrotrophic growth compared to control group.	Chong et al. 2015
Sliced (6 mm) kiwifruit (*Actinidia deliciosa*) cv. Hayward.	1. Control: Distilled water. 2. Alginate: 2% (w/v) + Calcium lactate (2%, w/v).	Immersion for 10 min at ambient temperature, followed by air drying for 15 min.	45 g (4 slices) were packed into polypropylene trays (450 ml) and sealed	12 days at 4 ± 1° C and 75% RH.	*Aloe vera* coating maintained the quality of kiwifruit slices by reducing	Benítez et al. 2015

(*Contd.*)

Sample	Treatments	Conditions	Storage	Results	Reference
	3. Chitosan: 1%, (w/v) in acetic acid solution (1%, v/v) or in citric acid solution (2%, w/v). 4. *Aloe vera* gel: 5%.	with a composite film (PP-PET, $PO_2 = 1.1 \times 10^{-3}$ ml/m²/24 h/Pa, $PCO_2 = 5 \times 10^{-3}$ ml/m2/24 h/Pa).		the microbial proliferation and preventing the tissue softening compared to chitosan and alginate edible coatings.	Ban et al. 2015
Segments of Honey pomelo (*Citrus grandis* L. Osbeck cv. Guanxi Pummelo at a fully ripening stage.	1. Control. 2. N,O-carboxy methyl chitosan (NOCC) (1%, w/v). 3. Sodium chlorite (SC): 600 mg/L. 4. SC (600 mg/L) + NOCC) (1% w/v).	2 fruit segments were packaged with an oriented polypropylene (OPP) film (20 μm thickness) bags and sealed. The O_2 and CO_2 permeance film were 8.7751×10^{-13} mol O_2 m²/s/Pa, and 3.2845×10^{-12} mol CO_2/m²/s/Pa at 23 °C and 0% RH. 2. Immersion for 2 min, then drained for 1 min, and air-dried for 20 min. 3. Immersion for 5 min, then drained for 1 min and air-dried for 20 min. 4. SC was applied in the same way and then NOCC.	19 days at 4 °C under passive and active MAP (10% O_2 + 10% CO_2).	The combined application of NOCC and SC under active MAP reduced weight loss and microbial growth, and best maintained the ascorbic acid content and sensory quality of freshcut pomelo.	
Plugs (4 cm long and 1.7 cm diameter) of Orange-flesh cantaloupe melons of ananas type, (*Cucumis melo*	1. Control: Uncoated. 2. Gelatin (1.5% w/v). 3. Chitosan (1.5% w/v). 4. Gelatin (1.5% w/v) + Chitosan (1.5% w/v). 5. Bilayered Films (LbL): Gelatin (1.5% w/v)/Chitosan w/v)/Chitosan (1.5% w/v).	5 plugs were packed in polyethylene terephthalate (PET) clamshell containers. 2 to 4. Individual plug immersion for 2 min and air-dried under aseptic conditions at room temperature for 30 min. 5. First immersion	14 days at 6 °C.	The LbL coating showed effective total microbial growth inhibition, was superior in preservation of fruit texture and also slightly reduced	Poverenov et al. 2014b

(Contd.)

Table 4. (*Contd.*)

Product	Treatment	Application of Edible Coating	Packaging	Storage Conditions	Benefits	Reference
L. subsp. *melo* var. *cantalupensis* Naudin), cv. Yaniv.		in a gelatin solution for 2 min and then in a chitosan solution for 2 min, before being air-dried at room temperature for 30 min.			fruit weight loss, and do not cause accumulation of off flavour volatiles.	
Cubed (15-mm side) melon (*Cucumis melo* L. var. Reticulatus Naud) group Inodorus, at a commercially ripe stage (10% soluble solid).	1. Control: Distilled water. 2. Multi-layer-by-layer coating of: a) Chitosan with glycerol (2% w/w) + Tween 20 (0.5% w/w), and the microencapsulated antimicrobial (trans-cinnamaldehyde-β-cyclodextrin) (2% w/w). b) Pectin (2% w/w). c) CaCl₂.	1. Immersion for 2 min in water. 2. Immersion into the first coating solution for 2 min and the residual solution was dripped off for 2 min. This process was repeated until the samples have been immersed into the three solutions (a, b and c).	Fifteen cubes (240 ± 10 g) packed in polypropylene trays (Ziploc Brand with Smart™ Snap Seal, 591 mL) Ziploc lid.	15 days at 4 °C.	The multilayered antimicrobial edible coating improved the microbiological and physicochemical quality of fresh-cut melon, extending the shelf life up to 15 days at 4 °C, compared with the control (<7 days). The coating maintained firmness, reduced weight loss, and vitamin C and total carotenoids losses.	Moreira et al. 2014

Cylinders (2.54 cm) of Cantaloupe melon at commercial ripening stage (8 to 10 °Brix).	1. Control: Distilled water. 2. Chitosan: 1.0, and 2.0% (w/w) + Pectin (1 or 2% w/w). 3. Multilayered edible coating with: a) Chitosan (0.5, 1, 2% w/w). b) beta-cyclodextrin encapsulated trans-cinnamaldehyde (1, 2, 3%, w/w). c) Pectin: 0.5, 1, 2% (w/w). Calcium chloride: 2% (w/w) was added to formulations 2 and 3.	Layer-by-layer procedure coating in the following order: Calcium chloride, chitosan and antimicrobial, pectin, and, calcium chloride. Immersion into each solution for 2 min, drip off for 2 min before dipping the samples into the next solution. Then 8 min of drying at room temperature.	Sixteen cylinders were packed in Ziploc® containers (Smart Snap® Seal, 591 ml, polypropylene) with a polyethylene lid.	15 days at 4 °C.	The best formulation of multilayered edible antimicrobial coating for fresh-cut cantaloupe was chitosan (2%) + pectin (1%) + 2 g/100 g encapsulated trans-cinnamaldehyde (2%). This coating maintained physicochemical and sensory quality attributes for longer (7 to 9 days versus 4 days for the uncoated controls), and also improved firmness.	Martiñon et al. 2014
Peeled white (*Opuntia albicarpa*) and red (*Opuntia ficus-indica*) cactus pear fruit.	1. Control: Uncoated. 2. Chitosan (1% w/v) with olive oil (0.6% v/v) in acetic acid (1 or 2.5% v/v).	Immersion twice in the chitosan solution for 1 min and then dried for 10 min at 25 °C in a food dehydrator.	Packed into clear polyethylene boxes (0.15 m × 0.15 m × 0.10 m).	16 days at 4 ± 1 °C and 85 ± 5% RH.	Chitosan coating with 1.0% of acetic acid maintained the best quality characteristics	Ochoa-Velasco and Guerrero-Beltrán 2014

(Contd.)

Table 4. (*Contd.*)

Product	Treatment	Application of Edible Coating	Packaging	Storage Conditions	Benefits	Reference
					and delayed the microbial growth in both varieties of cactus pear fruit.	
Wedges of pineapples (*Ananas comosus*) cv. MD2 Del Monte	1. Control: Distilled water. 2. Chitosan: 0.5% (w/v) in 0.5% acetic acid (v/v). 3. Sodium alginate: 1% (w/v) with glycerol (1.5%, w/v) and sunflower oil (0.025%, w/v). 4. Calcium lactate: 0.5% (w/v) or calcium ascorbate (1%, w/v).	Immersion for 5 min and then allowed to drain on cellulose paper for 15 min. For the alginate treatment, the wedges were dipped for 2 min in calcium lactate for alginate cross-linking.	110 g (four wedges) were packaged on polypropylene trays (255 mL), and heat sealed. The O_2 and CO_2 permeance of the PP-EVOH-PP film were 1.1×10^{-2} and 1 cm^3/ m^2/d/bar at 23 °C and 70% relative humidity, respectively.	15 days at 4 °C.	Calcium lactate dip combined with the use of a low permeable film may be useful in maintaining the quality of fresh-cut pineapple for up to 15 days of storage at 4 °C.	Benitez et al. 2014
Sliced (5 mm) mushrooms (*Agaricus bisporus*).	1. Control: Uncoated. 2. Chitosan: 0.3% in acetic acid (0.5%), pH 5.0. 3. Calcium chloride: 2.0%.	Spraying, and then drying using a fan at ambient temperature.	180 g were packaged in bags (22.0 × 30.0 cm) and heat-sealed. Film bags were: a) Polyvinyl chloride	7 days at 12 °C.	The polyethylene film (PE-2) packaging combined with coating treatment was the most	Ban et al. 2014

(Contd.)

	(PVC) wrap (28 μm thickness with a permeability to O_2 and CO_2 of 1.6×10^{-6} mL/m^2 s Pa and 4.8×10^{-6} mL/ m^2 s Pa at 23 °C and 0% RH, respectively). b) Polyethylene film (PE) 1: PE-1 film (25 μm thickness with a permeability to O_2 and CO_2 of 1.4×10^{-6} mL/m^2 s Pa and 5.6×10^{-6} mL/ m^2 s^1 Pa at 23 °C and 0% RH, respectively). c) PE-2: 30 μm thickness with a permeability to O_2 and CO_2 of 1.1×10^{-6} mL/m^2 s Pa and 2.5×10^{-6} mL/m^2 s Pa at 23 °C and 0% RH, respectively.		effective to lower maturity index (cap opening), to maintain fair colour, and to improve the preservation of mushrooms stored at 12 °C up to 7 days, as well as to satisfy consumer acceptance.	Xu et al. 2013
Sliced (4 mm thick) pears (*Pyrus communis* L.).	1. Control: Distilled water. 2. Chitosan (CH): 1.2% (w/v) in acetic acid (0.75%, v/v) + ascorbic acid (AA) (2%) + citric acid (CA) (1%).	1. and 2. Immersion for 5 min and then placed in a plastic sieve for 30 min. 3. Tissue paper (11 cm diameter) was dropped	150 g were packed in polypropylene film bags (20 × 25 cm).	15 days at 4 °C.	Combined application of chitosan coating (with AA and CA) and oil fumigation retarded the microbiological deterioration of

(Contd.)

Table 4. (*Contd.*)

Product	Treatment	Application of Edible Coating	Packaging	Storage Conditions	Benefits	Reference
	3. Oil fumigation. 4. CH + AA + CA + Oil fumigation.	with 0.10 mL of cinnamon oil and sealed. 4) Chitosan was applied as 2) and followed by treatments as 3).			fresh-cut pears, prevented product weight loss, reduced overall visual quality changes and loss of vitamin C and total phenolics.	
Arils of pomegranate, cv. Tarom at commercial mature stage.	1. Control: Distilled water with acetic acid (1% v/v). 2. Chitosan (0.25, 0.5 and 1% w/v) in acetic acid (1% v/v), pH 5.0.	Immersion for 1 min.	Packed in rigid polyethylene boxes (10 cm × 6 cm × 5 cm) with lids.	12 days at 4 °C and 95% RH.	Chitosan coating (1%) maintained acceptability, reduced fungi and bacteria growth. It also reduced the loss of anthocyanins, phenolic contents and antioxidant activity.	Ghasemnezhad et al. 2013
Florets of broccoli (*Brassica oleracea* L), cv Italica.	1. Control: Distilled water. 2. Medium molecular weight Chitosan (CH) with (98%	1., 2., 5. and 6. Immersion for 3 min at 20 °C. Then drying for 30 min with air at 30 °C and 60% RH.	Three florets (60 to 90 g) were packed in polymeric bags (PD960, Cryovac™) of 25 mm of thickness (with an O_2 permeability of	7 days at 5-7 °C.	CH coatings alone or enriched with BC/EO inhibited the growth of mesophilic and psychrotrophic bacteria.	

(*Contd.*)

deacetylation degree) (0.5, 1.0 and 2% w/v) in acetic acid (1%, v/v) with glycerol (Gly), (Gly/CH) weight ratio of 0.28. 3. Bioactive compounds (BC): Pomegranate and resveratrol (R) at 30, 60, 80, 100, 120 and 180 $\mu g/mL$, pollen and propolis (Pp) at 3, 6, 10 and 12 $\mu L/mL$. 4. Essential oils (EO): tea tree (TT) at 5, 10 and 15 $\mu L/mL$, Rosemary at 6, 12 and 18 $\mu L/mL$. 5. CH + BC: CH + R, CH + Pp 6. CH + EO: CH + TT	3) and 4): Hand-sprayed to a load of approximately 77 mL/m².	7000 cc/m²/d, CO₂ permeability of 20,000 cc/m²/d, and water vapour permeability of 1 g/m²/d).		The addition of BC to CH improved its antimicrobial action. CH coating alone and plus BC inhibited the enzymatic browning. CH coating alone and enriched with BC did not affect the texture and inhibited the florets' opening.	Alvarez et al. 2013
Shreds (2–3 mm thickness, and 35–40 mm length) of radish roots. 1. Control: Uncoated. 2. Purified chitosan (molecular weight 161 kDa and degree of deacetylation 85%): 0.2%.	Radish shreds were spread as a thin layer in polypropylene trays, and then chitosan powder	100 g radish shreds were packed in macro perforated LDPE resealable pouches (15 cm × 13 cm; 6 mm thickness)	10 days at 10 °C.	Purified chitosan and chitosan lactate reduced weight loss, respiration rate,	Pushkala et al. 2013

(Contd.)

Table 4. (*Contd.*)

Product	Treatment	Application of Edible Coating	Packaging	Storage Conditions	Benefits	Reference
	3. Chitosan lactate: 0.2%.	was applied over the surface of the shreds, and mixed.			titratable acidity, soluble solids. These treatments maintained the higher content of phytochemicals, better sensory acceptability, lower exudate volume, lesser browning and lower microbial load than the control.	
Sliced (5-mm-thick) nectarines (*Prunus persica* (L.) Batsch) cv. Big Top).	1. Control: Distilled water. 2. Chitosan: 2% (w/v) in citric acid 2% (v/v) + glycerol 50% (v/v) + Tween 20 (0.15% v/v). 3. Sodium alginate: 2% (w/v) in deionized water + glycerol 1.5% (v/v) + calcium chloride 2% (w/v). 4. Chitoplant®: 2% (w/v) in deionized water.	Immersion for 1 min at 20 °C and then dripped off.	Twenty-five slices were packaged in polypropylene plastic bags (20 cm × 30 cm size and 90 mm thickness) with 50 cm³ $O_2/m_2/bar/day$, 150 cm³ $CO_2/m^2/bar/$ day and 2.8 g/m²/ bar/day water vapour transmission.	Five days at 4 °C and 95% RH.	Alginate coating best maintained initial colour and acidity of nectarine slices, while Chitoplant® was the best coating to reduce browning and decay of fresh-cut nectarine.	Chiabrando and Giacalone 2013

(*Contd.*)

| Half sliced apples (*Malus domestica*), cv. Gala. | 1. Control: Acetic acid (1%) + inoculum.
2. Chitosan (0.2% w/v) in acetic acid (1% v/v) + inoculum.
3. Chitosan derivatives (0.2% w/v) in acetic acid (1% v/v) + inoculum: N-butylchitosan (ButChi), N-octylchitosan (OctChi), and N-dodecylchitosan (DodecChi).
4. Chitosan quaternary salt: N,N,N-trimethylchitosan (TMC), N-butyl-N,N-di-methylchitosan (ButDMC), N-oc-tyl-N,N-dimethylchi-tosan (OctDMC), and N-dodecyl-N,N-di-methylchitosan (DodecDMC).
Inoculum: Spore suspension of *Botrytis cinerea* and *Penicillium expansum* (1.0×10^6 spores/mL). | Immersion for 30 seconds, then drained and dried. | Seven days at 28 °C and 80% RH. | Water soluble quaternary salt of chitosan (TMC) was more effective in reducing browning, and having good antifungal activity against *Penicillium expansum*. | de Britto and Garrido Assis 2012 |

(Contd.)

Table 4. (*Contd.*)

Product	Treatment	Application of Edible Coating	Packaging	Storage Conditions	Benefits	Reference
Cubes (15 mm side) of papaya fruits (*Carica papaya* L.) cv. Maradol, at commercial ripe stage (9 °Brix).	1. Control: distilled water. 2. Multilayered edible coating with antimicrobial: a) Chitosan, medium MW (2% w/v) in acetic acid (1% v/v), with glycerol (5% v/v), Tween® 20 (0.5% w/w) and microencapsulated beta-cyclodextrin and *trans*-cinnamaldehyde complex (2% w/w). b) Pectin (from citrus peel) ((2% (w/w, galacturonic acid 74.0%). c) Calcium chloride: 2 g/100 g.	1. Immersion for 2 min. 2. Layer-by-layer deposition process (chitosan, pectin, CaCl₂). Immersion in the first coating solution for 2 min and then the residual solution was dripped off for 2 min. This process was repeated with the second and third solution.	Fifteen cubes (240 ± 10 g) were packed in polypropylene trays (Ziploc® with Smart Snap™ Seal, 591 mL).	15 days at 4 °C.	Polysaccharide-based multilayered antimicrobial edible coating reduced the losses of vitamin C and total carotenoids content, and extended the shelf life of fresh-cut papaya up to 15 days at 4 °C compared to the uncoated product (<7 days).	Brasil et al. 2012
Wedges (5 mm thick) and cubes (20 mm thick) of	1. Sodium chlorite (SC): 0, 300, 600 or 1000 mg/L. 2. Chitosan (CH) (1% w/v) in acetic acid (1% w/v), pH 4.0.	1. Immersion for 2 min then drained for 1 min, and dried in the air for 30 min.	Packed into unsealed plastic bags.	10 days at 4 °C.	SC combined with CMCH coating prevented tissue softening, and reduced cut-surface discolouration.	Xiao et al. 2011

(*Contd.*)

pears (*Pyrus communis* L.) cv. d'Anjou.	3. N, O-carboxy methyl chitosan (CMCH) (1% w/v) in water, pH 6.4.	2) and 3) first immersion into the solution 1. Then immersion into CH or CMCH solutions for 2 min, drained for 1 min and air-dried for 30 min.		
Cubes (1 cm) of apples, cv. Fuji at commercial maturity.	1. Control: Deionized water. 2. Chitosan (97% deacetylation degree) (1% w/v) in citric acid (1% w/v). 3. Ascorbic acid (AA, 2%) + CaCl$_2$ (0.5%). 4. AA (2%) + CaCl$_2$ (0.5%) + Chitosan (1%). pH of all solutions was 3.5-4.0.	Immersion for 2 min, then dripped off for 1 min at 5 °C.	Placed over plastic-coated wire racks inside plastic containers.	Eight days at 5 °C and 24 h at 20 °C.	Chitosan effectively inhibited browning and retarded tissue softening. Ascorbic acid and citric acid increased chitosan's inhibitory efficiency of browning.	Qi et al. 2011
Florets of (*Brassica oleracea* L. var. *Italica*).	1. Control: Distilled water with or without inoculum. 2. Chitosan (CH) (98% deacetylation degree): 2% (w/v) in acetic acid (1% v/v) + Glycerol (Gly), (Gly/CH) weight	Immersion for 3 min at 20 °C. Then dried by flowing air at 30 °C and 60% RH for 60 min.	Three broccoli florets (60-90 g) were packed in polymeric film bags (PD960, Cryovac™) of 25 mm of thickness (with an O$_2$ permeability of 7000 cc/m^2/d, CO$_2$	20 days at 5-7 °C.	Chitosan reduced total psychrotrophic and mesophilic bacteria counts. It also had a bactericidal effect on endogenous *E. coli* and decreased	Moreira et al. 2011a

(Contd.)

Table 4. (*Contd.*)

Product	Treatment	Application of Edible Coating	Packaging	Storage Conditions	Benefits	Reference
	ratio of 0.28, with or without inoculum. Inoculum: A spray of *E. coli* O157:H7 (3–4 log CFU/g).		permeability of 20,000 cc/m²/d, and water vapour permeability of 1 g/m²/d).		total *E. coli* counts (endogenous and exogenous). Chitosan inhibited the yellowing and opening of broccoli florets.	
Florets of broccoli heads (*Brassica oleracea* L. var. *Italica*).	1. Controls: a) Uncoated. b) Heat treatment (HT): 50 °C for 1.5 min. c) Distilled water and dried. d) HT and dried. 2. Chitosan: 2% (w/v) in acetic acid (1%) and glycerol (1%). 3. Carboxymethyl cellulose (CMC) (0.75%, w/v) in a water–ethyl alcohol mixture (31/11). 4. Heat treatment (HT). 5. CH + HT. 6. CMC + HT.	1 c), 2) and 3): Immersion for 3 min at 20 °C then drained and air dried for 60 min by flowing air at 30 °C and 60% RH. 4. Immersion in water for 1.5 min at 50 °C. Then cooled for 1 min in ice-water.	Three broccoli florets (90 g) were packed in polymeric film bags (multilayered polyolefin PD960, Cryovac) of 25 µm of thickness (with an O₂ permeability of 7000 cc/m²/d, CO₂ permeability of 20000 cc/m²/d, and water vapour permeability of 1 g/m²/d).	20 days at 5 °C.	Chitosan inhibited florets' opening, reduced mesophilic, psychrotrophic, yeast and molluscs, and coliforms counts, and extended the shelf life. The combination of chitosan and mild heat treatment (50 °C for 1.5 min) was the most effective treatment to maintain the quality attributes and reduced the microbial populations.	Moreira et al. 2011b

(*Contd.*)

Sliced apples (*Malus domestica*) cv. Gala.	1. Control: Uncoated. 2. Chitosan: 0.2% (w/v) in acetic acid (0.5 M), pH 4.5.	Immersion and draining off, afterwards dried at room temperature.		10 days at 25 ± 0.5 °C. Petri dishes containing fungi (mainly *Penicillium* sp. and *Alternaria* sp.) were allocated among the samples to allow inoculation.	Chitosan protected sliced apples against fungal contamination.	Garrido Asis and de Britto 2011
Carrot slices (0.5-cm thickness).	1. Control: uncoated. 2. Sodium caseinate (SC): with a protein concentration of Pro 2.5, w/v) + glycerol at a glycerol/protein weight ratio of 0.28. 3. Chitosan (CH): 2% (w/v) in acetic acid (1%, v/v) + glycerol at glycerol/chitosan weight ratio of 0.28, pH 4.4. 4. SC + CH: weight ratio CH/Pro = 0.8/1.	a) Coating. Immersion for 3 min at 20 °C, drained and then dried by flowing air at 30 °C and 50% RH for 50 min. b) Wrapping each slice in the film.	Non-packaged.	Five days at 10 °C and 65% RH.	CH and SC+/CH as a coating or wrapping reduced mesophilic, psychrotrophic, and yeasts and moulds count. Chitosan's antimicrobial properties were enhanced by the ionic interaction with SC.	Moreira et al. 2011c

(Contd.)

Table 4. (*Contd.*)

Product	Treatment	Application of Edible Coating	Packaging	Storage Conditions	Benefits	Reference
Florets of broccoli (*Brassica oleracea* L. var. *Italica*).	1. Control: a) Uncoated. b) Heat treatment (HT): 50 °C for 1.5 min. c) Distilled water and dried. d) HT and dried. 2. Chitosan (CH): 2% (w/v) in acetic acid (1%) and glycerol (1%). 3. Carboxymethylcellulose (CMC): 0.75% in water–ethyl alcohol mixture (31/11) + glycerol (1.9%). 4. HT + CH. 5. HT + CMC.	Immersion for 3 min at 20 °C, then drained and air dried by flowing air for 60 min at 30 °C and 60% RH.	Three florets (90 g) were packed in sealed bags (multilayered polyolefin PD960, Cryovac, of 25 μm of thickness with an O_2 permeability of 0.08 cm^3/m^2 s, CO_2 permeability of 0.2 cm^3/m^2 s, and water vapour permeability of 12 μg/m^2 s).	18 days at 5 °C.	Chitosan was superior to CMC in reducing total microbial counts, and improving ascorbic acid retention during storage. The combination of chitosan with HT was the most effective in preserving the quality attributes (reduced weight loss, and enzymatic browning, and delayed the yellowing) of fresh cut broccoli mainly reducing the mesophilic microbial counts during storage.	Ansorena et al. 2011

(*Contd.*)

Product	Treatments	Application method	Packaging	Storage	Results	Reference
Sliced (4 mm thick) lotus root (*Nelumbo nucifera* Gaerth).	1. Control: Uncoated. 2. Uncoated + MAP. 3. Chitosan: [1.2% (w/v) in acetic acid (1%) + Ascorbic acid (AA) (2%) + Citric acid (AC) (1%)] + MAP.	1. and 2. Immersion into sterilized distilled water for 5 min. 3) Immersion for 5 min, drained in a plastic sieve for 50 min with low speed air, and then dried with tissue paper.	100 g of fruit slices were packaged in trays (150×150 mm) and sealed with a microperforated polyethylene film (30 μm thickness).	10 days at 4 °C.	Chitosan + MAP treatment delayed browning, reduced polyphenol oxidase (PPO) activity and maintained the appearance of fresh cut lotus root, extending its shelf life.	Xing et al. 2010
Wedges of Huangguan pears (*Pyrus bretschneideri* Rehd) cv. Huangguan.	1. Control: Air + distilled water. 2. Air + Chitosan (CH) + Rosemary extract (ACHR): CH [(2%, w/v) in acetic acid (1%, v/v) at 4 °C + glycerol (1.5%, w/v) + Tween-80 (0.2%, v/v) pH 5.6] + rosemary extract (0.03%). 3. Pure oxygen pretreatment (PO) + CH (POCH): PO (100 kPa O_2 for 10 days at 2 ± 1 °C and 95% RH) + chitosan as 2. 4. PO + chitosan+ rosemary (POCHR).	Immersion for 2 min, and then drained on stainless steel screens by natural convection at room temperature.	130 g of fruit wedges were packed over plastic trays and covered with unsealed polyethylene film bags.	Three days at 20 °C and 60-70% RH.	The pure oxygen + chitosan + rosemary (POCHR) treatment was the most effective in reducing browning, softening and sensory changes of freshcut pear after three days at 20 °C. Additionally, this treatment reduced membrane permeability, vitamin C loss and weight loss.	Xiao et al. 2010a

(Contd.)

Table 4. (*Contd.*)

Product	Treatment	Application of Edible Coating	Packaging	Storage Conditions	Benefits	Reference
Wedges of Huangguan pears (*Pyrus bretschneideri* Rehd) cv. Huangguan.	1. Control: Distilled water. 2. Ascorbic acid + chitosan (AA+CH): Ascorbic acid (0.5% w/v) + chitosan (2%, w/v) in acetic acid (1%, v/v) at 4 °C + glycerol (1.5%, w/v) and Tween 80 (0.2%, v/v), pH 5.6. 3. Chitosan + rosemary extract (CH+R): Chitosan + rosemary extract (0.03%).	1. and 3. Immersion for 2 min and then drained off. 2) Immersion in AA for 3 min, and dried for 3 min with tissue paper, then immersion into CH for 2 min.	130 g of wedges were packed on foam trays, covered with unsealed polyethylene film bags.	Three days at 20 °C and 60-70% RH.	Chitosan + rosemary was more effective than ascorbic acid + chitosan, in reducing PPO activity, browning and the loss of phenolic content, maintained higher colour, firmness and visual appearance.	Xiao et al. 2010b
Cubes of Mango (*Mangifera indica*) cv 'Tommy Atkins' ripen to firmness and colour parameters of about 0.6 N,	1. Control: Distilled water. 2. Hot water dipping (HWD): 30 min at 50 °C, then cooled for 15 min in water at 17 °C. 3. Chitosan (CH) (0.25% w/v) in citric acid (0.5%, w/v) under	Immersion for 2 min.	100 g was packed in 0.5 L polyethylene terephthalate plastic tray.	Nine days at 6 °C.	HWD 50 °C for 30 min was more effective than chitosan to maintain firmness and colour during 9 days at 6 °C. Chitosan inhibited the microbial counts. But, its	Djioua et al. 2010

(*Contd.*)

L* = 57 and b* = 37, respectively.	heating (70 °C for 15 min). 4. HWD + CH.			antimicrobial effect was improved when combined with HWD.
Sliced carrots (20-mm diameter, 10 mm thick) (*Daucus carota*) cv. Nantesa.	1. Control: Acetic acid buffer (175 mM), pH 5.2. 2. Chitosan (high molecular weight) (CH): 1% (w/v) in acetic acid (1%, w/v) + Tween 80 (0.1%), pH 5.2. 3. CH + MC: Chitosan + methylcellulose (substitution degree of 1:9, 1% w/v), pH 5.2. 4. CH + OA: Chitosan + oleic acid (2% w/v), pH 5.2.	a) Immersion for 6 min at atmospheric pressure, dripped off and dried at 20 °C. b) Immersion for 6 min by applying a vacuum pulse (5 kPa for 4 min), and immersion 2 more min to restore atmospheric pressure, dripped off and dried at 20 °C.	10 slices were packed in 490 mL perforated PET packets.	Nine days at 5 °C.	Coating application with a vacuum pulse improved the resistance of water vapour transmission, maintained colour, and diminished the white blush values during storage.	Vargas et al. 2009
Sticks (6 cm) of Carrot (*Daucus carota* L.) cv. Tino.	1. Control: Uncoated. 2. Yam starch (1%, v/v) + Chitosan (0.5%, v/v, 75–85% deacetylated) + glycerol (1% v/v) + acetic acid (0.5%, v/v).	Immersion for 1 min and dried under forced air for up to 2.5 h at 6 °C and 90% RH.	Pack A, (10 kPa O$_2$ +10 kPa CO$_2$): 140 g were packaged in round trays (117 mm ∅) sealed with a micro perforated oriented polypropylene (OPP) films with 35 μm of thickness, and O$_2$	12 days of storage at 4 °C.	The combined application of the edible coating containing chitosan and moderate O$_2$ and CO$_2$ levels slightly reduced vitamin C and carotenoids, but	Simoes et al. 2009

(Contd.)

Table 4. (*Contd.*)

Product	Treatment	Application of Edible Coating	Packaging	Storage Conditions	Benefits	Reference
			transmission rate (PO_2) of 8.2×10^{-16} mol/m² day atm. Pack B, (2 kPa O_2 + 15–25 kPa CO_2): 200 g were packed in rectangular trays (118 mm×80 mm×60 mm) sealed with microperforated OPP of 35μm of thickness with $PO_2 = 3.5 \times 10^{-16}$ mol/m² day atm.		increased phenolic content, preserved the overall visual quality, reduced surface whiteness, and maintained sensory quality in carrot sticks.	
Slices (1 cm thick) Butternut squash (*Cucurbita moschata* Duch).	1. Control: a) Fresh slices, b) Distilled water and dried. 2. Sodium caseinate (SC) (5% w/w) + glycerol (Gly) (Gly: SC ratio of 0.25). 3. Chitosan (2% w/w) in acetic acid (1%) and glycerol (1%). 4. Carboxy methyl cellulose (CMC) (0.75% w/v) in a water–ethyl	Immersion, draining, and drying by convection air for 100 min at 20 °C, 50 min at 30 °C and 30 min at 50 °C.	Non-packaged.	Non-stored.	Chitosan reduced aerobic mesophilic bacteria count and drying this coating for 30 min at 50 °C enhanced the reduction effect. This drying condition preserved the ascorbic acid content of squash slices.	Moreira et al. 2009

(*Contd.*)

	alcohol mixture, (3:1) + glycerol (1.9% w/w).				
Cubes (13 mm) of Sweet potato (*Ipomoea batatas*), cv. Beauregard.	1. Control: Uncoated. 2. Chitosan (470 and 1110 kDa) (1% v/v) in acetic acid (1% w/v), pH 5.6.	Immersion for 2 min, and drained for 5 min.	Packed in sealed plastic bags.	17 days at 4 °C.	470 kDa chitosan coating maintained the surface colour, reduced total aerobic, and yeast and mould counts during storage.	Waimaleongora-Ek et al. 2008
Mushrooms (*Agaricus bisporus*) slices with a thickness of 4 mm.	1. Control: Malic acid (2%, w/v). 2. Chitosan (37% deacetylation): 0.5, 1.0 or 2.0% (w/v) in malic acid (2%).	Immersion for 1 min, and then dried in air for 30 min at room temperature.	Packed into a polyethylene film bag (0.02-mm-thick).	15 days at 4 °C.	Chitosan (2%) obtained the best results in reducing PPO, peroxidase (POD), catalase, phenylalanine ammonia lyase (PAL), and laccase activity. It also reduced total phenol content and total bacteria, yeasts and moulds counts.	Eissa 2008
Slices of Butternut squash (*Cucurbita moschata* Duch).	1. Control: a) No treatment, and b) Distilled water. 2. Sodium caseinate: 5% in distilled water +	Immersion for 3 min at 20 °C, drained and then dried for 50 min with flowing		Five days at refrigeration.	Edible coatings enriched with oleoresins did not enhance the reduction of mesophilic	Ponce et al. 2008

(Contd.)

Table 4. (*Contd.*)

Product	Treatment	Application of Edible Coating	Packaging	Storage Conditions	Benefits	Reference
	glycerol at a glycerol/protein ratio of 0.25. 3. Chitosan: 2% (w/w) in acetic acid (1%) with or without oleoresin (rosemary, garlic, capsicum, onion and cranberry). 4. Carboxymethyl cellulose (CMC): 0.75% in a water–ethanol mixture (31/11) + glycerol (1.9%). All formulations were with and without oleoresins (rosemary and olive).	air at 30 °C and 40–50% RH.			bacterial counts compared to oleoresins alone. However, chitosan enriched with oleoresins reduced PPO activity and prevented browning, while it did not affect the sensorial acceptability of butternut squash.	
Slices of Strawberries (*Fragaria* x *ananassa* Duch.).	1. Control: Distilled water. 2. Chitosan (low molecular weight chitosan, 50–160 kDa with a deacetylation	Strawberry slices were immersed for 5 min, drained and then dried.	Slices were packaged in high (HP) and low permeability (LP), 170 × 250 mm long, under passive modified atmosphere (PMAP) or active modified	4, 8, 12 and 15 °C.	Chitosan combined with the lowest temperature of storage (4 and 8 °C) and AMAP, especially those high in oxygen	Campaniello et al. 2008

(Contd.)

	degree of 75–85%): 1% (w/v) in citric acid (1%).				
	atmosphere (AMAP), MA1: 65% N2, 30% CO_2, 5% O_2 and MA2: 80% O_2, 20% CO_2 in LP bags.			(MA2) maintained the colour, delayed browning and inhibited the growth of microorganisms. These combined treatments maintained the quality and prolonged the shelf life of fresh cut strawberries.	
Slices of Red pitayas (*Hylocereus undatus*).	1. Control: Acetic acid (0.5%). 2. Chitosan (MW = 12.36 ± 0.17 kDa, 95–98% deacetylated and viscosity ≤30 mPas): 0.2, 0.5 and 1% (w/v) in acetic acid (0.5%, v/v), pH 5.0.	Immersion for 1 min, then air-dried for 20 min at 24 °C.	Slices were packed in plastic trays wrapped with 30 × 20 cm PVDC film.	Seven days at 8 °C.	Chitosan reduced weight loss, browning and total bacterial counts. It also maintained sensory quality.	Chien et al. 2007a

(Contd.)

Table 4. (*Contd.*)

Product	Treatment	Application of Edible Coating	Packaging	Storage Conditions	Benefits	Reference
Slices (5 × 4 × 1 cm) of Mango (*Mangifera indica* L.) cv. Irwin.	1. Control: Acetic acid (0.5%). 2. Chitosan (MW = 12.36 ± 0.17 kDa, 95–98% deacetylated and viscosity ≤30 mPas): 0.5, 1 and 2% (w/v) in acetic acid (0.5%, v/v), pH 5.0.	Immersion for 1 min, then air-dried for 30 min at 25 °C.	Slices were packed in plastic trays wrapped with 30 × 20 cm PVDC film.	Seven days at 6 °C.	Chitosan reduced weight and ascorbic acid losses, delayed browning, decreased total bacterial count, and maintained good sensory quality.	Chien et al. 2007b
Slices (5 mm thick of Carrot (*Daucus carota* L.).	1. Control: Sterile distilled water. 2. Starch: Yam starch (4% w/w) + glycerol (2% w/w). 3. Chitosan (CH) (0.5) + Starch: CH (0.5% w/v) in acetic acid (0.4, w/w) + Starch as (2). 4. CH (1.5%) + Starch: CH (1.5% w/v) in acetic acid (0.4%, w/w).	Immersion for 3 min, then air dried at 20 °C for 3 h.	Slices were packed in expanded polystyrene trays, wrapped with polyvinylchloride (PVC) film.	15 days at 10 °C.	Chitosan (1.5%) + starch coating was the most efficient treatment to reduce the growth of filamentous fungi and yeast, to inhibit the growth of total coliforms and lactic acid bacteria, and to reduce the growth of mesophilic aerobic bacteria during storage of sliced carrots.	Durango et al. 2006

(*Contd.*)

Peeled litchi (*Litchi chinesis* Sonn.) cv. Huaizhi at a commercially mature stage.	1. Control: Acetic acid (0.5%), pH 6.0. 2. Chitosan (90–95% of deacetylated degree and ≤100 mPa s of viscosity) (1, 2 and 3% w/v) in acetic acid (0.5% v/v), pH 6.0.	Immersion for 1 min then air-dried for 30 min at 25 °C.	Peeled fruit was packed onto plastic trays, and overwrapped with polyethylene film (0.02 mm thick).	6 days at –1 °C.	Chitosan decreased ascorbic acid loss and maintained higher values of sensory quality. Chitosan at 2 and 3% was the most effective to reduce weight loss and PPO and POD activity.	Dong et al. 2004
Sliced (4 mm thick) Chinese water chestnut (*Eleocharis tuberosa*) cv. Guilin.	1. Control: Acetic acid (0.5%, v/v), pH 5.0. 2. Chitosan (0.5, 1 or 2% w/v) in acetic acid (0.5% v/v), pH 0.6.	Immersion for 1 min, then air-dried for another 30 min.	Packed into trays over-wrapped with polyethylene films (0.02 mm thick).	15 days at 4 °C.	Chitosan coating at increasing concentration proved the most effect in reducing activities of PAL, PPO and POD, and delaying discolouration. It also decreased total phenolic content, eating quality and ascorbic acid loss.	Pen y Jiang 2003

Conclusion and Perspectives

Edible films and coatings have increased their applications in the food industry due to the wide range of mechanical (tensile strength, elongation at break, elastic modulus, etc.) and barrier properties (water vapor transmission rate and gas permeability) that they show and their use for solving several problems in the storage and transport of fresh produce. Several reviews have been written describing the properties and potential uses of traditional edible films and coatings. In this chapter, some of the most important aspects of edible films and coatings from marine sources (agar and chitosan) were described in detail. An extensive review of the application of chitosan as edible film and coating in fresh produce, clearly states that this biopolymer is widely used in fruits and vegetables due to antimicrobial characteristics. In contrast, agar is a biopolymer that has hardly been used as edible film or coating for fresh produce. However, this biopolymer has properties that could be very useful for these applications.

On the other hand, to have a rough idea about the perspective of these biopolymers as edible films and coatings, it is very important to understand the properties that these biopolymers exhibit individually. Almost all the drawbacks can be minimized or disappear when other materials or additives are added to the mixture or blend. The films or coatings ought to be characterized by the techniques described and evaluated in terms of performance using fresh produce. Moreover, some factors to be studied in order to understand their effects on coating performance for a particular produce are concentration, viscosity and composition of the coating solution; thickness, temperature and atmosphere of coating application; and the respiration rate, water activity and variety of the produce. The study of these materials is an interdisciplinary work because the information provided by analyses are related to particular issues for a specific coating and fresh produce. Another characteristic that should be studied is the interface adhesion between the surface of fresh produce and the coating, where the gas exchange process takes place, which has been discussed by few authors thus far.

Finally, we can expect that in the near future, a widely diverse range of fresh produce coated with an edible film of chitosan or agar will be marketed or consumed. Some research groups are performing interesting experiments with these polysaccharides and their blends.

References

Aday, M.S. and C. Caner. 2010. Understanding the effects of various edible coatings on the storability of fresh cherry. Packag. Technol. Sci. 23: 441–456.

Ali, A., M. T.M. Muhammad, K. Sijam and Y. Siddiqui. 2011. Effect of chitosan coatings on the physicochemical characteristics of Eksotika II papaya (*Carica papaya* L.) fruit during cold storage. Food Chem. 124: 620–626.

Ali, A., M. Maqbool, P.G. Alderson and N. Zahid. 2012. Efficacy of biodegradable novel edible coatings to control postharvest anthracnose and maintain quality of fresh horticultural produce. Acta Hort. 945: 39–44.

Ali, A., N. Zahid, S. Manickam, Y. Siddiqui and P.G. Alderson. 2014. Double layer coatings: A new technique for maintaining physico-chemical characteristics and antioxidant properties of dragon fruit during storage. Food Bioprocess Technol. 7: 2366–2374.

Alvarez, M.V., A.G. Ponce and M.R. Moreira. 2013. Antimicrobial efficiency of chitosan coating enriched with bioactive compounds to improve the safety of fresh cut broccoli. LWT – Food Sci Technol. 50: 78–87.

Ansorena, M.R., N.E. Marcovich and S.I. Roura. 2011. Impact of edible coatings and mild heat shocks on quality of minimally processed broccoli (*Brassica oleracea* L.) during refrigerated storage. Postharvest Biol. Tec. 59: 53–63.

Armisén, R. and F. Galatas. 2000. Agar. pp. 21–40. *In*: Phillips, G. and P. Williams (eds.). Handbook of Hydrocolloids. Woodhead Publishing, Grat Abington, Cambridge, UK.

Arnon, H., Y. Zaitseva, R. Porat and E. Poverenova. 2014. Effects of carboxymethyl cellulose and chitosan bilayer edible coating on postharvest quality of citrus fruit. Postharvest Biol. Tec. 87: 21–26.

Arnon, H., R. Granit, R. Porat and E. Poverenov. 2015. Development of polysaccharides-based edible coatings for citrus fruits: A layer-by-layer approach. Food Chem. 166: 465–472.

ASTM D882. 2012. Standard Test Method for Tensile Properties of Thin Plastic Sheeting, ASTM Book of Standards, D882–12.

ASTM E96. 2015. Standard Test Methods for Water Vapor Transmission of Materials, ASTM Book of Standards, E96–15.

ASTM D1434. 2015. Standard Test Method for Determining Gas Permeability Characteristics of Plastic Film and Sheeting. ASTM Book of Standards, D1434–15.

Baez-Sañudo, M., J. Siller-Cepeda, D. Muy-Rangel and J.B. Heredia. 2009. Extending the shelf-life of bananas with 1-Methylcyclopropene and a chitosan-based edible coating. J. Sci. Food Agr. 89: 2343–2349.

Bal, E. 2013. Postharvest application of chitosan and low temperature storage affect respiration rate and quality of plum fruits. J. Agr. Sci. Tech. 15: 1219–1230.

Baldwin E.A. 1994. Edible coatings for fresh fruits and vegetables: past, present and future. pp. 25–64. *In*: Krochta, J.M., E.A. Baldwin and M.O. Nisperos (eds.). Edible Coatings and Films to Improve Food Quality. CRC Press LLC, Boca Raton, Florida, USA.

Ban, Z., L. Li, J. Guan, J. Feng, M. Wu, X. Xu, et al. 2014. Modified atmosphere packaging (MAP) and coating for improving preservation of whole and sliced *Agaricus bisporus*. J. Food Sci. Tech. Mys. 51: 3894–3901.

Ban, Z., J. Feng, W. Wei, X. Yang, J. Li, J. Guan, et al. 2015. Synergistic effect of sodium chlorite and edible coating on quality maintenance of minimally processed *Citrus grandis* under passive and active MAP. J. Food Sci. 80: C1705–C1712.

Bangyekan, C., D. Aht-Ong and K. Srikulkit. 2006. Preparation and properties evaluation of chitosan-coated cassava starch films Carboh. Polym. 63: 61–71.

Benítez, S., L. Soro, I. Achaerandio, F. Sepulcre and M. Pujola. 2014. Combined effect of a low permeable film and edible coatings or calcium dips on the quality of fresh-cut pineapple. J. Food Process Eng. 37: 91–99.

Benítez, S., I. Achaerandio, M. Pujola and F. Sepulcre. 2015. *Aloe vera* as an alternative to traditional edible coatings used in fresh-cut fruits: A case of study with kiwifruit slices. LWT – Food Sci Technol. 61: 184–193.

Bill, M., D. Sivakumar, L. Korsten and A.K. Thompson. 2014. The efficacy of combined application of edible coatings and thyme oil in inducing resistance components in avocado (*Persea americana* Mill.) against anthracnose during post-harvest storage. Crop Prot. 64: 159–167.

Bixler, H.J. and H. Porse. 2011. A decade of change in the seaweed hydrocolloids industry. J. Appl. Phycol. 23: 321–335.

Bourtoom, T. 2008. Edible films and coatings: Characteristics and properties. Int. Food Res. J. 15: 237–248.

Brasil, I.M., C. Gomes, A. Puerta-Gomez, M.E. Castell-Perez and R.G. Moreira. 2012. Polysaccharide-based multilayered antimicrobial edible coating enhances quality of fresh-cut papaya. Lebensm. Wiss. Technol. 47: 39–45.

Campaniello, D., A. Bevilacqua, M. Sinigaglia and M.R. Corbo. 2008. Chitosan: Antimicrobial activity and potential applications for preserving minimally processed strawberries. Food Microbiol. 25: 992–1000.

Caro, Y. and J. Joas. 2005. Postharvest control of litchi pericarp browning (cv. Kwai Mi) by combined treatments of chitosan and organic acids II. Effect of the initial water content of pericarp. Postharvest Biol. Tec. 38: 137–144.

Chiabrando, V. and G. Giacalone. 2013. Effect of different coatings in preventing deterioration and preserving the quality of fresh-cut nectarines (cv. Big Top). CyTA-J. Food 11: 285–292.

Chien, P.J., F. Sheu and H.R. Lin. 2007a. Quality assessment of low molecular weight chitosan coating on sliced red pitayas. J. Food Eng. 79: 736–740.

Chien, P.J., F. Sheu and F.H. Yang. 2007b. Effects of edible chitosan coating on quality and shelf life of sliced mango fruit. J. Food Eng. 78: 225–229.

Chillo, S., S. Flores, M. Mastromatteo, A. Conte, L. Gerschenson and M.A. Del Nobile. 2008. Influence of glycerol and chitosan on tapioca starch-based edible film properties. J. Food Eng. 88: 159–168.

Chong, J.X., S. Lai and H. Yang. 2015. Chitosan combined with calcium chloride impacts fresh-cut honeydew melon by stabilizing nanostructures of sodium-carbonate-soluble pectin. Food Control 53: 195–205.

Cissé, M., J. Polidori, D. Montet, G. Loiseau and M.N. Ducamp-Collin. 2015. Preservation of mango quality by using functional chitosan-lactoperoxidase systems coatings. Postharvest Biol. Tec. 101: 10–14.

Contreras-Oliva, A., M.B. Pérez-Gago and C. Rojas-Argudo. 2012. Effects of chitosan coatings on physicochemical and nutritional quality of clementine mandarins cv. 'Oronules'. Food Sci. Technol. Int. 18: 303–315.

Craigie, J. 1990. Cell wall. pp. 221–258. *In*: Cole, K.M. and R.G. Sheath (eds.). Biology of the Red Algae. Cambridge University Press, Cambridge, UK.

Dang, Q.F., J.Q. Yan, Y. Li, X.J. Cheng, C.S. Liu and X.G. Chen. 2010. Chitosan acetate as an active coating material and its effects on the storing of *Prunus avium* L. J. Food Sci. 75: S125–S131.

de Aquino, A.B., A.F. Blank and L.C. Lins de Aquino Santana. 2015. Impact of edible chitosan-cassava starch coatings enriched with *Lippia gracilis* Schauer genotype mixtures on the shelf life of guavas (*Psidium guajava* L.) during storage at room temperature. Food Chem. 171: 108–116.

de Britto, D. and O.B. Garrido Assis. 2012. Chemical, biochemical, and microbiological aspects of chitosan quaternary salt as active coating on sliced apples. Cienc. Tecnol. Aliment. Campinas. 32: 599–605.

de Souza, E.L., C.V. Sales, C.E.V. de Oliveira, L.A.A. Lopes, M.L. da Conceicao, L.R.R. Berger, et al. 2015. Efficacy of a coating composed of chitosan from *Mucor circinelloides* and carvacrol to control *Aspergillus flavus* and the quality of cherry tomato fruits. Front. Microbiol. 6: 1–9.

Debeaufort, F. and A. Voilley. 1995. Effect of surfactants and drying rate on barrier properties of emulsified edible films. Int. J. Food Sci. Technol. 30: 183–189.

Debeaufort, F., N. Tesson and A. Voilley. 1995. Aroma compounds and water vapor permeability of edible and polymeric packagings. pp. 169–175. *In*: Ackermann, P., M. Jägerstad and T. Ohlsson (eds.). Food and Packaging Materials—Chemical Interactions. The Royal Society of Chemistry, Cambridge, UK.

Debeaufort, F. and A. Voilley. 1997. Methylcellulose-based edible films and coatings. II. Mechanical and thermal properties as a function of plasticizer content. J. Agric. Food Chem. 45: 685–694.

Debeaufort, F., J.A. Quezada-Gallo and A. Voilley. 1998. Edible films and coatings: Tomorrow's packagings: A review. Crit. Rev. Food Sci. 38: 299–313.

Di Pierro, P., B. Chico, R. Villalonga, L. Mariniello, A.E. Damiao and P. Masi. 2006. Chitosan–Whey protein edible films produced in the absence or presence of transglutaminase: Analysis of their mechanical and barrier properties. Biomacromol. 7: 744–749.

Djioua, T., F. Charles, M. Freire, Jr., H. Filgueiras, M.N. Ducamp-Collin and H. Sallanon. 2010. Combined effects of postharvest heat treatment and chitosan coating on quality of fresh-cut mangoes (*Mangifera indica* L.). Int. J. Food. Sci. Tech. 45: 849–855.

Dong, H.Q., L.Y. Cheng, J.H. Tan, K.W. Zheng and Y.M. Jiang. 2004. Effects of chitosan coating on quality and shelf life of peeled litchi fruit. J. Food Eng. 64: 355–358.

Duan, J., R. Wu, B.C. Strik and Y. Zhao. 2011. Effect of edible coatings on the quality of fresh blueberries (Duke and Elliott) under commercial storage conditions. Postharvest Biol. Tec. 59: 71–79.

Durango, A.M., N.F.F. Soares and N.J. Andrade. 2006. Microbiological evaluation of an edible antimicrobial coating on minimally processed carrots. Food Control 17: 336–341.

Eissa, H.A.A. 2007. Effect of chitosan coating on shelf life and quality of fresh-cut mushroom. J. Food Quality 30: 623–645.

Elsabee, M.Z. and E.S. Abdou. 2013. Chitosan based edible films and coatings: A review. Mat. Sci. Eng. C. 33: 1819–1841.

Eshghi, S., M. Hashemi, A. Mohammadi, F. Badii, Z. Mohammadhoseini and K. Ahmadi. 2014. Effect of nanochitosan-based coating with and without copper loaded on physicochemical and bioactive components of fresh strawberry fruit (*Fragaria x ananassa* Duchesne) during storage. Food Bioprocess Technol. 7: 2397–2409.

Falguera, V., J.P. Quintero, A. Jiménez, J.A. Muñoz and A. Ibarz. 2011. Edible films and coatings: Structures, active functions and trends in their use. Trends Food Sci. Technol. 22: 292–303.

Fernández, C.M., K. Milja, A. Sari, R. Jukka, K. Karin, H. Jyrki, et al. 2004. Physical stability and moisture sorption of aqueous chitosan–amylose starch films plasticized with polyols. Eur. J. Pharm. Biopharm. 58: 69–76.

Fishman, S., V. Rodov and S. Ben-Yehoshua. 1996. Mathematical model for perforation effect on oxygen and water vapor dynamics in modified-atmosphere packages. J. Food Sci. 61: 956–961.

Fisk, C.L., A.A. Silver, B.C. Strik and Y. Zhao. 2008. Postharvest quality of hardy kiwifruit (*Actinidia arguta* 'Ananasnaya') associated with packaging and storage conditions. Postharvest Biol. Tec. 47: 338–345.

Fonseca, S.C., F.A.R. Oliveira and J.K. Brecht. 2002. Modelling respiration rate of fresh fruits and vegetables for modified atmosphere packages: A review. J. Food Eng. 52: 99–119.

Friedman, M. and V.K. Juneja. 2010. Review of antimicrobial and antioxidative activities of chitosans in food. J. Food Protect. 73: 1737–1761.

Freile-Pelegrín, Y., T.J. Madera-Santana, D. Robledo, L. Veleva, P. Quintana and J.A. Azamar. 2007. Degradation of agar films in a humid tropical climate: Thermal, mechanical, morphological and structural changes. Polym. Degrad. Stabil. 92: 244–252.

Gao, P., Z. Zhu and P. Zhang. 2013. Effects of chitosan-glucose complex coating on postharvest quality and shelf life of table grapes. Carbohyd. Polym. 95: 371–378.

García, M., A. Casariego, R. Díaz and L. Roblejo. 2014. Effect of edible chitosan/zeolite coating on tomatoes quality during refrigerated storage. Emir. J. Food Agric. 26: 238–246.

García-Baldenegro, C.V., I. Vargas-Arispuro, J.A. Azamar-Barrios and T.J. Madera-Santana. 2014. Biodegradable agar films with natural semiochemical extract: Effect of extract concentration on physicochemical and release properties. J. Biob. Mat. Bioener. 8: 363–369.

Garrido Assis, O.B. and D. de Britto. 2011. Evaluation of the antifungal properties of chitosan coating on cut apples using a non-invasive image analysis technique. Polym. Int. 60: 932–936.

Gennadios, A., C.L. Weller and R.F. Testin. 1993. Property modification of edible wheat, gluten-based films. Trans. Am. Soc. Agric. Eng. 36: 465–475.

Geraldine R.M., N.F. Fereira Soares, D. Alvarenga Botrel and L.A. Goncalves. 2008. Characterization and effect of edible coatings on minimally processed garlic quality. Carboh. Polym. 72: 403–409.

Ghasemnezhad, M., S. Zareh, M. Rassa and R.H. Sajedi. 2013. Effect of chitosan coating on maintenance of aril quality, microbial population and PPO activity of pomegranate (*Punica granatum* L. cv. Tarom) at cold storage temperature. J. Sci. Food Agr. 93: 368–374.

Gimenez, B., A. Lopez de Lacey, E. Perez-Santín, M.E. Lopez-Caballero and P. Montero. 2013. Release of active compounds from agar and agar-gelatin films with green tea extract. Food Hydrocol. 30: 264–271.

Gol, N.B., P.R. Patel and T.V.R. Rao. 2013. Improvement of quality and shelf-life of strawberries with edible coatings enriched with chitosan. Postharvest Biol. Tec. 85: 185–195.

Gol, N.B., M.L. Chaudhari and T.V.R. Rao. 2015. Effect of edible coatings on quality and shelf life of carambola (*Averrhoa carambola* L.) fruit during storage. Food Sci. Tech. Mys. 52: 78–91.

Greener-Donhowe, I. and O. Fennema. 1994. Edible films and coatings: Characteristics, formation, definitions and testing methods. pp. 1–24. *In*: Krochta, J.M., E.A. Baldwin and M.O. Nisperos (eds.). Edible Coatings and Films to Improve Food Quality. CRC Press LLC, Boca Raton, Florida, USA.

Guilbert, S., N. Gontard and L.G.M. Gorris. 1996. Prolongation of the shelf life of perishable food products using biodegradable films and coatings. Lebensm. Wiss. Technol. 29: 1017–1023.

Guiseley, K.B. 1970. The relationship between methoxyl content and gelling temperature of agarose. Carbohyd. Res. 13: 247–256.

Hagenmaier, R.D. and P.E. Shaw. 1992. Gas permeability of fruit coating waxes. J. Amer. Soc. Hort. Sci. 117: 105–114.

Hamzah, H.M., A. Osman, C.P. Tan and F. Mohamad Ghazali. 2013. Carrageenan as an alternative coating for papaya (*Carica papaya* L. cv. Eksotika). Posth. Biol. Technol. 75: 142–146.

Han, C., Y. Zhao, S.W. Leonard and M.G. Traber. 2004. Edible coatings to improve storability and enhance nutritional value of fresh and frozen strawberries (*Fragaria x ananassa*) and raspberries (*Rubus ideaus*). Postharvest Biol. Tec. 33: 67–78.

Han, C., J. Zuo, Q. Wang, L. Xu, B. Zhai, Z. Wang, et al. 2014. Effects of chitosan coating on postharvest quality and shelf life of sponge gourd (*Luffa cylindrica*) during storage. Sci. Hortic-Amsterdam 166: 1–8.

Han, C.R., C. Lederer, M. McDaniel and Y.Y. Zhao. 2005. Sensory evaluation of fresh strawberries (*Fragaria ananassa*) coated with chitosan-based edible coatings. J. Food Sci. 70: S172–S178.

Han, J.H. 2000. Antimicrobial food packaging. Food Technol. 54: 56–65.

Han, J.H. 2003. Antimicrobial food packaging. pp. 50–70. *In*: Ahvenainen, R. (ed.). Novel Food Packaging Techniques. Woodhead Publishing, Cambridge, U.K.

Han, J.H. and A. Gennadios. 2005. Edible films and coatings: A review. pp. 239–262. *In*: Han, J.H. (ed.). Innovations in Food Packaging. San Diego: Elsevier Academic Press. New York, NY, USA.

Han, J.H. 2014. Edible films and coatings: A review. pp. 213–255. *In*: Han, J.H. (ed.). Innovations in Food Packaging. San Diego: Elsevier Academic Press. New York, NY. USA.

Hardenburg, R.E., A.E. Watada and C.Y. Wang. 1986. The Commercial Storage of Fruits, Vegetables, and Florist and Nursery Stocks. Agriculture Handbook 66. Agricultural Research Service. United States Department of Agriculture. Washington, D.C. USA.

Heredia-Guerrero, J.A., J.J. Benítez, E. Domínguez, I.S. Bayer, R. Cingolani, A. Athanassio, et al. 2014. Infrared and Raman spectroscopic features of plant cuticles: A review. Front. Plant Sci. 5: 1–14.

Hernández-Carmona, G., Y. Freile-Pelegrín and E. Hernández-Garibay. 2013. Conventional and alternative technologies for the extraction of algal polysaccharides. pp. 475–516. *In*: Domínguez, H. (ed.). Functional Ingredients from Algae for Foods and Nutraceuticals. Woodhead Publishing Series in Food Science, Technology and Nutrition, Cambridge, UK.

Hernández-Muñoz, P., E. Almenar, M.J. Ocio and R. Gavara. 2006. Effect of calcium dips and chitosan coatings on postharvest life of strawberries (*Fragaria x ananassa*). Postharvest Biol. Tec. 39: 247–253.

Holappa, J., M. Hjalmarsdottir, M. Masson, O. Runarsson, T. Asplund, P. Soininen, et al. 2006. Antimicrobial activity of chitosan N-bentainates. Carboh. Polym. 65: 114–118.

Holcroft, D. 2015. Water Relations in Harvested Fresh Produce. PEF White Paper No. 15-01. The Postharvest Education Foundation. P. 16.

Hong, K., J. Xie, L. Zhang, D. Sun and D. Gong. 2012. Effects of chitosan coating on postharvest life and quality of guava (*Psidium guajava* L.) fruit during cold storage. Sci. Hortic-Amsterdam 144: 172–178.

Howard, L.R. and A.R. Gonzalez. 2001. Food safety and produce operations: What is the future? Hort. Sci. 36: 3339–3351.

Imeson, A. 2009. Agar. pp. 31–49. *In*: Imeson, A. (ed.). Food Stabilisers, Thickeners and Gelling Agents. Wiley-Blackwell Publishing, West Sussex, UK.

ISO 2556:1974. Confirmation 2016. "Determination of the gas transmission rate of films and thin sheets under atmospheric pressure -- Manometric method". ISO. International Organization for Standardization.

Jacxsens, L., F. Devlieghhere and J. Debevere. 1999. Validation of a systematic approach to design equilibrium modified atmosphere packages for fresh-cut produce. LWT – Food Sci Technol. 32: 425–432.

Jiang, T., L. Feng and Y. Wang. 2013. Effect of alginate/nano-Ag coating on microbial and physicochemical characteristics of shiitake mushroom (*Lentinus edodes*) during cold storage. Food Chem. 141: 954–960.

Jin, T. and J.B. Gurtler. 2012. Inactivation of Salmonella on tomato stem scars by edible chitosan and organic acid coatings. J. Food Protect. 75: 1368–1372.

Joas, J., J. Caro, M.N. Ducamp and M. Reynes. 2005. Postharvest control of pericarp browning of litchi fruit (*Litchi chinensis* Sonn cv Kwai Mi) by treatment with chitosan and organic acids I. Effect of pH and pericarp dehydration. Postharvest Biol. Tec. 38: 128–136.

Kader, A. 2002. Postharvest biology and technology: An overview. pp. 39–47. *In*: Kader, A.A. (ed.). Postharvest Technology of Horticultural Crops. Third Edition. Publication 3311. University of California, Agriculture and Natural Resources. Oakland, CA, USA.

Kang, H., S. Kim, Y. You, M. Lacroix and J. Han. 2013. Inhibitory effect of soy protein coating formulations on walnut (*Juglans regia* L.) kernels against lipid oxidation. LWT – Food Sci. Tech. 51: 393–396.

Kanmani, P. and J.W. Rhim. 2014a. Antimicrobial and physical-mechanical properties of agar-based films incorporated with grapefruit seed extract. Carboh. Polym. 102: 708–716.

Kanmani, P. and J.W. Rhim. 2014b. Properties and characterization of bionanocomposite films prepared with various biopolymers and ZnO Nanoparticles. Carboh. Polym. 106: 190–199.

Karel, M., P. Issenberg, L. Ronsivalli and V. Jurin. 1963. Application of gas chromatography to the measurement of gas permeability of packaging materials. Food Technol. 3: 91–101.

Kays, S.J. 1997. Postharvest Physiology of Perishable Plant Products. Exon Press, Athens, GA. EUA.

Kester, J.J. and O.R. Fennema. 1989. An edible film of lipids and cellulose ethers barrier properties to moisture vapor transmission and structural evaluation. J. Food Sci. 54: 1383–1341.

Kou, X.H., W.I. Guo, R.Z. Guo, X.Y. Li and Z.-H. Xue. 2014. Effects of chitosan, calcium chloride, and pullulan coating treatments on antioxidant activity in pear cv. "Huang guan" during storage. Food Bioprocess. Tec. 7: 671–681.

Krochta, J.M. 2002. Proteins as raw materials for films and coatings: definitions, current status, and opportunities. pp. 1–41. *In*: Gennadios, A. (ed.). Protein-based Films and Coatings. CRC Press, New York, NY, USA.

Kumari, P., K. Barman, V.B. Patel, M.W. Siddiqui and B. Kole. 2015. Reducing postharvest pericarp browning and preserving health promoting compounds of litchi fruit by combination treatment of salicylic acid and chitosan. Sci. Hortic.-Amsterdam 197: 555–563.

Lahaye, M. and C. Rochas. 1991. Chemical structure and physico-chemical properties of agar. Hydrobiologia 221: 137–148.

Lara, I., B. Belge and L.F. Goulao. 2014. The fruit cuticle as a modulator of postharvest quality. Postharvest Biol. Tec. 87: 103–112.

Lara, I., B. Belge and L.F. Goulao. 2015. A focus on the biosynthesis and composition of cuticle in fruits. J. Agric. Food Chem. 63: 4005–4019.

Leceta, I., S. Molinaro, P. Guerrero, J.P. Kerry and K. de la Caba. 2015. Quality attributes of MAP packaged ready-to-eat baby carrots by using chitosan-based coatings. Postharvest Biol. Tec. 100: 142–150.

Lee, H.-B., B.S. Noh and S.C. Min. 2012. *Listeria monocytogenes* inhibition by defatted mustard meal-based edible films. Int. J. Food. Microbiol. 153: 99–105.

Letendre, M., G. D'Aprano, M. Lacroix, S. Salmieri and D. Sr-Gelais. 2002. Physicochemical properties and bacterial resistance of biodegradable milk protein films containing agar and pectin. J. Agric. Food Chem. 50: 6017–6022.

Lin, B., Y. Du, X. Liang, X. Wang, X. Wang and J. Yang. 2011. Effect of chitosan coating on respiratory behavior and quality of stored litchi under ambient temperature. J. Food Eng. 102: 94–99.

Liu, F., B. Qin and R. Son. 2009. Novel starch/chitosan blending membrane: Antibacterial, permeable and mechanical properties. Carboh. Polym. 78: 146–150.

López de Lacey, A.M., M.E. López-Caballero and P. Montero. 2014. Agar films containing green tea extract and probiotic bacteria for extending fish shelf-life. LWT - Food Sci. Techn. 55: 559–564.

López-Gálvez, G., R. El-Bassuoni, X. Nie and M. Cantwell. 1997. Quality of red and green fresh-cut peppers stored in controlled atmospheres. pp. 152–157. *In*: Gorny, J.R. (ed.). Proc. 7th Int. Contr. Atmosp. Res. Conf. Vol. 5. Davis, CA, USA.

Madera-Santana, T.J., M. Misra, L.T. Drzal, D. Robledo and Y. Freile Pelegrín. 2009. Preparation and characterization of biodegradable agar/poly(butylene adipate–co-terephatalate) composites. Polym. Eng. Sci. 49: 1117–1126.

Madera-Santana, T.J., D. Robledo, J.A. Azamar, C.R. Ríos-Soberanis and Y. Freile-Pelegrín. 2010. Preparation and characterization of low density polyethylene agar biocomposites: Torque-rheological, mechanical, thermal and morphological properties. Polym. Eng. Sci. 50: 585–591.

Madera-Santana, T.J., D. Robledo and Y. Freile-Pelegrín. 2011. Physicochemical properties of biodegradable polyvinyl alcohol-agar films from red algae *Hydropuntia cornea*. Marine Biotech. 13: 793–800.

Madera-Santana, T.J., Y. Freile-Pelegrín and J.A. Azamar-Barrios. 2014. Physicochemical and morphological properties of plasticized poly(vinyl alcohol)-agar biodegradable films. Int. J. Biol. Macromol. 69: 176–184.

Maneerat, C., A. Tongta, S. Kanlayanarat and C. Wongs-Aree. 1997. A transient model to predict O_2 and CO_2 concentrations in modified atmosphere packaging of bananas at various temperatures. pp. 191–197. *In*: Gorny, J.R. (ed.). Proc. 7th Int. Contr. Atmosp. Res. Conf. Vol. 5. Davis, CA, USA.

Maqbool, M., A. Ali, S. Ramachandran, D.R. Smith and P.G. Alderson. 2010. Control of postharvest anthracnose of banana using a new edible composite coating. Crop Prot. 29: 1136–1141.

Maqbool, M., A. Ali, P.G. Alderson, N. Zahid and Y. Siddiqui. 2011. Effect of a novel edible composite coating based on gum arabic and chitosan on biochemical and physiological responses of banana fruits during cold storage. J. Agric. Food Chem. 59: 5474–5482.

Marpudi, S. L., L.S.S. Abirami, R. Pushkala and N. Srividya. 2011. Enhancement of storage life and quality maintenance of papaya fruits using Aloe vera based antimicrobial coating. Indian J. Biotechnol. 10: 83–89.

Márquez, C.J., J.R.V. Cartagena and M.B. Pérez-Gago. 2009. Effect of edible coatings on Japanese loquat (*Eriobotrya japonica* T.). Postharvest Qual. Vitae 16: 304–310.

Martin, L.B.B. and J.K.C. Rose. 2014. There's more than one way to skin a fruit: formation and functions of fruit cuticles. J. Exp. Bot. doi:10.1093/jxb/eru301.

Martin-Polo, M.O. and A. Voilley. 1990. Comparative study of water vapor permeability of edible films composed of arabic gum and glycerol monostearate. Sci. Alim. 10: 473–482.

Martiñon, M.E., R.G. Moreira, M.E. Castell-Perez and C. Gomes. 2014. Development of a multilayered antimicrobial edible coating for shelf-life extension of fresh-cut cantaloupe (*Cucumis melo* L.) stored at 4 °C. Lebensm. Wiss. Technol. 56: 341–350.

Mayachiew, P. and S. Devahastin. 2010. Effects of drying methods and conditions on release characteristics of edible chitosan films enriched with Indian gooseberry extract. Food Chem. 118: 594–604.

McHugh, D.J. 2003. A Guide to the seaweed industry. FAO Fisheries Technical Paper, no. 441. Rome.

Medeiros, B.G.D.S., A.C. Pinheiro, M.G. Carneiro-da-Cunha and A.A. Vicente. 2012. Development and characterization of a nanomultilayer coating of pectin and chitosan - Evaluation of its gas barrier properties and application on 'Tommy Atkins' mangoes. J. Food Eng. 110: 457–464.

Meighani, H., M. Ghasemnezhad and D. Bakhshi. 2015. Effect of different coatings on post-harvest quality and bioactive compounds of pomegranate (*Punica granatum* L.) fruits. Food Sci. Tech. Mys. 52: 4507–4514.

Mitchell, F.G., R. Guillou and R.A. Parsons. 1972. Commercial Cooling of Fruits and Vegetables. Manual 43. Division of Agricultural Sciences. University of California, Berkeley, California, USA.

Mohammadi, A., M. Hashemi and S.M. Hosseini. 2015. Chitosan nanoparticles loaded with *Cinnamomum zeylanicum* essential oil enhance the shelf life of cucumber during cold storage. Postharvest Biol. Tec. 110: 203–213.

Moreira, M.D.R., A. Ponce, C.E. Del Valle and S.I. Roura. 2009. Edible coatings on fresh squash slices: Effect of film drying temperature on the nutritional and microbiological quality. J. Food Process. Pres. 33: 226–236.

Moreira, M.D.R., S.I. Roura and A. Ponce. 2011a. Effectiveness of chitosan edible coatings to improve microbiological and sensory quality of fresh cut broccoli. LWT - Food Science and Technology 44: 2335–2341.

Moreira, M.D.R., A. Ponce, R. Ansorena and S.I. Roura. 2011b. Effectiveness of edible coatings combined with mild heat shocks on microbial spoilage and sensory quality of fresh cut broccoli (*Brassica oleracea* L.). J. Food Sci. 76: M367–M374.

Moreira, M.D.R., M. Pereda, N.E. Marcovich and S.I. Roura. 2011c. Antimicrobial effectiveness of bioactive packaging materials from edible chitosan and casein polymers: Assessment on carrot, cheese, and salami. J. Food Sci. 76: M54–M63.

Moreira, S.P., W.M. de Carvalho, A.C. Alexandrino, H.C. Bezerra de Paula, M.D.C. Passos Rodrigues, R.W. de Figueiredo, et al. 2014. Freshness retention of minimally processed melon using different packages and multilayered edible coating containing microencapsulated essential oil. Int. J. Food. Sci. Tech. 49: 2192–2203.

Murano, E. 1995. Chemical structure and quality of agars from *Gracilaria*. J. Appl. Phycol. 7: 245–254.

Mustafa, M.A., A. Ali, S. Manickam and Y. Siddiqui. 2014. Ultrasound-assisted chitosan-surfactant nanostructure assemblies: Towards maintaining postharvest quality of tomatoes. Food Bioprocess Tec. 7: 2102–2111.

Muy Rangel, D., B. Espinoza Valenzuela, J. Siller Cepeda, J.A. Sañudo Barajas, B. Valdez Torres and T. Osuna Enciso. 2009. Effect of 1-methylcyclopropene (1-MCP) and an edible coating on enzyme activity and postharvest quality of mango 'Ataulfo'. Rev. Fitotec. Mex. 32: 53–60.

Noshad, M., M. Mohebbi, E. Ansarifar and B.A. Behbahani. 2015. Quantification of enzymatic browning kinetics of quince preserved by edible coating using the fractal texture Fourier image. Food Measure 9: 375–381.

Ochoa-Velasco, C.E. and J.A. Guerrero-Beltrán. 2014. Postharvest quality of peeled prickly pear fruit treated with acetic acid and chitosan. Postharvest Biol. Tec. 92: 139–145.

Oliveira, D.M., A. Kwiatkowski, C.I. Lourenzi Franco Rosa and E. Clemente. 2014. Refrigeration and edible coatings in blackberry (*Rubus* spp.) conservation. Food Sci. Tech. Mys. 51: 2120–2126.

Park, P.J., J.Y. Je, H.G. Byun, S.H. Moon and S.E. Kim. 2004. Antimicrobial activity of hetero-chitosan and their oligosaccharides with different molecular weights. J. Microbiol. Biotechnol. 14 317–323.

Park, S.I., S.D. Stan, M.A. Daeschel and Y.Y. Zhao. 2005. Antifungal coatings on fresh strawberries (*Fragaria x ananassa*) to control mold growth during cold storage. J. Food Sci. 70: M202–M207.

Paull, R.E. 1999. Effect of temperature and relative humidity on fresh commodity quality. Postharvest Biol. Tec. 15: 263–277.

Pen, L.T. and Y.M. Jiang. 2003. Effects of chitosan coating on shelf life and quality of fresh-cut Chinese water chestnut. Lebensm.-Wiss. U.-Technol. 36: 359–364.

Perdones, A., L. Sánchez-González, A. Chiralt and M. Vargas. 2012. Effect of chitosan-lemon essential oil coatings on storage-keeping quality of strawberry. Postharvest Biol. Tec. 70: 32–41.

Phan, D., F. Debeaufort, D. Luu and A. Voilley. 2005. Functional properties of edible agar-based and starch-based films for food quality preservation. J. Agric. Food Chem. 53: 973–981.

Pilon, L., P.C. Spricigo, M. Miranda, M.R. de Moura, O.B.G. Assis, L.H.C. Mattoso, et al. 2015. Chitosan nanoparticle coatings reduce microbial growth on fresh-cut apples while not affecting quality attributes. Int. J. Food. Sci. Tech. 50: 440–448.

Ponce, A.G., S.I. Roura, C.E. del Valle and M.R. Moreira. 2008. Antimicrobial and antioxidant activities of edible coatings enriched with natural plant extracts: *In vitro* and *in vivo* studies. Postharvest Biol. Tec. 49: 294–300.

Poverenov, E., Y. Zaitseva, H. Arnon, R. Granita, S. Alkalai-Tuvia, Y. Perzelan, et al. 2014a. Effects of a composite chitosan–gelatin edible coating on postharvest quality and storability of red bell peppers. Postharvest Biol. Tec. 96: 106–109.

Poverenov, E., S. Danino, B. Horev, R. Granit, Y. Vinokur and V. Rodov. 2014b. Layer-by-layer electrostatic deposition of edible coating on fresh cut melon model: Anticipated and unexpected effects of alginate–chitosan combination. Food Bioprocess Tec. 7: 1424–1432

Pranoto, Y., S.K. Rakshit and V.M. Salokhe. 2005. Enhancing antimicrobial activity of chitosan films by incorporating garlic oil, potassium sorbate and nisin. Food Sci. Technol. 38: 859–865.

Pushkala, R., P.K. Raghuram and N. Srividya. 2013. Chitosan based powder coating technique to enhance phytochemicals and shelf life quality of radish shreds. Postharvest Biol. Tec. 86: 402–408.

Qi, H., W. Hu, A. Jiang, M. Tian and Y. Li. 2011. Extending shelf-life of Fresh-cut 'Fuji' apples with chitosan-coatings. Innov. Food Sci. Emerg. 12: 62–66.

Raafat, D. and H.G. Sahl. 2009. Chitosan and its antimicrobial potential-a critical literature survey. Microb. Biotech. 2(2): 186–201.

Rabea, E.I., E.T. Badawy, C.V. Stevens, G. Smagghe and W. Steurbaut. 2003. Chitosan as antimicrobial agent: Applications and mode of action. Biomacromol. 4: 1457–1465.

Ramos-García, M., E. Bosquez-Molina, J. Hernández-Romano, G. Zavala-Padilla, E. Terrés-Rojas, I. Alia-Tejacal, et al. 2012. Use of chitosan-based edible coatings in combination with other natural compounds, to control *Rhizopus stolonifer* and *Escherichia coli* DH5α in fresh tomatoes. Crop Prot. 38: 1–6.

Ray, D., P. Roy, S. Sengupta, P. Sengupta, A.K. Mohanty and M. Misra. 2009. A study of physicomechanical and morphological properties of starch/poly(vinyl alcohol) based films. J. Polym. Environ. 17: 56–63.

Rhim, J.W., S.I. Hong, H.M. Park and P.K.W. Ng. 2006. Preparation and characterization of chitosan-based nanocomposite films with antimicrobial activity. J. Agric. Food Chem. 54: 5814–5822.

Rhim, J.W. 2013. Effect of PLA lamination on performance characteristics of agar/κ-carrageenan/clay bio-nanocomposite film. Food Res. Int. 51: 714–722.

Rhim, J.W. and L.F. Wang. 2014. Preparation and characterization of carrageenan-based nanocomposite films reinforced with clay mineral and silver nanoparticles. Appl. Clay Sci. 97–98: 174–18

Ribeiro, C., A.A. Vicente, J.A. Teixeira and C. Miranda. 2007. Optimization of edible coating composition to retard strawberry fruit senescence. Postharvest Biol. Tec. 44: 63–70.

Rivero, S., M.A. Garcia and A. Pinotti. 2009. Composite and bi-layer films based on gelatin and chitosan. J. Food Eng. 90: 531–539.

Robles-Sánchez, R.M., M.A. Rojas-Graü, I. Odriozola-Serrano, G. González-Aguilar and O. Martin-Belloso. 2013. Influence of alginate-based edible coating as carrier of antibrowning agents on bioactive compounds and antioxidant activity in fresh-cut Kent mangoes. LWT – Food Sci. Tech. 50: 240–246.

Rodríguez-Núñez, J.R., T.J. Madera-Santana, D.I. Sanchez-Machado, J. Lopez-Cervantes and H. Soto Valdez. 2014. Chitosan/hydrophilic plasticizer-based films: Preparation, physicochemical and antimicrobial properties. J. Environ. Polym. Degrad. 22: 41–51.

Rojas-Graü, M.A., R.J. Avena-Bustillos, C. Olsen, M. Friedman, P.R. Henika and O. Martín-Belloso. 2007. Effects of plant essential oils and oil compounds on mechanical, barrier and antimicrobial properties of alginate–apple puree edible films. J. Food Eng. 81: 634–641.

Romanazzi, G., E. Feliziani, M. Santini and L. Landi. 2013. Effectiveness of postharvest treatment with chitosan and other resistance inducers in the control of storage decay of strawberry. Postharvest Biol. Tec. 75: 24–27.

Saltveit, M.E. 2014. Respiratory metabolism. pp. 68. *In:* Gross, K.C., C.Y. Wang and M.E. Saltveit (eds.). The Commercial Storage of Fruits, Vegetables, and Florist and Nursery Stocks. Agriculture Handbook Number 66. United States Department of Agriculture. Agricultural Research Service. http://www.ba.ars.usda.gov/hb66/contents.html.

Sánchez, C., F.C. Lidon, M. Vivas, P. Ramos, M. Santos and M.G. Barreiro. 2015. Effect of chitosan coating on quality and nutritional value of fresh-cut 'Rocha' pear. Emir. J. Food Agric. 27: 206–214.

Sánchez-González, L., M. Vargas, C. González-Martínez, A. Chiral and M. Chafer. 2011a. Use of essential oils in bioactive edible coatings. Food Eng. Rev. 3: 1–16.

Sánchez-González, L., C. Pastor, M. Vargas, A. Chiralt, C. González-Martínez and M. Chafer. 2011b. Effect of hydroxypropylmethylcellulose and chitosan coatings with and without bergamot essential oil on quality and safety of cold-stored grapes. Postharvest Biol. Tec. 60: 57–63.

Serrano, M., D. Martínez-Romero, F. Guillen, J.M. Valverde, P.J. Zapata, S. Castillo, et al. 2008. The addition of essential oils to MAP as a tool to maintain the overall quality of fruits. Trends Food Sci. Technol. 19: 464–471.

Seymour, G.B., L. Østergaard, N.H. Chapman, S. Knapp and C. Martin. 2013. Fruit development and ripening Annual Rev. Plant Biol. 64: 219–241.

Shiri, M.A., D. Bakhshi, M. Ghasemnezhad, M. Dadi, A. Papachatzis and H. Kalorizou. 2013a. Chitosan coating improves the shelf life and postharvest quality of table grape (*Vitis vinifera*) cultivar Shahroudi. Turk. J. Agric. For. 37: 148–156.

Shiri, M.A., M. Ghasemnezhad, D. Bakhshi and H. Sarikhani. 2013b. Effect of postharvest putrescine application and chitosan coating on maintaining quality of table grape cv. "Shahroudi" during long-term storage. J. Food Process. 37: 999–1007.

Simoes, A.D.N., J.A. Tudela, A. Allende, R. Puschmann and M.I. Gil. 2009. Edible coatings containing chitosan and moderate modified atmospheres maintain quality and enhance phytochemicals of carrot sticks. Postharvest Biol. Tec. 51: 364–370.

Simonaitiene, D., I. Brink, A. Sipailiene and D. Leskauskaite. 2015. The effect of chitosan and whey proteins-chitosan films on the growth of *Penicillium expansum* in apples. J. Sci. Food Agr. 95: 1475–1481.

Skurtys, O., C. Acevedo, F. Pedreschi, J. Enronoe, F. Osorio and J.M. Aguilera. 2010. Food hydrocolloid edible films and coatings. Nova Science Publishers, Inc. New York, NY, USA.

Soares N.F.F., D.F.P. Silva, G.P. Camilloto, C.P. Oliveira, N.M. Pinheiro and E.A.A. Medeiros. 2011. Antimicrobial edible coating in post-harvest conservation of guava. Rev. Bras. Frutic. Jaboticabal E: 281–289.

Sothomvit, R. and J.M. Krochta. 2005. Plasticizers in edible films and coatings. pp. 403–433. *In*: Han, J.H. (ed.). Innovations in Food Packaging. Elsevier Academic Press, San Diego, CA.

Souza, M.P., A.F.M. Vaz, M.A. Cerqueira, J.A. Texeira, A.A. Vicente and M.G. Carneiro-da-Cunha. 2015. Effect of an edible nanomultilayer coating by electrostatic self-assembly on the shelf life of fresh-cut mangoes. Food Bioprocess Tec. 8: 647–654.

Srinivasa, P.C., R. Baskaran, M.N. Ramesh, K.V.H. Prashanth and R.N. Tharanathan. 2002. Storage studies of mango packed using biodegradable chitosan film. Eur. Food Res. Technol. 215: 504–508.

Suseno, N., E. Savitri, L. Sapei and K.S. Padmawijaya. 2014. Improving shelf-life of Cavendish banana using chitosan edible coating. pp. 113–120. *In*: Sugih, A.K., H. Muljana, A.A. Arie, L. Janssen and J.K. Lee (ed.). International Conference and Workshop on Chemical Engineering Unpar 2013. Vol. 9.

Suyatma, E., L. Tighzert and A. Copinet. 2005. Effect of hydrophilic plasticizers on mechanical, thermal, and surface properties of chitosan films. J. Agric. Food Sci. Chem. 53: 3950–3967.

Talasila, P.C. 1992. Modeling of heat and mass transfer in a modified atmosphere package. Ph.D. dissertation, University of Florida, Gainesville, FL, USA.

Tang, X.G., P. Kumar, S. Alavi and K.P. Sandeep. 2012. Recent advances in biopolymers and biopolymer-based nanocomposites for food packaging materials. Crit. Rev. Food Sci. Nutr. 52: 426–442.

Tavassoli-Kafrani, E., H. Shekarchizadeh and M. Masoudpour-Behabadi. 2016. Development of edible films and coatings from alginates and carrageenans. Carboh. Polym. 137: 360–374.

Tezotto-Uliana, J.V., G.P. Fargoni, G.M. Geerdink and R.A. Kluge. 2014. Chitosan applications pre- or postharvest prolong raspberry shelf-life quality. Postharvest Biol. Tec. 91: 72–77.

Thompson, J.F., G. Mitchell, T.R. Rumsey, R.F. Kasmire and C.H. Crisosto. 2002. Commercial Cooling of Fruits, Vegetables, and Flowers. Agriculture and Natural Resources Publication 21567. University of California, Oakland, CA, USA.

Trevino-Garza, M.Z., S. García, M.S. Flores-González and K. Arevalo-Niño. 2015. Edible active coatings based on pectin, pullulan, and chitosan increase quality and shelf life of strawberries (*Fragaria x ananassa*). J. Food Sci. 80: M1823–M1830.

Varasteh, F., K. Arzani, M. Barzegar and Z. Zamani. 2012. Changes in anthocyanins in arils of chitosan-coated pomegranate (*Punica granatum* L. cv. Rabbab-e-Neyriz) fruit during cold storage. Food Chem. 130: 267–272.

Vargas, M., A. Albors, A. Chiralt and C. González-Martínez. 2006. Quality of cold-stored strawberries as affected by chitosan-oleic acid edible coatings. Postharvest Biol. Tec. 41: 164–171.

Vargas, M., A. Chiralt, A. Albors and C. González-Martínez. 2009. Effect of chitosan-based edible coatings applied by vacuum impregnation on quality preservation of fresh-cut carrot. Postharvest Biol. Tec. 51: 263–271.

Vasconcelos de Oliveira, C.E., M. Magnani, C.V. de Sales, A.L. de Souza Pontes, G.M. Campos-Takaki, T.C. Montenegro Stamford, et al. 2014a. Effects of chitosan from *Cunninghamella elegans* on virulence of post-harvest pathogenic fungi in table grapes (*Vitis labrusca* L.). Int. J. Food Microbiol. 171: 54–61.

Vasconcelos de Oliveira, C.E., M. Magnani, C.V. de Sales, A.L. de Souza Pontes, G.M. Campos-Takaki, T.C. Montenegro Stamford, et al. 2014b. Effects of post-harvest treatment using chitosan from *Mucor circinelloides* on fungal pathogenicity and quality of table grapes during storage. Food Microbiol. 44: 211–219.

Velickova, E., E. Winkelhausen, S. Kuzmanova, V.D. Alves and M. Moldao-Martins. 2013. Impact of chitosan-beeswax edible coatings on the quality of fresh strawberries (*Fragaria x ananassa* cv. Camarosa) under commercial storage conditions. Lebensm. Wiss. Technol. 52: 80–92.

Vu, K.D., R.G. Hollingsworth, E. Leroux, S. Salmieri and M. Lacroix. 2011. Development of edible bioactive coating based on modified chitosan for increasing the shelf life of strawberries. Food Res. Int. 44: 198–203.

Waimaleongora-Ek, P., A.J.H. Corredor, H.K. No, W. Prinyawiwatkul, J.M. King, M.E. Janes, et al. 2008. Selected quality characteristics of fresh-cut sweet potatoes coated with chitosan during 17-day refrigerated storage. J. Food Sci. 73: S418–S423.

Wang, J., B. Wang, W. Jiang and Y. Zhao. 2007. Quality and shelf life of mango (*Mangifera indica* l. cv. 'Tainong') coated by using chitosan and polyphenols. Food Sci. Technol. Int. 13: 317–322.

Wang, S.Y. and H. Gao. 2013. Effect of chitosan-based edible coating on antioxidants, antioxidant enzyme system, and postharvest fruit quality of strawberries (*Fragaria x aranassa* Duch.). Lebensm. Wiss. Technol. 52: 71–79.

Xiao, C., L. Zhu, W. Luo, X. Song and Y. Deng. 2010a. Combined action of pure oxygen pretreatment and chitosan coating incorporated with rosemary extracts on the quality of fresh-cut pears. Food Chem. 121: 1003–1009.

Xiao, C., W. Luo, M. Liu, L. Zhu, M. Li, H. Yang, et al. 2010b. Quality of fresh-cut pears (*Pyrus bretschneideri* Rehd cv. Huangguan) coated with chitosan combined with ascorbic acid and rosemary extracts. Philipp. Agric. Scientist 93: 66–75.

Xiao, Q., L.T. Lim and Q. Tong. 2012. Properties of pullulan-based blend films as affected by alginate content and relative humidity. Carboh. Polym. 87(1): 227–234.

Xiao, Z., Y. Luo, Y. Luo and Q. Wang. 2011. Combined effects of sodium chlorite dip treatment and chitosan coatings on the quality of fresh-cut d'Anjou pears. Postharvest Biol. Tec. 62: 319–326.

Xing, Y., X. Li, Q. Xu, Y. Jiang, J. Yun and W. Li. 2010. Effects of chitosan-based coating and modified atmosphere packaging (MAP) on browning and shelf life of fresh-cut lotus root (*Nelumbo nucifera* Gaerth). Innov. Food Sci. Emerg. 11: 684–689.

Xing, Y., X. Li, Q. Xu, J. Yun, Y. Lu and Y. Tang. 2011. Effects of chitosan coating enriched with cinnamon oil on qualitative properties of sweet pepper (*Capsicum annuum* L.). Food Chem. 124: 1443–1450.

Xu, Q., Y. Xing, Z. Che, T. Guan, L. Zhang, Y. Bai, et al. 2013. Effect of chitosan coating and oil fumigation on the microbiological and quality safety of fresh-cut pear. J. Food Safety 33: 179–189.

Xu, W.T., K.L. Huang, F. Guo, W. Qu, J.J. Yang, Z.H. Liang, et al. 2007. Postharvest grapefruit seed extract and chitosan treatments of table grapes to control *Botrytis cinerea*. Postharvest Biol. Tec. 46: 86–94.

Xu, Y.X., K.M. Kim, M.A. Hanna and D. Nag. 2005. Chitosan–starch composite film: preparation and characterization. Ind. Crop. Prod. 21: 185–192.

Yang, G., J. Yue, X. Gong, B. Qian, H. Wang, Y. Deng, et al. 2014. Blueberry leaf extracts incorporated chitosan coatings for preserving postharvest quality of fresh blueberries. Postharvest Biol. Tec. 92: 46–53.

Yang, H., J. Zheng, C. Huang, X. Zhao, H. Chen and Z. Sun. 2015. Effects of combined aqueous chlorine dioxide and chitosan coatings on microbial growth and quality maintenance of fresh-cut bamboo shoots (*Phyllostachys praecox f. prevernalis.*) during storage. Food Bioprocess Technol. 8: 1011–1019.

Ye, M., H. Neetoo and H. Chen. 2008. Control of *Listeria monocytogenes* on ham steaks by antimicrobials incorporated into chitosan-coated plastic films. Food Microbiol. 25: 260–268.

Younas, M.S., M.S. Butt, I. Pasha and M. Shahid. 2014. Development of zinc fortified chitosan and alginate based coatings for apricot. Pak. J. Agri. Sci. 51: 1033–1039.

Younes, I., S. Sellimi, M. Rinaudo, K. Jellouli and M. Nasri. 2014. Influence of acetylation degree and molecular weight of homogeneous chitosans on antibacterial and antifungal activities. Int. J. Food Microbiol. 185: 57–63.

Younes, I. and M. Rinaudo. 2015. Chitin and chitosan preparation from marine sources. Structure, properties and applications. Mar. Drugs. 13: 1133–1174.

Zheng, L.Y. and J.F. Zhu. 2003. Study on antimicrobial activity of chitosan with different molecular weights. Carbohyd. Polym. 54: 527–530.

Zhou, R., Y. Mo, Y. Li, Y. Zhao, G. Zhang and Y. Hu. 2008. Quality and internal characteristics of Huanghua pears (*Pyrus pyrifolia* Nakai, cv. Huanghua) treated with different kinds of coatings during storage. Postharvest Biol. Tec. 49: 171–179.

Ziani, K., J. Oseas, V. Coma and J.I. Mate. 2008. Effect of the presence of glycerol and Tween 20 on the chemical properties of films base on chitosan with different degree of deacetylation. LWT Food Sci. Tec. 41: 2159–2165.

6

Agroindustrial Biomass: Potential Materials for Production of Biopolymeric Films

Delia R. Tapia-Blácido*, Bianca C. Maniglia, Milena Martelli-Tosi, and Vinícius F. Passos

Departamento de Química, Faculdade de Filosofia, Ciências e Letras de Ribeirão Preto, Universidade de São Paulo, Av. Bandeirantes, 3900, CEP 14040-901, Ribeirão Preto, SP, Brazil Ribeirão Preto/SP, Brazil

Introduction

The global production of plastic reached 299 million tonnes in 2013. Unfortunately, less than 5% of all this plastic was recycled, leading to rapid accumulation of plastic waste and consequent environmental pollution (Plastics Europe 2015). In this context, the use of biopolymers in multiple food-packaging applications has emerged as an alternative and environmentally friendly technology to produce biodegradable films (Espitia et al. 2014). Application of these materials will only be feasible if they are competitive in terms of cost and functionality as compared to synthetic plastics. Initially, many researchers considered producing biodegradable films from polysaccharides and/or proteins, but the use of pure polymers yielded expensive materials. In an attempt to find cheaper raw materials to produce biodegradable films, current research has focused on agroindustrial biomass as a source of natural polymers such as carbohydrates, proteins and fibres. Agroindustrial biomass may also contain bioactive compounds like antioxidants and antimicrobials, allowing the production of bioactive films that offer extra benefits in relation to conventional materials. For example, fruit and vegetable processing generates a large amount of residues and is

---

*Corresponding author: delia@ffclrp.usp.br

therefore a serious environmental concern. However, this agroindustrial matter could be a potential source of natural antioxidants and serve as an additive to produce biopackaging film, providing nutritional and technological advantages (Oliveira et al. 2015; Schieber et al. 2001).

Harvesting and industrial processes generate agroindustrial biomass. For example, soybean harvesting gives rise to stalks, stems and leaves, collectively designated soybean straw. The ratio between the soybean straw weight and the soybean weight varies from 120 to 150%. The soybean grain processing industry generates soybean hulls as a byproduct, which represents approximately 8% of the whole seed (Tapia-Blácido et al. 2016). The residue remaining from the extraction of turmeric dye is another interesting example of agroindustrial biomass. Turmeric is widely used in the food, cosmetic and pharmaceutical industries (Sigrist et al. 2011; Cecilio-Filho et al. 2000). Maniglia (2012) simulated the industrial extraction of curcumin with ethanol/isopropanol by the Soxhlet method and obtained a yield of 11.3% of turmeric dye. The process furnished large amounts of a residue that had an interesting composition.

Biodegradable films from agroindustrial biomass combine the film-forming properties of biopolymers (mainly polysaccharides such as pectin and cellulose derivatives) with additional properties like antioxidant and antimicrobial activity and have therefore attracted increasing attention from researchers worldwide (Maniglia et al. 2014, 2015; Azeredo et al. 2009; Kaya and Maskan 2003). Agroindustrial biomass such as cassava bagasse, rice flour, canola meal, turmeric flour and orange waste flour are examples of residues that are rich in polysaccharides like starch and pectin, which present filmogenic capacity (Oliveira et al. 2015; Maniglia et al. 2014, 2015; Chang and Nickerson 2014; Suppakul et al. 2013; Dias et al. 2010; Wanyo et al. 2009). Agroindustrial biomass rich in fibres; e.g. cassava and sugarcane bagasse, have been explored as filler and reinforcement material, to improve film properties. These materials have to be treated before they are applied in films (Slavutsky and Bertuzzi 2014; Wicaksono et al. 2013). On the other hand, residues from the extraction of turmeric and annatto dyes still contain antioxidant and antimicrobial compounds, which make them potentially applicable in the production of bioactive biopackaging.

In this context, agroindustrial biomass or waste with high contents of interesting polymers, such as polysaccharides and proteins, represent an interesting alternative in the area of biofilms. This chapter focuses on the potential use of some types of agroindustrial biomass as a source of biopolymers to produce films or as a source of fibres to act as filler in biopackaging.

Biopolymers in Agroindustrial Biomass

The agroindustrial sector comprises economic activities related to manufacturing sectors that are primarily engaged in the processing of raw materials and intermediate agricultural, fishery and forestry products. Depending on the agroindustrial source and on processing, the agroindustrial residue contains several polymers such as polysaccharides (e.g. starch, pectin and chitosan), protein (e.g., zein, gluten and gelatin) and lignocellulosic material (Table 1).

Cassava bagasse, an agroindustrial residue of the cassava starch industry, consists mainly of starch and fibres (Pasquini et al. 2010). Isolation of starch from 250-300 t of fresh cassava gives about 280 t of a wet solid residue, or bagasse, composed of fibrous material (10.61-18.35% (w/w dry basis)) and starch (61.84-79.90% (w/w dry basis)) (Teixeira et al. 2009, 2012; Srinorakutara et al. 2005; Bertol and Lima 1999; Leonel et al. 1999).

Starch is a storage polysaccharide present in the cells of tubers, such as potatoes and cassava, and seeds, like corn and wheat. Starch contains two types of glucose polymers, namely amylose and amylopectin. Amylose consists of long chains of unbranched D-glucose units connected by links (α 1-4) and has excellent film-forming ability, rendering strong, isotropic, odorless, tasteless and colorless films (Campos et al. 2011). The molecular weight of amylose chains ranges from a few thousand to over one million. Amylopectin also has high molecular weight (to 100 million). However, unlike amylose, amylopectin is highly branched. The glycosidic linkages between the glucose units of amylopectin chains are α 1-4, but the branch points (about one every 24 to 30 units) contain α 1-6 glycosidic linkages (Nelson and Cox 2002). Native starches are semi-crystalline and contain crystalline and non-crystalline regions in alternating layers formed by hydrogen bonding between the structures of amylose and amylopectin (Jenkins et al. 1993). Therefore, the dispersion of starch requires heating in water or another solvent, which disrupts the crystalline structure of starch and allows the solvent molecules to interact with the hydroxyl groups in amylose and amylopectin, leading to partial solubilization of starch (Hoover 2001).

Starch is relatively easy to separate from agroindustrial biomass by alkaline, acid or enzymatic separation methods (Arvanitoyannis and Kassaveti 2009). Because most isolation methods affect the properties of the final material, it is essential to identify the most suitable isolation technique in terms of the purity, yield and properties of starch (Correia and Beirão-da-Costa 2012). Commercial isolation of starch involves milling or grating the biomass, separating the fibres, and suspending starch in water, followed by centrifugation, purification, dehydration and drying (Agama-Acevedo et al. 2014; Santos et al. 2013).

Rice harvesting and processing gives rise to straw, hull and flour. Rice harvesting generates between 1 and 1.5 kg of straw for each kilogram of rice grain (Maiorella 1985). Rice flour represents 14% of the total grain mass and is used to produce beer and animal food (Nabeshima and El-Dash 2004). Rice straw is rich in fibre (66%) and consists of 43-49% of cellulose, 26-32% of hemicellulose and 16-21% of lignin (Garay et al. 2009). The disposal of

Table 1. Chemical composition of different types of agroindustrial biomass (g/100 g of material on dry basis).

Biomass type	Cellulose	Hemicellulose	Lignin	Other polysaccharides	Protein	Lipid	Ash	References
Cassava Bagasse		17.5		Starch: 82	-	-	-	Slavutsky and Bertuzzi 2014
Rice flour		0.6		Starch: 80	6.8	0.54	0.22	Wanyo et al. 2009
Rice straw	43-49	23-28	12-16	-	-	-	12-20	Garay et al. 2009
Rice hulls	35.47	13.35	7.24	70.95	4.65	0.28	14.74	Jacometti et al. 2015
Corn bran		73.4		12.54	5.50	5.49	1.46	Oliveira Junior et al. 2014
Corn stover	35.5	36.9	4.9		-	-	6.67	Barten 2013
Wheat bran		4.51		Starch: 69.04	11	1.25	0.53	Wanyo et al. 2009
Wheat straw	30	50	15	-	-	-	-	Sun and Cheng 2002
Sugarcane bagasse	38	27	20	-	-	-	-	Saxena et al. 2009
Soybean straw	34-35	16-17	22	-	-	-	-	Wan et al. 2011, Cabrera et al. 2015

(Contd.)

Table 1. (*Contd.*)

Biomass type	Cellulose	Hemicellulose	Lignin	Other polysaccharides	Protein	Lipid	Ash	References
Soybean hulls	35.8	23.1	9.1	-	15.4	-	4.0	Rojas et al. 2014
Oat hulls	48.0	25.5	14.5	90.02	1.85	1.94	3.27	Jacometti et al. 2015
Orange waste	9.2	10.5	-	Pectin: 42.5	-	-	-	Oliveira et al. 2016
Canola meal	12				36	3.5	6.1	Newkirk 2009
Cottonseed Flour	-	-	-	-	34.4	38.8	1.8	Marquié et al. 1995
Shrimp waste ("*Litopenaeus vannamei*")	-	-	-	Chitin: 9.3	54.4	11.9	21.2	Trung and Phuong 2012
Fish waste "Tilapia"	-	-	-	-	57.9	19.10	21.8	Ghaly et al. 2013
Turmeric Flour	49.4	2.9	6.9	Starch: 24.8	9.2	0.9	5.8	Maniglia et al. 2014
Annatto flour		28.4		Starch: 42.2	11.5	2.2	5.2	Valério et al. 2015

rice straw is limited because it has low bulk density, is slow to degrade in soil, harbors rice stem diseases and displays high mineral content (Binod et al. 2010). Rice hulls represent approximately 20% of the dry weight of the rice harvest (Dagnino et al. 2013) and contain 36-40 g of cellulose/100 g of hull and 12-19 g of hemicelluloses/100 g of hull (Banerjee et al. 2009). They also contain fats, gums, alkaloids, resins, essential oils and other cytoplasmic components (extractives) as well as ash composition of approximately 12 g of ash/100 g of hull. The ash consists primarily of silica (80 to 90 g of silica/ 100 g of ash) (Jacometti et al. 2015; Dagnino et al. 2013; Banerjee et al. 2009).

Along with wheat and rice, corn is one of the most widely cultivated cereals in the world. The corn milling industry generates corn bran, which consists of fibre and starch as well as a small amount of sulfite ion (SO_3^{2-}) (Zhang and Zheng 2015; Oliveira Junior 2014). Corn bran is usually employed to feed animals, but increased production of corn (~28%) requires that other uses for this bran be found in order to avoid problems related to the disposal of this residue. On the other hand, corn harvesting produces corn stover, which consists of a complex mixture composed mainly of cellulose, hemicellulose and lignin (~77% fibres) (Barten 2013). Corn stover can be burned on the field, discarded or used as ground cover after the harvest (Portal do Agronegócio 2009; Ereno 2007; Wang et al. 2005).

Wheat grain harvesting generates straw, which corresponds to 50% of the plant weight and has fibres as the major constituents (95%). Wheat straw is usually employed as soil fertilizer or bedding for animals (Sun and Cheng 2002). Straw is an abundant and low-cost (about 25 euros/ton but with high price volatility) agricultural residue that is primarily composed of cellulose microfibril conglomerates bound by an intercellular matrix consisting of hemicelluloses, lignin and pectins (Hornsby et al. 1997).

Sugarcane processing to extract juice produces a fibrous residue (85%- see Table 1) is called bagasse (Hofsetz and Silva 2012). Brazil is the world's largest producer (25% of the total sugar in the world) and exporter (60% of the total sugar exports in the world) of sugar. Until recently, Brazil was the world's largest producer of ethanol and exporter of bagasse (Drabik et al. 2015). Sugarcane bagasse contains cellulose (38-46%), hemicelluloses (23-27%) and lignin (19-32%), whereas sugarcane straw presents lower cellulose (33-36%), higher lignin (26-40%) and similar hemicellulose content (18-29%) as compared to sugarcane bagasse. Sugarcane bagasse can be used as raw material to produce paper, particle board, furfural and other products. In Brazil, sugarcane bagasse is usually employed as fuel to produce steam and electric energy in sugarcane mills (Saxena et al. 2009; Sosa-Arnao et al. 2009).

Soybean harvesting generates stalks, stems and leaves, collectively designated soybean straw. Soybean hulls, which represent approximately 8% of the whole seed, also constitute a residue of the soybean grain processing industry (Merci et al. 2015; Gnanasambandam and Proctor 1999). According to estimates, the 2015/2016 harvest will provide 116.3 million ton of soybean straw, which consists of cellulose (34-35%), hemicelluloses (16-17%; 11.4%

xylan, 1.8% galactan, 1.0% arabinan, and 1.8% mannan), acid insoluble lignin (21%), acid soluble lignin (1%), extractives (6-11%), ash (5-11%) and other non-identified compounds (10-12%; e.g., protein, pectin, acetyl groups and glucuronic acid substitutes) (Cabrera et al. 2015; Wan et al. 2011). Traditionally, soybean straw has been used for low-value purposes like livestock feeding, mulch and bedding materials for animals (Mo et al. 2005). Because soybean straw has low nutritional value, application of the entire residue is difficult. The hulls contain cellulose (29-51%), hemicelluloses (10-25%), lignin (1-4%), pectins (4-8%), proteins (11-15%) and minor extractives (Mielenz et al. 2009; Rojas et al. 2014; Yoo et al. 2011).

Oat hulls are a byproduct of oat groat milling and are discarded during processing, which makes them an environmental pollutant. Oat hulls contain approximately 90 g of fibre/100 g of hull, which is higher than the fibre content in wheat (47 g of fibre/100 g of wheat) or corn bran (62 g of fibre/100 g of corn bran) (Jacometti et al. 2015; Galdeano and Grosmann 2006).

Orange bagasse resulting from extraction of juice from the fruit is another residue with high polysaccharide content, mainly pectin. Brazil is the largest producer and exporter of orange juice, contributing with 50% of the world production. Of the total production of orange in Brazil, 70% is intended for the production of juice (Neves et al. 2012). Considering that 100 kg of orange gives 55 kg of juice, the remaining 45 kg consists of residues such as discarded oranges, zest, seed, and waste resulting from the extraction of essential oil and washed pulp (Cavichiolo 2010; Sánchez-Sáenz et al. 2015). Orange bagasse has high contents of pectin (42.5%) and other compounds such as soluble sugars (16.9%), celullose (9.21%) and hemicellulose (10.5%) (Rezzadori and Benedetti 2009). Pectin corresponds to poly α1–4-galacturonic acids with varying degree of methylation of the carboxylic acid residues and/or amidated polygalacturonic acids (Mishra et al. 2012; White et al. 1999). According to the degree of esterification (DE) with methanol (esterified galacturonic acid groups/total galacturonic acid groups ratio) (Farris et al. 2009; Sila et al. 2009), pectin can be classified as high methoxyl pectin (HMP) or low methoxyl pectin (LMP). Over 50% and less than 50% of the carboxyl groups in HMP (DE > 50) and LMP (DE < 50) are esterified, respectively. Pectin has higher gelification power and ability to form films, so orange bagasse has potential application in the production of biopackaging.

As mentioned previously, agroindustrial biomass such as corn, wheat, rice, soybean straw, sugarcane bagasse and orange bagasse are sources of lignocellulosic materials; that is, cellulose, hemicellulose and lignin, which are the major components of plant cell walls (~90% of dry biomass) (Gibson 2012; Harris and Stone 2008; Pauly and Keegstra 2008). Cellulose is a linear macromolecule consisting of D-anhydroglucose ($C_6H_{11}O_5$) repeating units joined by β-1,4-glycosidic linkages with a degree of polymerization (DP) of around 10,000. Each repeating unit contains three hydroxyl groups. These hydroxyl groups and their ability to form hydrogen bonds play a major role in directing crystalline packing and governing the physical properties of cellulose materials (Bismarck et al. 2005; Tapia-Blácido et al. 2016). Cellulose is resistant to strong alkali (17.5%), but acids easily hydrolyze it to water-soluble sugars. Hemicelluloses are polysaccharides composed of linear and branched D-xylose, L-arabinose, D-galactose, D-glucose and D-mannose heteropolymers. The polymer chains in hemicelluloses are much shorter (DP from around 50 to 300), branched, and noncrystalline, which makes hemicelluloses more susceptible to hydrolysis by acids (Chang and Holtzapple 2000; Maurya et al. 2015; Oh et al. 2015; Tapia-Blácido et al. 2016). Additionally, hemicellulose is very hydrophilic and soluble in alkali (Kubo et al. 2005). Lignin is a racemic heteropolymer consisting of three monomers of hydroxycinnamyl alcohol (p-coumaryl, coniferyl and sinapyl alcohols) that differ in terms of their degree of methoxylation (Cazacu et al. 2012). Lignin is amorphous and hydrophobic in nature. Acids cannot hydrolyze lignin, which is soluble in hot alkali, readily oxidized and easily condensable with ethanol (Kubo et al. 2005).

Proteins are also present in agroindustrial biomass but at lower concentration as compared to starch. They have received much attention for the production of biopolymeric film. Extraction of oil from canola affords a protein-rich residue (36%) called canola meal, which consists mainly of proteins (Table 1). Canola meal is typically used for animal feeding (Kimber and McGregor 1998; Newkirk 2009), but it has limited application due to the presence of anti-nutritional compounds, such as glucosinolates, phenolics and phytates (Mccurdy 1990; Manamperi et al. 2011). Therefore, canola meal is mainly applied as low-value animal feed (Tan et al. 2011), but better use and greater value could be found for this material.

Extraction of cotton oil also generates a solid residue called cottonseed flour, which has high protein content (34%). This residue is mainly used as animal feed, but it can also be employed to produce biopolymeric films (Marquié et al. 1995, 1997, 1998).

In the seafood industry, shrimp is usually processed to obtain shrimp meat for export. Between 35 and 45% of the leftovers are shells and heads, considered as byproducts (Gildberg and Stenberg 2011). As a result, shrimp processing produces massive amounts of shrimp biowaste, estimated at more than 200,000 metric tons (wet weight) per year. The major components of the shrimp byproducts are protein (54%), chitin (deacetylated chitosan) (9%), lipid and minerals (Trung and Phuong 2012). Chitin is insoluble in water, organic solvents, and diluted acid and alkali. It presents three polymeric forms: α-quitin, β-quitin, and γ-quitin depending on the crystalline structure, arrangement of the chains and presence of water molecules. Chitin deacetylation in alkaline medium gives chitosan (Abou et al. 2007), a copolymer consisting of β-(1–4)-2-acetamido-D-glucose and β-(1–4)-2-amino-D-glucose units with the latter usually exceeding 60% of chitosan. Chitosan is important for its antimicrobial properties, cationicity and film-forming properties (Coma et al. 2002, 2003; Durango et al. 2006; Han et al. 2004; Park et al. 2004; Ribeiro et al. 2007). Chitosan nanoparticles have served as fillers in polymer films based on hydroxypropyl methylcellulose (Moura et al. 2009) and banana over-ripe puree (Martelli et al. 2013). Chitosan nanoparticles can be produced by ionically cross-linking cationic chitosan with specific polyanions, like tripolyphosphate or by methacrylic acid polymerization (Moura et al. 2008).

Fish processing for food production requires that bones, skin, head and viscera (byproducts), which represent approximately 60–70% of the total weight of the fish, be removed (Fogaça et al. 2015; Taskaya and Jaczynski 2009). The majority of fish waste is discharged into the ocean. In the presence of oxygen, aerobic bacteria present in water break the organic matter down, considerably reducing the water oxygen content (Ghaly et al. 2013). Filleting byproducts can be a source of proteins such as collagen (Hosseini et al. 2015; Mellinas et al. 2012; Tao et al. 2015), whose conversion to gelatin occurs upon heating of collagen in acid or alkaline medium. Two types of gelatin commercially known as type-A gelatin (isoelectric point at pH ~ 8 e 9) and type-B gelatin (isoelectric point at pH ~ 4 e 5) can be obtained under acid and alkaline pre-treatment conditions, respectively (Gómez-Guillén 2011). The physical properties of gelatin influence its quality and potential application because these properties are related to the structure of the gelatin (Yang and Wang 2009). In recent years, fish gelatin has aroused interest as an alternative to bovine and porcine gelatin because of social and health reasons, such as the bovine spongi form of encephalopathy (Mellinas et al. 2016). Protein recovery from the filleting byproducts offers many benefits to the production of human food and biopackaging and reduces the environmental concerns associated with the disposal of processing byproducts (Jaczynski 2008). Canada is the major world exporter of seafood and marine products. This country exports 75% of its fish products to more than 80 countries (595, 615, 738 metric tonnes in 2012) (FAO 2013).

Carrot residues and residues from the extraction of turmeric and annatto dyes are among the agroindustrial biomass containing additional compounds with antioxidant and antimicrobial activity. Carrot residues bear a large amount of β-carotene and other carotenoids, which act as antioxidants (Hiranvarachat and Devahastin 2014). Besides, carrot residues are a source of dietary fibre. Chantaro et al. (2008) produced fibre from carrot residues with up to 73% total dietary fibre (g/100 g dry weight). Therefore, this material displays both fibres and antioxidant properties and represents an excellent option for incorporation into biodegradable films (Oliveira et al. 2015).

Turmeric (*Curcuma longa L.*) belongs to the family Zingiberaceae. It is mainly harvested in India. The world production of turmeric is around 1,100,000 ton/year (Nair 2013). Turmeric provides the yellow pigment the food industry uses to prepare pickles, mayonnaise and mustard; coat frozen fish fillets and meat products; and pasta, juices, gelatin, cheese and butter (Freund et al. 1998; Martins and Rusig 1992). Extraction of the turmeric oleoresin with solvents generates a residue that is rich in fibres and starch (25 and 52%, respectively–see Table 1; Fig. 1A) and contains curcuminoids (15.02 mg/L of curcumin, 25.50 mg/L of demethoxicurcumin, and 22.19 mg/L of bisdemethoxicurcumin). Curcumin accounts for the yellow color of turmeric (Maniglia et al. 2014).

Annatto (*Bixaorellana* L.) seeds are an important raw material to obtain the pigment bixin, norbixin, and norbixinate, whose levels vary depending on the maturation of the seed (Dornelas et al. 2005; Mantovani et al. 2013; Valério et al. 2015). These pigments can be separated from annatto seeds by many processes, including immersion of the seeds in hot vegetable oil, diluted alkaline aqueous solutions and solvents. In the first case, abrasion of the exocarp submerged in warm vegetable oil (70 °C) affords the pigment (Taham et al. 2015). Brazil is one of the largest producers and exporters of natural dye extracted from this plant (Anselmo et al. 2008; Mantovani et al. 2013; Valério et al. 2015). The residue generated after dye is extracted from annatto seeds and presents high protein (11.5%), starch (42.2%) and fibre (28.4%) contents (Valério et al. 2015) and is still red, which indicates that it displays a residual carotenoid content (Fig. 1B).

(A)   (B)

Fig. 1. Photograph of (A) turmeric dye extraction residue, and (B) annatto dye extraction residue.

Biopolymeric Film from Agroindustrial Biomass

Because agroindustrial biomass contains several polymers such as starch, pectin, chitosan, proteins and lignocellulosic material (hemicellulose, cellulose and lignin) and since this type of biomass displays antioxidant and antimicrobial compounds, they are potentially applicable in biopolymeric film technology both as a source of biopolymers for the formation of film matrixes and as a source of lignocellulosic materials for use as filler in films. Process conditions (temperature, pH, polymer concentration, etc.), type and concentration of the plasticizer, method of preparation (casting or extrusion) and type of polymer can influence the properties of the resulting biopackaging film. Therefore, evaluating all these parameters is mandatory to meet the requirements of biopackaging films (Kader et al. 1989), which should:

- allow slow but controlled respiration (reduced O_2 absorption) of the commodity;
- constitute a selective barrier to gases (CO_2) and water vapour;
- create a modified atmosphere with respect to internal gas composition, thus regulating the ripening process and extending the shelf-life;
- diminish the migration of lipids for use in the confectionery industry;
- maintain structural integrity (to delay chlorophyll loss) and improve mechanical handling;
- serve as a vehicle to incorporate food additives (flavour, colours, antioxidants and antimicrobial agents); and
- prevent (or reduce) microbial spoilage during extended storage.

Agroindustrial Biomass as Source of Biopolymers for the Formation of Films

Table 2 lists the properties of some films produced from agroindustrial biomass. In general, starch films exhibit poor water barrier and mechanical properties because they are greatly hygroscopic, fragile and brittle (Acosta et al. 2013). Table 2 summarizes research works on starch film from agroindustrial biomass.

Suppakul et al. (2013) produced films from cassava bagasse flour and evaluated the effect of the concentration of sorbitol, used as plasticizer, on the mechanical properties of the film. Increasing concentration of sorbitol in the films reduced the mechanical resistance and improved the flexibility of the film while the water vapour permeability (WVP) remained unaffected. Dias et al. (2010) assessed the filmogenic capacity of rice flour residue in the presence of different plasticizers (glycerol and sorbitol). Films with sorbitol were less permeable to water (0.12 g.mm/m².h.kPa) and more rigid (248.57 MPa), whereas films with glycerol were more plasticized (66.43%) and had poorer water vapour barrier properties (0.64 g.mm/m².h.kPa). Maniglia et al. (2014) optimized the process conditions (heating temperature and pH) for the production of films from the residue remaining after extraction of turmeric dye plasticized with glycerol. The optimal conditions were T = 86.7 °C and pH = 8.5, which afforded films with high mechanical strength (18 MPa), low solubility (36%) and low WVP (0.167 g.mm/m².h.kPa). These films also had antioxidant activity due to the presence of curcuminoids (0.0037 mg.mL⁻¹ bisdemethoxycurcumin, 0.0008 mg.mL⁻¹ demethoxycurcumin, and 0.0001 mg.mL⁻¹ curcumin).

Chang and Nickerson (2015) isolated canola protein from canola meal residue and evaluated how glycerol and sorbitol impacted the mechanical, optical and water vapour barrier properties of edible films. The canola protein films containing glycerol were more transparent (78% opacity),

Table 2. Film properties produced from biopolymers obtained from agroindustrial residues.

Polymer	Plasticizer	Thickness (μm)	Tensile Strength (MPa)	Young's Modulus (MPa)	Elongation at break (%)	WVP*	References
Cassava flour (5 g/100 g of solution)	Sorbitol (30 g /100 g of flour)	98	28.65	2629.72	5.20	0.002	Suppakul et al. 2013
	Sorbitol (40 g /100 g of flour)	99	9.25	971.20	28.24	0.002	
Rice flour (5 g/100 g of solution)	Glycerol (30 g /100 g of flour)	130	1.34	22.21	66.43	0.64	Dias et al. 2010
	Sorbitol (30 g /100 g of flour)	120	7.23	248.57	4.32	0.12	
Canola meal protein isolate (5 g/100 g of solution)	Glycerol (50 g /100 g of protein)	70	1.2	-	10.2	1.2	Chang and Nickerson 2015
	Sorbitol (50 g /100 g of protein)	90	10.2	-	3.9	0.5	
Flour from fruit residue (FRV) (8 g/100 g of solution)	-	176	0.020	0.001	29.40	-	Ferreira et al. 2015

(Contd.)

Table 2. (*Contd.*)

Polymer	Plasticizer	Thickness (μm)	Tensile Strength (MPa)	Young's Modulus (MPa)	Elongation at break (%)	WVP*	References
Flour from fruit residue (FRV) (8 g/100 g of solution) + Potato peel (4 g/100 g of solution)	-	263	0.092	0.003	36.10	-	
Flour from fruit residue (FRV) (8 g/100 g of solution)	-	242	27.00	3	31.38	2.45	Barbosa et al. 2011
Flour from fruit residue (FRV) (10 g/100 g of solution)	-	242	28.00	3	30.51	2.48	
Flour from fruit residue (FRV) (8 g/100 g of solution)		190	0.13	0.07	17.44	-	Fai et al. 2015
Gelatin capsule residue (40 g /100 g of solution)		133	2.57	202.21	282.75	0.566	Iahnke et al. 2015

(*Contd.*)

Material	Plasticizer						Reference
Gelatin capsule residue (54 g /100 g of solution) + Carrot residue (8.5 g /100 g of solution)		298	2.03	1066.94	106.11	0.594	
Cottonseed flour (6 g/100 g of solution)	Glycerol (10 g /100 g of flour)	124		Puncture strength (N): 0.5 Solubility (%): 100			Marquié et al. 1995
	Glycerol (30 g /100 g of flour)	124		Puncture strength (N): 0.10 Solubility (%): 100			
Gelatin extracted from fish skin waste (4 g/100 g of solution)	Glycerol (30 g /100 g of gelatin)	51	7.44	287.03	102.04	1.42	Hosseini et al. 2015
Gelatin extracted from fish skin waste (4 g/100 g of solution) + Chitosan nanoparticles (8 g/100 g of gelatin)	Glycerol (30 g /100 g of gelatin)	65	11.28	467.20	32.73	0.88	
"Tilapia" fish skin gelatin (3 g/100 g of solution) + Nanoclay (1 g/100 g of gelatin)	Glycerol (25 g /100 g of gelatin)	49	43.99	1144.85	9.20	0.08	Nagarajan et al. 2015

(Contd.)

Table 2. (*Contd.*)

Polymer	Plasticizer	Thickness (μm)	Tensile Strength (MPa)	Young's Modulus (MPa)	Elongation at break (%)	WVP*	References
"Tilapia" fish skin gelatin (3 g/100 g of solution) + Nanoclay (1 g/100 g of gelatin) + extract from coconut husk (0.4 g/100 g of gelatin)	Glycerol (25 g /100 g of gelatin)	52	38.82	1044.60	7.19	0.07	Nagarajan et al. 2015
Flour extracted from Pectin orange waste (6 g/100 g of solution)	Glycerol (25 g /100 g of pectin)	-	2.42	189.53	6.55	2.52	Oliveira et al. 2016
Chitosan extracted from shrimp waste (1 g /100 g of solution)	-	30	0.5	-	2.47	-	Martínez-Camacho et al. 2010
	Glycerol (20 g /100 g of chitosan)	20	0.25	-	5.06	-	

*WVP: Water vapour permeability (g mm h^{-1} m^{-2} kPa^{-1}).

less permeable to water vapour (1.2 g.mm/m^2.h.kPa)and more flexible (elongation at break of 10.2%) as compared to the films containing sorbitol (90% opacity, WVP of 0.5 g.mm/m^2.h.kPa and elongation at break of 3.9%).

Solid residues from fruit and vegetables present potential benefits for use as packaging material: they are a source of bioactive ingredients that can be incorporated into the packaging system and consumed with the food, supplementing functional attributes (Benbettaïeb et al. 2012; Ferreira et al. 2015; Pascall and Lin 2013). Few studies have dealt with the use of residues from the processing of fruit and vegetables to prepare edible films (Andrade et al. 2016; Ferreira et al. 2015; Fai et al. 2015; Iahnke et al. 2015; Barbosa et al. 2011). Ferreira et al. (2015) developed edible films from the flour of fruit residue (FRV) by processing whole fruits and vegetables used to prepare isotonic drink and then adding flour made from potato peels to FRV. These authors evaluated how the concentration of FRV and potato peel flour affected the properties and solubility of the film. Incorporation of the potato peel flour into FRV improved the tensile strength (from 0.020 until 0.092 MPa) because it increased the starch content. Without plasticizer, the FRV/potato peel flour films were highly hydrophilic, water soluble (87 to 95%) and flexible (elongation at break of 29 to 36%) due to the high content of soluble compounds such as sugar and globular proteins in FRV. These films performed well as film packaging for acerolas—they increased the shelf life by 50% as compared to control fruits. Andrade et al. (2016) also developed edible films from FRV by using two concentrations of flour: 8 and 10% (w/w). These films presented good visual appearance and a bright yellow colour due to the high content of carotenoids in the fruit waste. The films also had pronounced fruit flavour and visually smooth surface. The complex composition of FVR—fibre (48%; 9.6% soluble and 39% insoluble), carbohydrates (26%), proteins (9.5%) and lipids (5%) influenced the mechanical properties and the WVP of the FRV films. Regardless of the concentration of FRV, these films had poor mechanical resistance but higher flexibility due to the presence of sugars such as fructose and sucrose, which might play an important role in providing flexibility to the films. The WVP of FRV films was higher than the WPV of synthetic films containing high-density polyethylene (0.00079 g.mm/m^2.h.kPa) or low-density polyethylene (0.00310 g.mm/m^2.h.kPa). Fai et al. (2015) evaluated how the FRV film performed as coating applied to fresh-cut carrots by immersion and spraying. The FRV coating effectively delayed weight loss. Immersion and spray coating of shredded carrots led to 12 and 25% less weight loss as compared to the control sample. The FRV coating also prevented carrots from whitening during storage.

Residues of gelatin capsule and minimally processed carrot were used as raw materials to produce films and to evaluate the effect of the concentration of these residues on the antioxidant, physicochemical, barrier, optical and mechanical properties of the resulting films (Iahnke et al. 2015). The increasing of concentration of carrot residue improved the antioxidant activity, moisture content, water solubility, and WVP of the gelatin film, but elongation at break decreased. Meanwhile, films based on residues from gelatin capsules only had the greatest tensile strength (2.57 MPa) and elongation at break (282.75%). Because the carrot residue is rich in fibres, films containing this residue presented rough, irregular top surface morphology and yielded films with lower elongation at break (~68%).

Cottonseed flour is rich in proteins. Marquié et al. (1995) studied the filmogenic capacity of this residue by using different concentrations of glycerol. Content of glycerol below 10% w/w dry basis gave excessively brittle films, whereas a glycerol content of 30% made films sticky (Marquié et al. 1995). Decreasing glycerol content lowered the puncture strength of the films (glycerol 10%: 0.5 N and glycerol 30%: 0.1 N). These films were soluble in water independent of the concentration of glycerol (100%).

Fish waste is a valuable source of gelatin to produce films (Badii and Howell 2006). Gelatin from fish waste has very good film-forming characteristics and gives transparent and flexible films that act as effective barriers of UV light and oxygen (Gennadios et al. 1994; Jongjareonrak et al. 2008; Pérez-Gago 2001). Due to their hydrophilic nature, these films have poor water vapour barrier and mechanical properties, which limits their use as packaging materials (Hosseini et al. 2013).

Nowadays, one of the most effective alternatives to improve the barrier and mechanical properties of synthetic and natural packaging materials is to produce nanocomposites (Bae et al. 2009; Farahnaky et al. 2014; Nagarajan et al. 2015). Hosseini et al. (2013) studied how the addition of different amounts of chitosan nanoparticles (CSNPs) affected the properties of gelatin films. The barrier and mechanical properties of the resulting composite films improved after formation of gelatin/CSNPs nanocomposites. Subsequently, films based on gelatin/CSNPs were incorporated with oregano essential (0.4, 0.8 and 1.2% w/v) oil, and the effect of this incorporation on the polymer matrix and antimicrobial activity was assessed. Addition of oregano oil to the films markedly reduced film WVP with a maximum permeability reduction of 32% in the film containing 0.8% (w/v) of incorporated oil. Addition of an antimicrobial agent increased the flexibility (elongation at break rose from 44.7% until 151.8%) but decreased the tensile strength (from 10.6 until 3.3 MPa) of the film. A minimum concentration of 1.2% (w/v) of oregano essential oil was necessary to ensure antibacterial efficacy.

Nagarajan et al. (2015) evaluated how the addition of ethanolic extract from coconut husk (EECH) at 0–0.4 % (w/w, on protein basis) impacted the properties of nanocomposite films consisting of tilapia skin gelatin and gelatin/Cloisite Na$^+$. The nanocomposite film incorporated with 0.4% EECH (w/w) showed the lowest WVP (from 2.19 until 1.94 10^{-11} g.m.m^{-2}.s^{-1}.Pa^{-1}), possibly due to the larger number of interactions between the functional group of gelatin and phenolics. However, these films showed lowest tensile strength (38.8 MPa) and elongation at break (7.2%) because incorporation of EECH afforded a film with heterogeneous structure and discontinuous areas.

Pectin films from orange waste developed by Oliveira et al. (2016) showed poor elongation (6.65%) and good water vapour barrier property (2.52g.mm/m^2.h.kPa).

Chitosan edible films produced from shrimp waste have antibacterial and antifungal properties that qualify them for use in food protection. However, their use is limited by their weak mechanical properties and gas and water vapour permeability (Elsabee and Abdou 2013; Martínez-Camacho et al. 2010). Martínez-Camacho et al. (2010) produced films using commercial chitosan and chitosan obtained from shrimp waste as raw material.

Lignocellulosic Biomass as Filler for Biopolymeric Film

Pure polysaccharide or protein films generally exhibit adequate mechanical and overall optical properties, but they are highly sensitive to moisture and have high WVP. Recently, there has been growing interest in adding reinforcement materials like polymers with hydrophobic character; for example, fibres and nanoparticles, to films in order to improve their mechanical and water vapour barrier properties (Montaño-Leyva et al. 2013; Satyanarayana et al. 2009; Reddy and Yang 2005). The use of natural fibres as additives in materials can potentially enhance their performance and technological application due to their low cost, large abundance, excellent biodegradability and high specific strength (Tserki et al. 2009). Furthermore, the low density of natural fibres reduces composite mass (Bledzki and Gassan 1999). Many kinds of fibres exist for this application, but cellulose fibres are preferable due to their low cost and satisfactory rigidity and biodegradability, which make these fibres a suitable raw material to reinforce films. Most works have focused on how the addition of fibre affects the mechanical properties of composite films and have reported a similar behavior—incorporation of nano/microfibre increases the tensile strength and elasticity modulus of composite films and decreases their elongation capacity (Avérous et al. 2001; Curvelo et al. 2001; Dufresne and Vignon 1998; Müller et al. 2009, among others).

To obtain cellulose micro- or nanocrystals, it is usually necessary to apply several pretreatments to remove hemicelluloses and lignin from the structure of the fibre. Physical, biological, chemical and physicochemical methods can be used separately or in combination to produce cellulose micro/nanofibres for later addition to starch or protein films (Andrade-Mahecha et al. 2015; De Campos et al. 2013; Flauzino Neto et al. 2013; Hendriks and Zeeman 2009; Mosier et al. 2005; Oh et al. 2015; Singh et al. 2015; Sun and Cheng 2002; Taherzadeh and Karimi 2008). Table 3 shows some polymeric films incorporated with nanofibres obtained from agroindustrial residues.

Wicaksono et al. (2013) produced cellulose nanofibres from cassava bagasse and added them to tapioca films. Cellulose nanofibres were obtained from cassava bagasse through a series of chemical (alkali treatment, bleaching and acid hydrolysis) and mechanical (high-velocity mixer) treatments. These

Table 3. Mechanical properties of films added with lignocellulosic material from agroindustrial residues.

Polymer	Plasticizer	Lignocellulosic material	Fibres	Tensile Strength (MPa)	Young's Modulus (MPa)	Elongation at break (%)	References
Tapioca starch (4 g/100 g of solution)	Sorbitol (62.5 g/100 g of starch)	Cellulose Nanofibres from Cassava Bagasse 5-8 nm and length of several micrometers	0 wt% (dry basis) 4 wt% (dry basis)	1.45 2.75	- -	33.89 19.57	Wicaksono et al. 2013
Cassava starch (7 g/100 g of solution)	Glycerol/sorbitol mixture (1:1) (30 g/100 g of starch)	Cellulose Nanofibrils from Cassava Bagasse Thickness: 2–11 nm Length: 360–1700 nm	0 wt% (dry basis) 20 wt% (dry basis)	4.1 4.2	44.5 49.1	83.3 92.4	Teixeira et al. 2009
Poly(vinyl alcohol) (5 g/100 g of solution)	-	Nanocellulose from sugarcane bagasse Width: 71 nm	0 wt% (dry basis) 7.5 wt% (dry basis)	40 60	- -	135 185	Mandal and Chakrabarty 2014
Cornstarch (4 g/100 g of solution)	Glycerol (20 g/100 g of starch)	Cellulose nanocrystals from sugarcane bagasse length: 247 nm diameter: 10 nm	0 wt% (dry basis) 3 wt% (dry basis)	2.8 17.4	112 520	44.9 9.1	Slavutsky and Bertuzzi 2013
Corn starch (5 g/100 g of solution)	Glycerol (60 g/100 g of starch)	Cellulose nanocrystals from rice straw diameter from 10 to 12 nm	0 wt % (dry basis) 15 wt % (dry basis)	10.0 26.0	327 896	33.1 3.6	Agustin et al. 2014

treatments partially removed hemicelluloses and lignin from the structure of the fibre and improved its crystallinity by 14.52 and 39.37% in the case of cassava bagasse and cellulose nanofibres, respectively. Addition of cellulose nanofibres to tapioca films increased the tensile strength (~90%) and decreased elongation at break (70%).

Teixeira et al. (2009) produced cellulose nanofibrils from the residue of cassava bagasse by acid hydrolysis and reinforced the thermoplastic cassava starch with the nanofibrils, to obtain starch with decreased hydrophilic character (~15%). The presence of sugar in the suspension of cassava bagasse nanofibrils probably increased elongation at break, resulting in a weak reinforcement effect. Films added with 20 wt% (dry basis) of nanofibrils had unaltered tensile strength but, as mentioned above, elongation at break increased by ~11% as compared to nonreinforced cassava starch films.

Mandal and Chakrabarty (2014) reinforced poly(vinyl alcohol) with nanocellulose obtained by acid hydrolysis of cellulose from sugarcane bagasse. Reinforced poly(vinyl alcohol) films with 7.5 wt% (dry basis) nanocellulose had 50% higher tensile strength and 30% higher elongation at break as compared to the control poly(vinyl alcohol) films. Slavutsky and Bertuzzi (2013) also produced cellulose nanocrystals by acid hydrolysis of sugarcane bagasse residue aiming to use them to reinforce commercial corn starch films. These authors noted that starch films reinforced with 20 wt% (dry basis) of cellulose nanocrystals had 6.2 times higher tensile strength and 4.9 times lower elongation at break as compared to nonreinforced starch films.

To reinforce corn starch films, Agustin et al. (2014) produced cellulose nanocrystals from rice straw residue after treatments like delignification, sulfuric acid hydrolysis and sonication. The nanocrystals reinforced the starch matrix (the reinforced film had 2.6 times higher tensile strength than the control film), although the resulting films were less thermally stable. Addition of cellulose nanocrystals shifted the temperature at maximum weight loss to lower value (323.75 and 277.36 °C for nonreinforced and reinforced films, respectively). The decreased thermal stability of the reinforced film can be ascribed to the inherent low thermal stability of nanocrystals obtained by sulfuric acid hydrolysis.

At the Chemistry Department of FFCLRP, University of Sao Paulo (Ribeirão Preto, SP, Brazil), we have studied two types of agroindustrial biomass: residue from the extraction of turmeric dye, as a source of polymers for film formation, and soybean straw, as a filler in starch films. This chapter describes some of our results.

Biopolymeric Film from the Residue Remaining from Extraction of Turmeric Dye

Dye industries extract oleoresin from dried turmeric rhizomes with solvents such as acetone, methanol, ethanol or isopropanol. The residues obtained after extraction consist mainly of starch (52.4%, with starch corresponding to 37% amylose and 63% amylopectin) and fibres (25.4%) (Table 1). Residues may also contain residual levels of curcuminoids (15 mg.L^{-1} curcumin, 25 mg.L^{-1} demethoxycurcumin and 22 mg.L^{-1} bisdemethoycurcumin) (Maniglia et al. 2014, 2015). The turmeric oleoresin contains 30 to 40% of dyes expressed as curcumin as well as 15 to 25% of volatile oil. This oleoresin is a highly viscous orange-brown product that acquires a bright yellow color after dilution (Pereira and Stringheta 1998).

In the laboratory of the Chemistry Department of FFCLRP-USP, our research group used the Soxhlet method to extract turmeric dye (oleoresin) with ethanol/isopropanol (1:1, v/v) as solvent. This procedure was employed so as to simulate an industrial process and obtain a residue to develop biodegradable films. Initially, we tried to extract starch from this residue, but the milling and sieving processes were not successful—starch present in the residue obtained after extraction of turmeric dye may have undergone pregelatinization due to the temperature and duration of the process (~47 °C and 3 h, respectively), in agreement with observations made by Kuttigounder et al. (2011) for cured-dried turmeric powder. Therefore, we decided to obtain the flour from the turmeric residue by the wetting milling method. Three of the fractions were achieved by wet milling in water and sieving through 80- (F1), 200- (F2), and 270- (F3) mesh screens, which was followed by drying for 24 h in an oven at 35 °C. The liquid fraction that passed through the sieves was collected and centrifuged at 6000 x g and 10 °C. The separated solids were dried at 35 °C for 24 h (F4) (Maniglia et al. 2014, 2015). Table 4 depicts the chemical composition of the turmeric flour fractions. Because fraction F2 showed higher starch content and lower cellulose, hemicellulose and lignin contents than fractions F1, F3 and F4, we selected this fraction to produce films.

Furthermore, fraction F2 also contained curcuminoids (curcumin: 10 mg.L^{-1}, demethoxycurcumin: 17 mg.L^{-1}, and bisdemethoxycurcumin: 15 mg.L^{-1}) and antioxidant activity (66.6%), but at a lower level than the turmeric residue.

Figure 2 shows the SEM micrographs of the turmeric flour fractions. All the flour fractions contained starch granules (indicated by an arrow). The starch granules formed agglomerates in all fractions; there were no free starch granules in the flour. These agglomerates hindered the access of water inside the granule and prevented gelatinization of starch, which required higher temperature (85-90 °C) and longer heating time (4 h) for gelatinization and formation of the film (Maniglia 2012). To produce biopolymer films, it was necessary to add a plasticizer to render biopolymer-based films more flexible, processable and extensible. The type and amount of plasticizer affect the mechanical properties, oxygen and water vapour barrier properties, optical clarity, degree of crystallinity and glass transition temperature (Tg) of the film.

Table 4. Chemical composition (g/100 g of turmeric on dry basis) of the different turmeric flour fractions obtained from the residues remaining after extraction of turmeric dye with ethanol/isopropanol by the Soxhlet extraction method.

Material	Moisture*	Protein	Lipid	Ashes	Starch	Cellulose	Hemicellulose	Soluble Lignin	Non-soluble lignin
Residue	5.44 ± 0.09^d	9.33 ± 0.01^b	1.35 ± 0.03^b	11.51 ± 0.14^b	52.37	8.75 ± 0.05^c	5.37 ± 0.29^d	2.18 ± 0.04^a	9.14 ± 0.59^c
F1	7.52 ± 0.28^a	8.56 ± 0.11^d	0.68 ± 0.02^e	2.54 ± 0.09^f	48.16	10.93 ± 0.34^a	15.14 ± 0.19^a	2.23 ± 0.08^a	11.76 ± 0.50^a
F2	7.73 ± 0.42^a	9.25 ± 0.11^b	0.88 ± 0.02^d	5.80 ± 0.26^d	69.02	2.61 ± 0.24^e	3.87 ± 0.95^e	1.83 ± 0.05^c	6.75 ± 0.57^e
F3	6.93 ± 0.35^b	9.14 ± 0.09^c	0.72 ± 0.02^e	5.40 ± 0.06^e	60.76	3.43 ± 0.36^d	10.28 ± 1.35^b	1.97 ± 0.07^b	8.30 ± 0.38^d
F4	6.48 ± 0.25^c	10.60 ± 0.02^a	4.10 ± 0.10^a	11.97 ± 0.09^a	47.38	7.93 ± 0.32^b	6.92 ± 0.47^c	1.99 ± 0.03^b	9.10 ± 0.30^c

*Moisture content expressed on wet basis (g/100 g of wet turmeric).
Turmeric residue flour fractions F1 (sieved through 80-mesh screen), F2 (sieved through 200-mesh screen), F3 (sieved through 270-mesh screen) and F4 (centrifuged fraction).

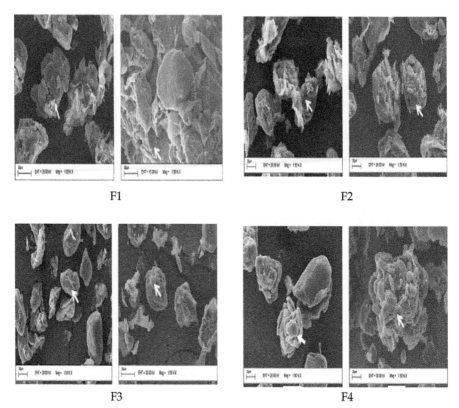

Fig. 2. Microstructure of the turmeric flour fractions obtained from the residues remaining after turmeric dye extraction with ethanol/isopropanol by Soxhlet. The arrows show starch granules (Maniglia 2012; Maniglia et al. 2015).

Therefore, we evaluated the effect of the type (glycerol and sorbitol) and concentration of plasticizer on the mechanical and water vapour barrier properties as well as on the solubility and antioxidant activity of turmeric films. Concentrations of glycerol of 19, 22 and 30 g/100 g of flour were used, and concentrations of sorbitol of 20, 25 and 30 g/100 g of flour were employed. The turmeric flour films were prepared by the casting method, by using a suspension of 5 g of flour/100 g, heating for 4 h and 12,000-rpm cycles at every hour. After addition of the plasticizer (glycerol or sorbitol), the suspension was heated for 20 min and poured onto acrylic plates (a weight of 0.15 g.m^{-2} was maintained).

Finally, the plates containing the filmogenic solution were dried at 35 °C, for 7 h, in an oven with forced circulation. All the turmeric flour films were yellowish and did not present bubbles or cracks (Maniglia et al. 2014). Table 5 lists the mechanical properties, solubility in water, moisture content and WVP of the turmeric flour films plasticized with different concentrations of glycerol or sorbitol.

Table 5. Mechanical properties and functional properties of turmeric residue flour films plasticized with glycerol (Maniglia 2012).

Films	Thickness (µm)	Elongation at break (%)	Tensile Strength (MPa)	Young's modulus (MPa)	Solubility (%)	WVP* (*10⁻¹⁰g/m.s.Pa)
Glycerol 19%	91.3 ± 5.9[bA]	1.4 ± 0.1[cB]	24.0 ± 0.0[cB]	1178 ± 23[aB]	9.9 ± 0.6[bB]	1.6 ± 0.1[cA]
Glycerol 22%	103.5 ± 2.5[aA]	2.5 ± 0.1[bA]	19.5 ± 0.1[bB]	816 ± 28[bB]	29.7 ± 1.9[aB]	1.9 ± 0.1[bA]
Glycerol 30%	111.3 ± 13.3[aA]	4.1 ± 0.0[aA]	15.1 ± 0.1[aB]	752 ± 11[cB]	33.8 ± 3.6[aB]	2.7 ± 0.1[aA]
Sorbitol 20%	65.9 ± 1.2[cB]	1.8 ± 0.2[cA]	38.5 ± 0.2[aA]	2718 ± 19[aA]	35.1 ± 2.3[bA]	0.9 ± 0.1[aB]
Sorbitol 25%	87.9 ± 2.6[aB]	2.8 ± 0.2[bA]	20.8 ± 0.1[bA]	886 ± 25[bA]	37.1 ± 2.1[aA]	0.6 ± 0.1[bB]
Sorbitol 30%	77.3 ± 1.3[bB]	4.1 ± 0.3[aA]	16.2 ± 0.3[cA]	835 ± 2[cA]	43.1 ± 5.2[aA]	0.3 ± 0.1[cB]

*WVP (water vapour permeability).

a, b, c: Different small letters in the same column represent a significant difference between the various concentrations for the same plasticizer (p<0.05).

A, B: Different caps letters represent a significant difference between films with different plasticizers and close concentrations (p<0.05).

For both plasticizers, increasing plasticizer concentration augmented elongation at break and decreased the tensile strength and Young's modulus. However, regardless of the type (glycerol or sorbitol) and concentration of plasticizer, the turmeric flour films were less elongable than other flour films, such as cellulose fibre flour films (4.0%) (Dias et al. 2011), amaranth flour film of *caudatus* and *cruentus* species (83.7 and 51.9%) (Tapia-Blácido et al. 2011), achira flour films (22%) (Andrade-Mahecha et al. 2012) and banana flour films (24.2%) (Pelissari et al. 2013). Thus, the complex structural configuration of the macromolecules (cellulose, hemicellulose, lignin, protein and starch) present in the flour and which formed the film matrix minimized the plasticizing effect of glycerol and sorbitol.

On the other hand, the type and concentration of plasticizer influenced the solubility of the turmeric flour film. Films plasticized with sorbitol were more soluble in water than films plasticized with glycerol. Furthermore, increasing plasticizer content (glycerol or sorbitol) increased the solubility of the film. Turmeric flour films plasticized with 19% glycerol were less soluble as compared to other films. The smaller solubility of the films plasticized with glycerol was possibly due to the lower molecular weight of this plasticizer. The smaller size of glycerol allowed for its better incorporation inside the film matrix and improved its interaction with the structures present therein, making the plasticizer less available for binding with water molecules (Tapia-Blácido et al. 2016). In contrast to glycerol, which contains three hydroxyls, sorbitol bears six hydroxyls, which made the latter plasticizer more hydrophilic as compared to glycerol and afforded more soluble films. However, these films were less permeable to water vapour than the films plasticized with glycerol because sorbitol probably filled the turmeric film matrix, decreasing pore size and water permeation through the films.

The WVP of turmeric films plasticized with glycerol increased with increasing plasticizer content, whereas the opposite trend occurred in the case of turmeric films plasticized with sorbitol. High glycerol content can disrupt the matrix, increasing mobility and decreasing the density between molecules, to facilitate the transport of water and gases through the film (McHugh and Krochta 1994).

Figure 3 illustrates the microstructure of turmeric flour films plasticized with different concentrations of glycerol and sorbitol as obtained by SEM. All the films had a rough surface. The presence of pores, agglomerates and swollen starch granules on the surface of the film was evident. The starch-fibre, protein-fibre and fibre-fibre interactions in the film matrix may have led to a rougher surface and culminated in a discontinuous film matrix with lower elongation at break. The micrographs of the cross-section of the films plasticized with lower concentration of glycerol and sorbitol revealed a more compact and cohesive structure than the structure observed at higher concentrations of plasticizer. This explains the lower elongation and higher tensile strength of these films. Comparison of the effect of plasticizers showed that the turmeric film plasticized with sorbitol displayed a more compact

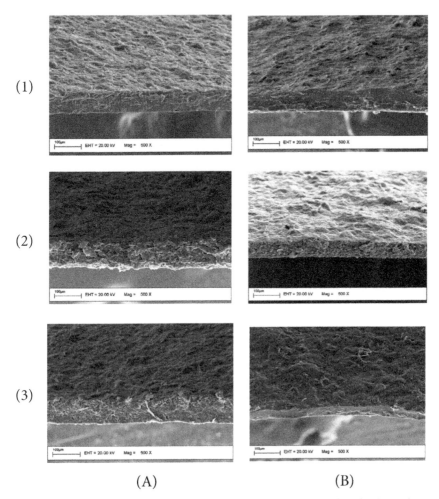

(A)   (B)

Fig. 3. SEM microstructure of turmeric flour film plasticized with glycerol
(A): (1) 19%, (2) 22% and (3) 30%, and plasticized with sorbitol
(B): (1) 20%, (2) 25% and (3) 30%.

structure than the turmeric film plasticized with glycerol, which explained the higher mechanical resistance and lower WVP of the former film.

Maniglia et al. (2014, 2015) optimized the conditions to process turmeric flour films plasticized with 22% of glycerol and 30% of sorbitol. The optimized conditions for the flour film obtained from turmeric residue plasticized with sorbitol and glycerol were heating temperature of 86.7 and 85.1°C and pH of 8.5 and 8.1, respectively (Maniglia et al. 2014, 2015). These optimized films presented high antioxidant activity (51% and 46% for sorbitol and glycerol, respectively) and contained curcuminoids as observed in Table 6. The concentration of curcuminoids in the films was lower as compared with the concentration of curcuminoids in the turmeric residue and in the

Table 6. Curcuminoids concentration in the residue obtained after extraction of turmeric dye, of the turmeric residue flour (F2) and of the optimized films plasticized with glycerol and sorbitol.

Samples	Curcumin (mg/L)	Demethoxycurcumin (mg/L)	Bisdemethoxycurcumin (mg/L)
Turmeric residue	15.02 ± 1.11	25.50 ± 2.47	22.19 ± 0.31
Turmeric residue flour (F2)	10.26 ± 0.10	16.75 ± 0.31	14.86 ± 0.91
Turmeric flour film plasticized with glycerol	2.18 ± 0.02	1.41 ± 0.00	5.81 ± 0.11
Turmeric flour film plasticized with sorbitol	3.70 ± 0.04	0.8 ± 0.13	0.10 ± 0.00

Maniglia et al. 2014, 2015.
Calibration curve: Curcumin: 17.19–0.54 (10^{-3} g.L^{-1}), R^2 = 0.992; Bisdemethoxycurcumin: 35.00–0.10. (10^{-3} g/L^{-1}), R^2 = 0.999; Demethoxycurcumin: 48.40–0.19 (10^{-3} g. L^{-1}), R^2 = 0.999.

F2 turmeric residue flour, indicating that the film production process caused loss of curcuminoids. The turmeric residue obtained by curcumin extraction exhibited interesting filmogenic capacity and antioxidant properties. These films could serve as food packaging sensitive to oxidation due to their antioxidant activity.

Pretreated Soybean Straw as Filler in Cassava Starch

In our laboratory, soybean straw residues (SR: straw and pods) underwent different chemical pretreatments, aimed at their further addition to cassava starch (CS) films and evaluation of their effect on the properties of the film. Four different SR chemical pretreatments were used, including alkali followed by bleaching. Table 7 summarizes the sequence of pre-treatments. Depending on the temperature of the reaction, the treatments were classified as severe or mild:

(i) Severe: SR treatment with NaOH at 5% (T1) or 17.5% (T2) (w/v) at 90 °C for 1 h, repeated twice. The solution was then brought to room temperature and rinsed to neutralization. The fibres were bleached with 0.7% acetic acid and 3.3% sodium chlorite ($NaClO_2$) aqueous solution under stirring at 75 °C for 4 h;

(ii) Mild: SR treatment with NaOH at 5% (T3) or 17.5% (T4) (w/v) at 30 °C for 15 h. After this period, the solution was brought to room temperature and rinsed to neutralization. The resulting fibres were bleached in a mixture of 4% H_2O_2 (w/v) and 2% NaOH (w/v) at 90 °C for 3 h. $MgSO_4$ $7H_2O$ at 0.3% (w/v) was added as stabilizer. The solution was cooled to room temperature, and the fibres were filtered and washed with distilled water until neutral pH, followed by rinsing with ethanol and acetone.

Treated and nontreated SR were sieved through 100-mesh sieves (Tyler series, 150 µm) and re-suspended in water 24 h before film processing. The film solution was prepared by using 5% (w/w) cassava starch, 25% (g/100 g of starch) glycerol and 1% (g/100 g of starch) SR fibres (nontreated and

Table 7. Chemical pretreatments of soybean residues.

Treatment		Alkali	Bleaching
Severe	T1	NaOH 5% (1 h at 90 °C) - 2x	$NaClO_2$ 3.3% + CH_3COOH 0.7% (3 h at 75 °C)
	T2	NaOH 17.5% (1 h at 90 °C) - 2x	$NaClO_2$ 3.3% + CH_3COOH 0.7% (3 h at 75 °C)
Mild	T3	NaOH 5% (15 h at 30 °C)	H_2O_2 4% + NaOH 2% + $MgSO_4$ $7H_2O$ 0.3% (3 h at 90 °C)
	T4	NaOH 17.5% (15 h at 30 °C)	H_2O_2 4% + NaOH 2% + $MgSO_4$ $7H_2O$ 0.3% (3 h at 90 °C)

treated). The surface micrograph of the CS/SR fibre films obtained by SEM showed that the SR fibres were homogeneously distributed into the CS film matrix, regardless of the type of SR fibre, nontreated (Fig. 4.1A) or chemically treated via T1 (Fig. 4.2A), T2 (Fig. 4.3A), T3 (Fig. 4.4A), or T4 (Fig. 4.5A). The micrographs of the cross-section of the CS films (Figs. 4.1B, 4.2B, 4.3B, 4.4B and 4.5B) revealed that the SR fibres were included in the matrix. However, the SR fibres disrupted the matrix because they prevented amylose-amylose, amylopectin-amylopectin and amylose-amylopectin interactions.

Table 8 summarizes the mechanical properties (tensile strength and elongation at break), moisture, solubility in water and WVP of CS films added with nontreated or chemically treated SR. Concerning thickness, there were no statistical differences among the samples. CS films added with nontreated SR presented lower tensile strength values as compared to control films (CS), whereas elongation at break was similar for both films. Meanwhile, elongation of the films containing chemically treated SR fibres improved by about 28 to 62%, whereas the tensile strength decreased regardless of the chemical treatment. We had expected that incorporation of natural fibres into cassava starch films would increase the tensile strength and reinforce the material, as mentioned by Versino and García (2014). The SR chemical treatments induced production of sugar monomers, which possibly acted as plasticizer and reduced the mechanical resistance while increasing the flexibility of the CS/SR fibre films. The interaction of sugar with the amylose and amylopectin in CS enhanced the mobility of the chains, increasing the elongation at break of the CS/SR fibre films.

Concerning functional properties, CS films added with SR fibre treated via T3 and T4 presented lower solubility values and were thus less hydrophilic. Compared to the other treatments, T3 and T4 (bleaching with H_2O_2) generated SR fibres with lower lignin content, so cellulose OH-groups were more available to interact with glycerol, amylose and amylopectin, resulting in fewer sites available for binding with water. The various film formulations did not differ in terms of WVP: $2.4 \pm 0.4 \times 10^{-10}$ g/m.s.Pa. Therefore, nontreated and chemically treated SR fibres were well incorporated into the CS/SR fibre film, as verified by SEM (Fig. 4).

Figure 5 corresponds to the DSC thermograms of CS films added with nontreated or chemically treated SR and of the control CS film. No visible glass transition occurred in the studied temperature range for any of the samples, which indicated good inclusion of SR fibres into the CS/SR fibre film matrix. Furthermore, the thermograms evidenced an endothermic peak from 69.2 to 84.2 °C, which could be related to gelatinization of starch. The temperature of the control CS film peaked at 84.2 °C. The addition of SR fibres to CS decreased this value, pointing to the lower thermal stability of the CS/SR fibre film. This reduction was less pronounced for films reinforced with nontreated SR (81.6 °C) and SR treated via T4 (79.4 °C) as compared to the control CS film (84.2 °C). Savadekar and Mhaske (2012) observed the same trend when they incorporated nanofibres at a concentration of 0.5% into starch films.

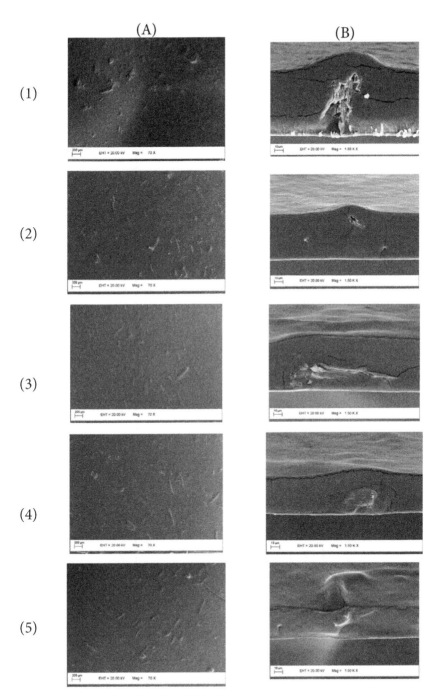

Fig. 4. SEM surface (A) and cross section (B) of cassava starch film added with (1) nontreated soybean straw and soybean chemically treated via (2) T1, (3) T2, (4) T3 and (5) T4.

Table 8. Properties of cassava starch (CS) films containing 5% of CS, 25% of glicerol (g/100g of CS) and 1% of soybean residue (SR) (g SR/100g of CS), nontreated and chemically treated with T1, T2, T5 and T6, as described in Table 7.

Film	Thickness (μm)	Tensile strength (MPa)	Elongation at break (%)	Moisture (%)	Solubility in water (%)	WVP (10^{-10} g/m.s.Pa)
CS 5%	77 ± 7[a]	4.8 ± 0.4[a]	28 ± 6[b]	15.1 ± 0.3[a]	18.0 ± 2.5[a]	2.72 ± 0.16[a]
CS 5%/SR 1%	89 ± 13[a]	2.5 ± 0.1[bc]	23 ± 6[b]	14.2 ± 0.5[a]	17.1 ± 1.1[a]	2.02 ± 0.06[a]
CS 5%/SR-T1 1%	70 ± 8[a]	2.2 ± 0.2[cd]	58 ± 6[a]	14.5 ± 0.4[a]	17.4 ± 2.9[a]	2.39 ± 0.18[a]
CS 5%/SR-T2 1%	70 ± 11[a]	2.7 ± 0.3[b]	38 ± 29[ab]	14.2 ± 1.3[a]	13.9 ± 0.5[ab]	2.76 ± 0.24[a]
CS 5%/SR-T3 1%	79 ± 2[a]	2.5 ± 0.2[bc]	55 ± 7[a]	12.7 ± 1.7[a]	11.7 ± 1.1[b]	2.33 ± 0.31[a]
CS 5%/SR-T4 1%	78 ± 14[a]	1.8 ± 0.2[d]	62 ± 8[a]	13.9 ± 1.6[a]	12.1 ± 1.5[b]	2.43 ± 0.64[a]

[a-e] Means with different superscript letters in the same column are statistically different ($p < 0.05$).

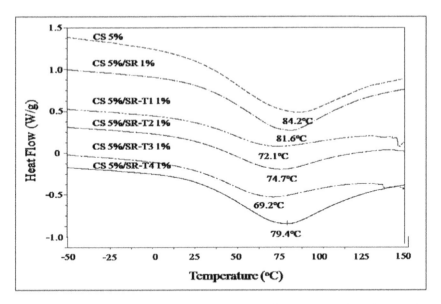

Fig. 5. DSC of cassava starch films added with 1% of soybean residues fibres (g/100 g of starch), nontreated or chemically treated. Control (CS 5%), CS film added with nontreated fibres (CS 5%/SR 1%) and CS films added with chemically treated fibres (T1, T2, T3 and T4).

Starch granules are sensitive to retrogradation during film storage because they contain high level of amylose (17%). The incorporation of fibres into starch films does not play a role in the retrogradation of amylose. However, the chemical treatment of cellulosic materials usually changes the physical and chemical structure of fibres, making more OH groups available to interact with amylose and amylopectin in the film matrix. This decreases the retrogradation of starch during storage, thereby reducing the gelatinization temperature.

Conclusions

Agroindustrial biomass produced during harvesting or large-scale agroindustrial processes are an important source of natural polymers and bioactive compounds with antimicrobial and antioxidant properties for application in film biopackaging. Polysaccharides and proteins present in agroindustrial residues can serve as film-forming matrix, whereas the lignocellulosic material can act as film matrix filler. Compared to conventional materials, the presence of antimicrobial and antioxidant compounds in agroindustrial biomass provides extra benefits to biopackaging. Furthermore, application of agroindustrial biomass as raw material to produce biopolymeric film is advantageous over the use of petroleum-based materials because they are biodegradable, inexpensive, abundant and renewable, contributing to the reduction of environmental pollution. However, improving the water vapour barrier and mechanical properties of films based on agroindustrial biomass is necessary to make the use of these residues in the packaging industry viable. Hence, there still is a lot of room for research in this area to evaluate the potential of other unexplored residues and propose strategies to improve the performance of known residues for film production.

Acknowledgments

The authors would like to thank Fundação de Amparo à Pesquisa do Estado de São Paulo (São Paulo Research Support Foundation–FAPESP) and CAPES (Coordenacão de Aperfeiçoamento de Pessoal de Nível Superior) for financial support. The authors would like to express their gratitude to EMBRAPA Soja (Brazil), for providing soybean residues (straw and pods).

References

Abdou, E.S., K.S.A. Nagy and M.Z. Elsabee. 2007. Extraction and characterization of chitin and chitosan from local sources. Bioresource Technol. 99(5): 1359–1367.

Acosta, S., A. Jiménez, A. Chiralt, C. González-Martinez and M. Cháfer. 2013. Mechanical, barrier and microstructural properties of films, based on cassava starch gelatin blends: Effect of aging and lipid addition. Inside Food Symposium, Leuven, Belgium, 1–6.

Agama-Acevedo, E., S.L. Rodriguez-Ambriz, F.J. García-Suárez, F. Gutierrez-Méraz, G. Pacheco-Vargas and L.A. Bello-Pérez. 2014. Starch isolation and partial characterization of commercial cooking and dessert banana cultivars growing in Mexico. Starch-Stärke 66(3–4): 337–344.

Agustin, M.B., B. Ahmmad, S.M.M. Alonzo and F.M. Patriana. 2014. Bioplastic based on starch and celulose nanocrystals from rice straw. J. Reinf. Plast. Compos. 33(24): 2205–2213.

Andrade, R.M.S, M.S.L. Ferreira and E.C.B.A. Gonçalves. 2016. Development and characterization of edible films based on fruit and vegetable residues. J. Food Sci. 81(2): E412–E418.

Andrade-Mahecha, M.M., D.R. Tapia-Blácido and F.C. Menegalli. 2012. Development and optimization of biodegradable films based on achira flour. Carbohydr. Polym. 88(2): 449–458.

Andrade-Mahecha, M.M., F.M. Pelissari, D.R. Tapia-Blácido and F.C. Menegalli. 2015. Achira as a source of biodegradable materials: Isolation and characterization of nanofibers. Carbohydr. Polym. 123: 406–415.

Anselmo, G.C.S., M.E.R.M.C. Mata and E. Rodrigues. 2008. Comportamento Higroscópico do extrato seco de urucum (*Bixa orellana L.*). Ciência Agrotécnica. 32(6): 1888–1892.

Arvanitoyannis, I.S. and A. Kassaveti. 2009. Starch–cellulose blends. pp. 19–50. In: Yu, Long (ed.). Biodegradable Polymer Blends and Composites from Renewable Resources. John Wiley & Sons. Hoboken, NJ, USA.

Avérous, L., C. Fringant and L. Moro. 2001. Plasticized starch-cellulose interactions in polysaccharides composites. Polymer 42(15): 6565–6572.

Azeredo, H.M.C., L.H. Mattoso, D. Wood, T.G. Williams, R.J. Avena-Bustillos and T.H. McHugh. 2009. Nanocomposite edible films from mango puree reinforced with cellulose nanofibers. J. Food Sci. 74(5): N31–N35.

Badii, F. and N.K. Howell. 2006. Fish gelatin: Structure, gelling properties and interaction with egg albumen proteins. Food Hydrocolloid. 20(5): 630–640.

Bae, H.J., H.J. Park, S.I. Hong, Y.J. Byun, D.O. Darbya, R.M. Kimmela, et al. 2009. Effect of clay content, homogenization RPM, pH, and ultrasonication on mechanical and barrier properties of fish gelatin /montmorillonite nanocomposite films. LWT - Food Sci. Technol. 42(6): 1179–1186.

Banerjee, S., R. Sen, R.A. Pandey, T. Chakrabarti, D. Satpute, B.S. Giri, et al. 2009. Evaluation of wet air oxidation as a pretreatment strategy for bioethanol production from rice husk and process optimization. Biomass Bioenerg. 33(12): 1680–1686.

Barbosa, H.R., D.P.R. Ascheril, J.L.R. Ascheri and C.W.P. de Carvalho. 2011. Permeabilidade, estabilidade e funcionalidade de filmes biodegradáveis de amido de caroço de jaca (*Artocarpus heterophyllus*). Revista Agrotec. 2: 73–88.

Barten, T.J. 2013. Evaluation and prediction of corn stover biomass and composition from commercially available corn hybrids in Master Thesis in Crop Production and Physiology, Iowa State University, USA.

Benbettaïeb, N., M. Kurek, S. Bornaz and F. Debeaufort. 2014. Barrier, structural and mechanical properties of bovine gelatin–chitosan blend films related to biopolymer interactions. J. Sci. Food Agri. 94: 2409–2419.

Bertol, T.M. and G.J.M.M. Lima. 1999. Níveis de resíduos industrial de fécula da mandioca na alimentação de suínos em crescimento e terminação. Pesquisa Agropecuária Brasileira 34(2): 243–248.

Binod, P., R. Sindhu, R.R. Singhania, S. Vikram, L. Devi, S. Nagalakshmi, et al. 2010. Bioethanol production from rice straw: An overview. Bioresource Technol. 101(13): 4767–4774.

Bismarck, A., S. Mishra and T. Lampke. 2005. Plant fibers as reinforcement for green composites. pp. 39–97. *In*: Mohanty, A.K., M. Misra and L.T. Drzal (eds.). Natural Fibers, Biopolymers and their Biocomposites. CRC Press, Boca Raton, USA.

Bledzki, A.K. and J. Gassan. 1999. Composites reinforced with cellulose-based fibres. Progress Polymer Sci. 24(2): 221–274.

Cabrera, E., M.J. Muñoz, R. Martín, I. Caro, C. Curbelo and A.B. Díaz. 2015. Comparison of industrially viable pretreatments to enhance soybean straw biodegradability. Bioresource Technol. 194: 1–6.

Campos, R.P., A. Kwiatkowski and E. Clemente. 2011. Post-harvest conservation of organic strawberries coated with cassava starch and chitosan. Revista Ceres. 58(5): 554–560.

Cavichiolo, J.R. 2010. Secagem do bagaço de laranja em secador tipo flash in Master Thesis in Agricola Engineer, Universidade de Campinas, Campinas, 83 pages.

Cazacu, G., M. Capraru and V.I. Popa. 2012. Advances concerning lignin utilization in new materials in advances. pp. 255–318. *In*: Thomas, S., P.M. Visakh and A.P. Mathew (eds.). Natural Polymers: Composite and Nanocomposites. Springer Berlin Heidelberg, Germany

Cecilio-Filho, A.B., R.J. Souza, L.T. Braz and M. Tavares. 2000. Cúrcuma: Planta medicinal, condimentar e de outros usos potenciais. Ciência Rural Santa Maria. 30(1): 171–177.

Chang, C. and M.T. Nickerson. 2014. Effect of plasticizer-type and genipin on the mechanical, optical, and water vapor barrier properties of canola protein isolate-based edible films. Eur. Food Res. Technol. 238(1): 35–46.

Chang, C. and M.T. Nickerson. 2015. Effect of protein and glycerol concentration on the mechanical, optical, and water vapor barrier properties of canola protein isolate-based edible films. Food Sci. Technol. Intern. 21(1): 33–44.

Chang, V.S. and M.T. Holtzapple. 2000. Fundamental factors affecting biomass enzymatic reactivity. Appl. Biochemi. Biotechnol. 84–86: 5–37.

Chantaro, P., S. Devahastin and N. Chiewchan. 2008. Production of antioxidant high dietary fiber powder from carrot peels. LWT - Food Sci Technol. 41(10): 1987–1994.

Coma, V., A. Martial-Gros, S. Garreau, A. Copinet, F. Salin and A. Deschamps. 2002. Edible antimicrobial films based on chitosan matrix. J. Food Sci. 67(3): 1162–1169.

Coma, V., A. Deschamps and A. Martial-Gros. 2003. Bioactive packaging materials from edible chitosan polymer: Antimicrobial activity assessment on diary related contaminants. J. Food Sci. 68(9): 2788–2792.

Correia, P.R. and M.L. Beirão-da-Costa. 2012. Starch isolation from chestnut and corn flours through alkaline and enzymatic methods. Food Bioprod. Process. 90(2): 309–316.

Curvelo, A.A.S., A.J.F de Carvalho and J.A.M. Agnelli. 2001. Thermoplastic starch-cellulosic fibers composites: Preliminary results. Carbohydr. Polym. 45(2): 183–188.

Dagnino, E., E.R. Chamorro, S.D. Romano, F.E. Felissia and M.C. Area. 2013. Optimization of the acid pretreatment of rice hulls to obtain fermentable sugars for bioethanol production. Ind. Crop Prod. 42: 363–368.

De Campos, A., A.A. Correa, D. Cornella, E.M. Teixeira, J.M. Marconcini, A. Dufresne, et al. 2013. Obtaining nanofibers from curaua and sugarcane bagasse fibers using enzymatic hydrolysis followed by sonication. Cellulose. 20: 1491–1500.

Dias, A.B., C.M.O. Müller, F.D.S. Larotonda and J.B. Laurindo. 2011. Mechanical and barrier properties of composite films based on rice flour and cellulose fibers. LWT - Food Sci Technol. 44(2): 535–542.

Dias, A.M., C.M.O. Müller, F.D.S. Larotonda and J.B. Laurindo. 2010. Biodegradable films based on rice starch and rice flour. J. Cereal Sci. 51(2): 213–219.

Dornelas, C.S.M., F.A.C. Almeida, A.F. Neto, D.M.M. Sousa and A.P. Evangelista. 2015. Desenvolvimento na maturação de frutos e sementes de Urucum (*Bixa orellana L.*). Scientia Plena. 11(1): 1–8.

Drabik D., H. De Gorter, D. R. Just and G. R. Timilsina. 2015. The economics of Brazil's ethanol-Sugar markets, mandates and tax exemptions. Am. J. Agr. Econ. 97(5): 1433–1450.

Dufresne, A. and M.R. Vignon. 1998. Improvement of starch film performances using cellulose microfibrils. Macromolecules 31(8): 2693–2696.

Durango, A.M., N.F.F. Soares and N.J. Andrade. 2006. Microbiological evaluation of an edible antimicrobial coating on minimally processed carrots. Food Control 17(5): 336–341.

Elsabee, M.Z. and E.S. Abdou. 2013. Chitosan based edible films and coatings: A review. Mat. Sci. Eng. C. 33(4): 1819–1841.

Ereno, D. 2007. Álcool de celulose. Revista FAPESP, Edição Impressa 133.

Espitia, P.J.P., W-X Du, R.J. Avena-Bustillos, N.F.F. Soares and T.H. McHugh. 2014. Edible films from pectin: Physical-mechanical and antimicrobial properties - A review. Food Hydrocolloid. 35: 287–296.

Fai, A.E.C., M.R.A. de Souza, N.V. Bruno and É.C.B.A. Gonçalves. 2015. Produção de revestimento comestível à base de resíduo de frutas e hortaliças: aplicação em cenoura (*Daucus carota L.*) minimamente processada. Scientia Agropecuaria, Trujillo. 6(1): 59–68.

FAO. 2013 Trade: Canadian Trade Exports. Available in: http://www.fao.org/fishery/facp/CAN/en Accessed: January 4, 2016.

Farahnaky, A., S.M.M. Dadfar and M. Shahbazi. 2014. Physical and mechanical properties of gelatin-clay nanocomposite. J. Food Eng. 122: 78–83.

Farris, S., K.M. Schaich, L. Liu, L. Piergiovanni and K.L. Yam. 2009. Development of polyion-complex hydrogels as an alternative approach for the production of bio-based polymers for food packaging applications: A review. Trends Food Sci. Tech. 20(8): 316–332.

Ferreira, M.S.L., M.C.P. Santos, T.M.A. Moro, G.J. Basto, R.M.S. Andrade and E.C.B.A. Gonçalves. 2015. Formulation and characterization of functional foods based on fruit and vegetable residue flour. J. Food Sci. Technol. 52(2): 822–830.

Flauzino Neto, W.P., H.A. Silverio, N.O. Dantas and D. Pasquini. 2013. Extraction and characterization of cellulose nanocrystals from agro-industrial residue – Soy hulls. Ind. Crop Prod. 42: 480–488.

Fogaça, F.H., L.S. Sant'ana, J.A.F. Lara, A.C.G. Mai and D.J. Carneiro. 2015. Restructured products from tilapia industry byproducts: The effects of tapioca starch and washing cycles. Food Bioprod. Process. 94: 482–488.

Freund, R.P., J.C. Washaw and M. Maggion. 1998. Natural color for use in foods. Cereal Foods World 33(7): 553–556.

Galdeano, M.C. and M.V.E. Grossamnn. 2006. Oat hulls treated with alkaline hydrogen peroxide associated with extrusion as fiber source in cookies. Ciência e Tecnologia de Alimentos. 26(1): 123–126.

Garay, R.M.M., M.B. Rallo, R.C. Carmona and J.C. Araya. 2009. Characterization of anatomical, chemical, and biodegradable properties of fibers from corn, wheat, and rice residues. Chil. J. Agric. Res. 69(3): 406–415.

Gennadios, A., T.H. Mchugh, C.L. Weller and J.M. Krochta. 1994. Edible coating and films based on protein. pp. 201–278. *In*: Krochta, J.M., E.A. Baldwin and M.O. Nisperos-Carriedo (eds.). Edible Coatings and Films to Improve Food Quality. CRC, New York, NY, EUA.

Ghaly, A.E., V.V. Ramakrishnan, M.S. Brooks, S.M. Budge and D. Dave. 2013. Fish processing wastes as a potential source of proteins, amino acids and oils: A critical review. J. Microb. Biochem. Technol. 5(4): 107–129.

Gibson, L.G. 2012. The hierarchical structure and mechanics of plant materials. J. R. Soc. Interface. 9(76): 2749–2766.

Gildberg, A. and E. Stenberg. 2011. A new process for advanced utilization of shrimp waste. Process Biochemi. 36(8–9): 809–812.

Gnanasambandam, R. and A. Proctor. 1999. Preparation of soy hull pectin. Food Chem. 65(4): 461–467.

Gómez-Guillén, M.C., B. Giménez, M.E. López-Caballero and M.P. Montero. 2011. Functional and bioactive properties of collagen and gelatin from alternative sources: A review. Food Hydrocolloid. 25(8): 1813–1827.

Han, C.Y. Zhao, S.W. Leonard and M.G. Traber. 2004. Edible coatings to improve storability and enhance nutritional value of fresh and frozen strawberries (Fragaria × ananassa) and raspberries (Rubus ideaus). Postharvest Biol. Technol. 33(1): 67–78.

Harris, P.J. and B.A. Stone. 2008. Chemistry and molecular organization of plant cell walls. pp. 60. Himmel, M.E. (ed.). Biomass Recalcitrance. Blackwell, Oxford.

Hendriks, A.T.W.M. and G. Zeeman. 2009. Pretreatments to enhance the digestibility of lignocellulosic biomass. Bioresource Technol. 100(1): 10–18.

Hiranvarachat, B. and S. Devahastin. 2014. Enhancement of microwave-assisted extraction via intermittent radiation: Extraction of carotenoids from carrot peels. J. Food Eng. 126: 17–26.

Hofsetz, K. and M.A. Silva. 2012. Brazilian sugarcane bagasse: Energy and non-energy consumption. Biomass Bioenerg. 46: 564–573.

Hoover, R. 2001. Composition, molecular structure, and physicochemical properties of tuber and root starches: A review. Carbohydr. Polym. 45(3): 253–267.

Hornsby, P., E. Hinrichsen and K. Tarverdu. 1997. Preparation and properties of polypropylene composites reinforced with wheat and flax starw fibres: Part II, Analysis of composite microstructure and mechanical properties. J. Mat Sci. 32(4): 1009–1015.

Hosseini, S.F., M. Rezaei, M. Zandi and F. Farahmand. 2013. Preparation and functional properties of fish gelatin–chitosan blend edible films. Food Chem. 136(1–4): 1490–1495.

Hosseini, S.F., M. Rezaei, M. Zandi and F. Farahmand. 2015. Fabrication of bio-nanocomposite films based on fish gelatin reinforced with chitosan nanoparticles. Food Hydrocolloid. 44: 172–182.

Iahnke, A.O.S, T.M.H. Costa, A.O. Rios and S.H. Flôres. 2015. Residues of minimally processed carrot and gelatin capsules: Potential materials for packaging films. Ind. Crop Prod. 76: 1071–1078.

Jacometti, G.A., L.R.P.F. Mello, P.H.A. Nascimento, A.C. Sueiro, F. Yamashita and S. Mali. 2015. The physicochemical properties of fibrous residues from the agro industry. LWT - Food Sci. Technol. 62(1): 138–143.

Jaczynski, J. 2008. Protein, lipid recovery from fish-processing by-products. pp. 1–32. *In*: Papadopoulos, Konstantinos N. (ed.). Food Chemistry Research Developments. Nova Science Publishers, Inc. New York.

Jenkins, D., M.G. Richard and G.T. Daigger. 1993. Manual on the Causes and Control of Activated Sludge Bulking and Foaming, 3rd Edition. Lewis Publishers, New York.

Jongjareonrak, A., S. Benjakul, W. Visessanguan and M. Tanaka. 2008. Antioxidative activity and properties of fish skin gelatin films incorporated with BHT and α-tocopherol. Food Hydrocolloid. 22: 449–458.

Kader, A.A, D. Zagory and E.L. Kerbel. 1989. Modified atmosphere packaging of fruits and vegetables. Crit. Rev. Food Sci. Nutr. 28(1): 1–3.

Kaya, S. and A. Maskan. 2003. Water vapour permeability of pestil (a fruit leather) made from boiled grape juice with starch. J. Food Eng, Essex. 57(3): 295–299.

Kimber, D.S. and D.I. McGregor. 1995. The species and their origin, cultivation and world production. pp. 1–7. *In*: Kimber, D.S. and D.I. McGregor (eds.). Brassica Oil Seed; Production and Utilization. Centre for Agriculture and Biosciences International, University Press, Cambridge.

Kubo, S., R.D. Guilbert and J.F. Kadla. 2005. Lignin-based polymer blends and biocomposite Materials. pp. 672–695. *In*: Mohanty, A.K., M. Misra and L.T. Drzal (eds.). Natural Fibers, Biopolymers and Biocomposites. CRC Press, Boca Raton, USA.

Kuttigounder, D., L. Rao and S. Bhattacharya. 2011. Turmeric powder and starch: selected physical, physicochemical, and microstructural properties. J. Food Sci. 76: C1284–C1291.

Leonel, M., M.P. Cereda and X. Roau. 1999. Aproveitamento de resíduo da produção de etanol a partir de farelo de mandioca, como fonte de fibras dietéticas. Ciência e Tecnologia de Alimentos. 19(2): 241–245.

Maiorella, B.L. 1985. Ethanol. pp. 861. *In*: Moo-Young, M. (ed.). Comprehensive Biotechnology. Pergamon Press, Oxford, USA.

Manamperi, W.A., D.P. Wiesenborn, S.K. Chang and S.W. Pryor. 2011. Effects of protein separation conditions on the functional and thermal properties of canola protein isolates. J. Food Sci. 76(3): E266–E273.

Mandal, A. and D. Chakrabarty. 2014. Studies on the mechanical, thermal, morphological and barrier properties of nanocomposites based on poly(vinyl alcohol) and nanocellulose from sugarcane bagasse. J. Ind. Eng. Chem. 20(2): 462–473.

Maniglia, B.C. 2012. Elaboração de filmes biodegradáveis a partir do resíduo da extração do pigmento de Cúrcuma. Master Thesis of Chemistry, University of São Paulo, Ribeirão Preto, Brazil.

Maniglia, B.C., J.R. Domingos, R.L. de Paula and D.R. Tapia-Blácido. 2014. Development of bioactive edible film from turmeric dye solvent extraction residue. LWT - Food Sci. Technol. 56(2): 269–277.

Maniglia, B.C., J.R. Domingos, R.L. de Paula and D.R. Tapia-Blácido. 2015. Turmeric dye extraction residue for use in bioactive film production: Optimization of turmeric film plasticized with glycerol. LWT - Food Sci. Technol. 64(2): 1187–1195.

Mantovani, N.C., M.F. Grando, A. Xavier and W.C. Otoni. 2013. Avaliação de genótipos de urucum (*Bixa orellana L.*) por meio da caracterização morfológica de frutos, produtividade de sementes e teor de bixina. Ciência Florestal. 23(2): 355–362.

Marquié, C., C. Aymard, J.L. Cuq and S. Guilbert. 1995. Biodegradable packaging made from cottonseed flour: Formation and improvement by chemical treatments with gossypol, formaldehyde and glutaraldehyde. J. Agr. Food Chem. 43(10): 2762–2767.

Marquié, C., A.-M. Tessier, C. Aymard and S. Guilbert. 1997. HPLC determination of the reactive lysine content of cottonseed protein films to monitor the extent of crosslinking by formaldehyde, glutaraldehyde, and glyoxal. J. Agr. Food Chem. 45(3): 922–926.

Marquié, C., A.M. Tessier, C. Aymard and S. Guilbert. 1998. How to monitor the protein cross-linking by formaldehyde, glutaraldehyde or glyoxal in cotton-seed protein based films? Nahrung. 42(3–4): 264–265.

Martelli, M.R., T.T. Barros, M.R. de Moura, L.H. Mattoso and O.B. Assis. 2013. Effect of chitosan nanoparticles and pectin content on mechanical properties and water vapor permeability of banana puree films. J. Food Sci. 78(1): N98–104.

Martínez-Camacho, A.P., M.O. Cortez-Rocha, J.M. Ezquerra-Brauer, A.Z. Graciano-Verdugo, F. Rodriguez-Félix, M.M. Castillo-Ortega, et al. 2010. Chitosan composite films: Thermal, structural, mechanical and antifungal properties. Carbohydr. Polym. 82(2): 305–315.

Martins, M.C. and O. Rusig. 1992. Utilização de açafrão (*Curcuma longa L.*) como corante natural para alimentos. *Boletim da SBCTA*, Campinas, Brazil, 26(1): 56–65.

Maurya, D.P., A. Singla and S. Negi. 2015. An overview of key pretreatment processes for biological conversion of lignocellulosic biomass to bioethanol. Biotechnol. 3: 1–144.

Mccurdy, S.M. 1990. Effect of processing on the functional properties of canola/rapeseed protein. J. Am. Oil Chem. Soc. 67(5): 281–284.

McHugh, T.H. and J.M. Krochta. 1994. Sorbitol- vs glycerol-plasticized whey protein edible films: integrated oxygen permeability and tensile property evaluation. J. Agri. Food Chemi. 42(4): 841–845.

Mellinas, C., A. Valdés, M. Ramos, N. Burgos, M.C. Garrigos and A. Jimenéz. 2016. Active edible films: Current state and future trends. J. Appl. Polymer Sci. 133(2): 1–15.

Merci, A., A. Urbano, M.V.E. Grossmann, C.A. Tischera and S. Mali. 2015. Properties of microcrystalline cellulose extracted from soybean hulls by reactive extrusion. Food Res. Int. 73: 38–43.

Mielenz, J.R., J.S. Bardsley and C.E. Wyman. 2009. Fermentation of soybean hulls to ethanol while preserving protein value. Bioresource Technol. 100(14): 3532–3539.

Mishra, R. K., A.K. Banthia and A.B.A. Majeed. 2012. Pectin based formulations for biomedical applications: A review. Asian J. Pharm. Clin. Res. 5(4): 1–7.

Mo, X., D. Wang and X. Sun. 2005. Straw-based and biocomposites. pp. 450–496. *In*: Mohanty, A.K., M. Misra and L.T. Drzal (eds.). Natural Fibers, Biopolymers, and Biocomposites. CRC Taylor & Francis Press, Boca Raton, USA.

Montaño-Leyva, B., G.G.D. da Silva, E. Gastaldi, P. Torres-Chávez, N. Gontard and H. Angellier-Coussy. 2013. Biocomposites from wheat proteins and fibers: Structure/mechanical properties relationships. Ind. Crop Prod. 43: 545–555.

Mosier, N., C. Wyman, B. Dale, R. Erlander, Y.Y. Lee, M. Holtzapple et al. 2005. Features of promising technologies for pretreatment of lignocellulosic biomass. Bioresource Technol. 96(6): 673–686.

Moura, M.R., F.A. Aouada and L.H.C. Mattoso. 2008. Preparation of chitosan nanoparticles using methacrylic acid. J. Colloid Interf. Sci. 321(2): 477–483.

Moura, M.R., F.A. Aouada, R.J. Avena-Bustillos, T.H. McHugh, J.M. Krochta and L.H.C. Mattoso. 2009. Improved barrier and mechanical properties of novel hydroxypropyl methylcellulose edible films with chitosan/tripolyphosphate nanoparticles. J. Food Eng. 92(4): 448–453.

Müller, C., J. Laurindo and F. Yamashita. 2009. Effect of cellulose fibers addition on the mechanical properties and water vapor barrier of starch-based films. Food Hydrocolloid. 23: 1328–1333.

Nabeshima, E.H. and A.A. EL-Dash. 2004. Modificação química da farinha de arroz como alternativa para o aproveitamento dos subprodutos do beneficiamento do arroz. *Boletim do Centro de Pesquisa de Processamento de Alimentos*. 22(1): 107–120.

Nagarajan, M., S. Benjakul, T. Prodpran and P. Songtipya. 2015. Properties and characteristics of nanocomposite films from tilapia skin gelatin incorporated with ethanolic extract from coconut husk. J. Food Sci. Technol. 52(12): 7669–7682.

Nair, K.P.P. 2013. The agronomy and economy of turmeric and ginger. pp. 1–5. *In*: Nair, K.P.P. (ed.). The Invaluable Medicinal Spice Crops. Elsevier, London.

Nelson, D.L. and M.M. Cox. 2002. Princípios de Bioquímica, 3th ed., Sarvier, Brazil.

Neves, M.F., V.G. Trombin, P. Milan, F.F. Lopes, F. Cressoni and R. Kalaki. 2012. O retrato da citricultura brasileira. Ribeirão Preto: FEA-RP e FUNDACE. Available in: <http://www.citrusbr.com.br/download/biblioteca/o_retrato_da_citricultura_brasileira_baixa.pdf>. Accessed in January 23, 2016.

Newkirk, R. 2009. Canola meal feed industry guide. Canadian International Grains Institute, 4th Edition, Winnipeg, Canada.

Oh, Y.H., I.Y. Eom, J.C. Joo, J.H. Yu, B.K. Song, S.H. Lee, et al. 2015. Recent advances in development of biomass pretreatment technologies used in biorefinery for the production of bio-based fuels, chemical and polymers. Korean J. Chem. Eng. 32(10): 1945–1959.

Oliveira, A., S. Iahnke, T.M.H. Costa, A.O. Rios and S.H. Flôres. 2015. Residues of minimally processed carrot and gelatin capsules: Potential materials for packaging films. Ind. Crop Prod. 76: 1071.

Oliveira Junior, G.I., N.M.B. Costa, H.S.D. Duarte and M.C.D. Paes. 2014. Chemical composition and effects of micronized corn bran on iron bioavailability in rats. Food Sci. Technol. 34(3): 616–622.

Oliveira, T.I.S., L. Zea-Redondo, G.K. Moates, N. Wellner, K. Cross, K.W. Waldron, et al. 2016. Pomegranate peel pectin films as affected by montmorillonite. Food Chem. 198: 107–112.

Park, S.-I., M.A. Daeschel and Y. Zhao. 2004. Functional properties of antimicrobial lysozyme-chitosan composite films. J. Food Sci. 69(8): M215–M221.

Pascall, M.A. and S.J. Lin. 2013. The application of edible polymeric films and coatings in the food industry. J. Food Process. Technol. 4: 116.

Pasquini, D., E.M. Teixeira, A.A.S. Curvelo, M.N. Belgacem and A. Dufresne. 2010. Extraction of cellulose whiskers from cassava bagasse and their applications as reinforcing agent in natural rubber. Ind. Crop Prod. 32(3): 486–490.

Pauly, M. and K. Keegstra. 2008. Cell wall carbohydrates and their modification as a resource for biofuels. Plant J. 54(4): 559–568.

Pelissari, F.M., M.M. Andrade-Mahecha and P.J.A. Sobral. 2013. Optimization of process conditions for the production of films based on the flour from plantain bananas (*Musa paradisíaca*). LWT - Food Sci Technol. 52(1): 1–11.

Pereira, A.S. and P.C. Stringheta. 1998. Considerações sobre a cultura e processamento do açafrão. Horticultura Brasileira, Brasília. 16(2): 102–105.

Pérez-Gago, M.B. 2012. Protein-based films and coatings. pp. 14–58. *In*: Baldwin, A.E., R. Hagenmaier and J. Bai (eds.). Edible Coatings and Films to Improve Food Quality. 2nd Ed. CRC Press, Taylor and Francis Group, Boca Raton, USA.

Plastics Europe. Plastics – the Facts 2014/2015. An analysis of European plastics production, demand and waste data. Available in: <http://www.plasticseurope.org/>. Accessed in April 18, 2016.

Portal do Agronegócio. 2009. Available in <http://www.portaldoagronegocio.com.br> Accessed: December 14, 2015.

Reddy, N. and Y. Yang. 2005. Biofibers from agricultural byproducts for industrial applications. Trends Biotechnol. 23(1): 22–27.

Rezzadori, K. and S. Benedetti. 2009. Proposições para valorização de resíduos do processamento do suco de laranja. *In*: International Workshop Advances in Cleaner Production. São Paulo, Brazil.

Ribeiro, C., A.A. Vicente, J.A. Teixeira and C. Miranda. 2007. Optimization of edible coating composition to retard strawberry fruit senescence. Postharvest Biol. Technol. 44(1): 63–70.

Rojas, M.J., P.F. Siqueira, L.C. Miranda, P.W. Tardioli and R.L.C. Giordano. 2014. Sequential proteolysis and cellulolytic hydrolysis of soybean hulls for oligopeptides and ethanol production. Ind. Crop Prod. 61: 202–210.

Sánchez-Sáenz, C.M., V.R.G. Nascimento, J.D. Biagi and R.A. Oliveira. 2015. Mathematical modeling of the drying of orange bagasse associating the convective method and infrared radiation. Revista Brasileira de Engenharia Agrícola e Ambiental 19(12): 1178–1184.

Santos, L.S., R.C.F. Bonomo, R.C.I. Fontam, P. Bonomo, C.X.S. Leite and D.O. Santos. 2013. Efeito dos metodos de extração na composição, rendimento e propriedades da pasta do amido obtido da semente de jaca. Revista Brasileira de Produtos Agroindustriais. 15(3): 255–261.

Satyanarayana, K.G., G.G.C. Azizaga and F. Wypych. 2009. Biodegradable composites based on lignocellulosic fibers – An overview. Progress Polymer Sci. 34(9): 982–1021.

Savadekar, N.R and S.T. Mhaske. 2012. Synthesis of nano cellulose fibers and effect on thermoplastics starch based films. Carbohydr. Polym. 89(1): 146–151.

Saxena, R.C., D.K. Adhikari and H.B. Goyal. 2009. Biomass-based energy fuel through biochemical routes: A review. Renew. Sust. Energ. Rev. 13(1): 167–178.

Schieber, A., F.C. Stintzing and R. Carle. 2001. By-products of plant food processing as a source of functional compounds - recent developments. Trends Food Sci. Tech. 12(11): 401–413.

Sigrist, M.S., J.B. Pinheiro, J.A.A. Filho and M.I. Zucchi. 2011. Genetic divergence among Brazilian turmeric germplasm using morpho-agronomical descriptors. Crop Breeding Appl. Biotechnol. 11: 70–76.

Sila, D.N., S.V. Buggenhout, T. Duvetter, I. Fraeye, A. De Roeck, A. Van Loey, et al. 2009. Pectins in processed fruits and vegetables: Part II—Structure–function relationships. Comp. Rev. Food Sci. F. 8(2): 86–104.

Singh, J., M. Suhag and A. Dhak'a. 2015. Augmented digestion of lignocellulose by steam explosion, acid and alkaline pretreatment methods: A review. Carbohydr. Polym. 117(6): 624–631.

Slavutsky, A.M. and M.A. Bertuzzi. 2013. Water barrier properties of starch films reinforced with cellulose nanocrystals obtained from sugarcane bagasse. Carbohydr. Polym. 110: 53–61.

Sosa-Arnao, J.H. and S.A. Nebra. 2009. Bagasse dryer role in the energy recovery of water tube boilers. Dry Technol. 27: 587–594.

Srinorakutara, T., C. Suesat, B. Pitiyont, W. Kitpreechavanit and S. Cattithammanit. 2004. Utilization of waste from cassava starch plant for ethanol production. Proceedings of The Joint International Conference on "Susta": 1–6.

Sun, Y. and J. Cheng. 2002. Hydrolysis of lignocellulosic materials for ethanol production: A review. Bioresource Technol. 83: 1–11.

Suppakul, P., B. Chalernsooka, B. Ratisuthawata, S. Prapasitthia and N. Munchukangwana. 2013. Empirical modeling of moisture sorption characteristics and mechanical and barrier properties of cassava flour film and their relation to plasticizing-antiplasticizing effects. LWT - Food Sci. Technol. 50(1): 290–297.

Taham, T., F.A. Cabral and M.A.S. Barrozo. 2015. Extraction of bixin from annatto seeds using combined technologies. J. Supercritic Fluid. 100: 175–183.

Taherzadeh, M.J. and K. Karimi. 2008. Pretreatment of lignocellulosic wastes to improve ethanol and biogas production: A review. Int. J. Mol. Sci. 9: 1621–1651.

Tan, S.H., R.J. Mailer, C.L. Blanchard and S.O. Agbool. 2011. Extraction and characterization of protein fractions from Australian canola meals. Food Res. Int. 44(4): 1075–1082.

Tao, Z, W.Y. Weng, M.J. Cao, G.M. Liu, W.J. Su, K. Osako, et al. 2015. Effect of blend ratio and pH on the physical properties of edible composite films prepared from silver carp surimi and skin gelatina. J. Food Sci. Technol. 52(3): 1618–1625.

Tapia-Blácido, D.R., P.J.A. Sobral and F.C. Menegalli. 2011. Optimization of amaranth flour films plasticized with glycerol and sorbitol by multi- response analysis. LWT - Food Sci. Technol. 44(8): 1731–1738.

Tapia-Blácido, D.R., B.C. Maniglia and M.R. Martelli. 2016. Sustainable polymers from biomass. pp. 227–253. *In*: Tang, C. (ed.). Biopolymers from Sugarcane and Soybean Biomass. Wiley, New Jersey, EUA. (In press)

Taskaya, L. and J. Jaczynski. 2009. Flocculation-enhanced protein recovery from fish processing by-products by isoelectric solubilization and precipitation. LWT - Food Sci. Technol. 42(2): 570–575.

Teixeira, E.M., D. Pasquini, A.A.S. Curvelo, E. Corradini, M.N. Belgacem and A. Dufresne. 2009. Cassava bagasse cellulose nanofibrils reinforced thermoplastic cassava starch. Carbohydr. Polym. 78(3): 422–431.

Teixeira, E.M., A.A.S. Curvelo, A.C. Corrêa, J.M. Marconcini, G.M. Glenn and L.H.C. Mattoso. 2012. Properties of thermoplastic starch from cassava bagasse and cassava starch and their blends with poly (lactic acid). Ind. Crop Prod. 37(1): 61–68.

Trung, T.S. and P.T.D. Phuong. 2012. Bioactive compounds from by-products of shrimp processing industry in Vietnam. J. Food Drug Anal. 20(1): 194–197.

Tserki, V., P. Matzinos and C. Panayiotou. 2006. Novel biodegradable composites based on treated lignocellulosic waste flour as filler Part II. Development of biodegradable composites using treated and compatibilized waste flour. Composites Part A – Appl. Sci. Manufact., Kidlington. 37(9): 1231–1238.

Valério, A., M.I.L. Ramos, J.A.B. Neto and M.L.R. Macedo. 2015. Annatto seed residue (*Bixa orellana L.*): Nutritional quality Melissa. Food Sci. Technol. 35(2): 326–330.

Versino, F. and M.A. García. 2014. Cassava (*Manihot esculenta*) starch films reinforced with natural fibrous filler. Ind. Crop Prod. 58: 305–314.

Wan, C., Y. Zhou and Y. Li. 2011. Liquid hot water and alkaline pretreatment of soybean straw for improving cellulose digestibility. Bioresource Technol. 102: 6254–6259.

Wang, J.S., J. Wang and M. Gulfraz. 2005. Efficient cellulase production from corn straw by *Trichoderma Reesei* LW1 through solid state fermentation process, Ethnobotanical leaflets: 1–8. Available in: http://www.ethnoleaflets.com//leaflets/wang.htm. Accessed: January 22, 2016.

Wanyo, P., C. Chomnawang and S. Siriamornpun. 2009. Substitution of wheat flour with rice flour and rice bran in flake products: Effects on chemical, physical and antioxidant properties. World Appl. Sci. J. 7(1): 49–56.

White, G. W., T. Katona and J.P. Zodda. 1999. The use of high-performance size exclusion chromatography (HPSEC) as a molecular weight screening technique for polygalacturonic acid for use in pharmaceutical applications. J. Pharm. Biomed. Anal. 20(6): 905–912.

Wicaksono, R., K. Syamsu, I. Yuliasih and M. Nasir. 2013. Cellulose nanofibers from cassava bagasse: Characterization and application on tapioca film. Chem. Mater. Res. 3(13): 79–87.

Yang, H. and Y. Wang. 2009. Effects of concentration on nanostructural images and physical properties of gelatin from channel catfish skins. Food Hydrocolloid. 23(3): 577–584.

Yoo, J., S. Alavi, P. Vadlani and V. Amanor-Boadu. 2011. Thermo-mechanical extrusion pretreatment for conversion of soybean hulls to fermentable sugars. Bioresource Technol. 102(16): 7583–7590.

Zhang, J. and P. Zheng. 2015. A preliminary investigation of the mechanism of hexavalent chromium removal by corn-bran residue and derived chars. RSC Adv. 5: 17768–17774.

Vegetable Nanocellulose in Food Packaging

C. Gómez H.[1], A. Serpa[1], J. Velásquez-Cock[1], C. Castro[2], B. Gómez H.[3],
L. Vélez[1], P. Gañán[3] and R. Zuluaga[1]*

[1] Facultad de Ingeniería Agroindustrial, Universidad Pontificia Bolivariana,
    Circular 1° N° 70-01, Medellín, Colombia, Zip Code 050031
[2] Facultad de Ingeniería Textil, Universidad Pontificia Bolivariana,
    Circular 1° N° 70-01, Medellín, Colombia, Zip Code 050031
[3] Facultad de Ingeniería Química, Universidad Pontificia Bolivariana,
    Circular 1° N° 70-01, Medellín, Colombia, Zip Code 050031

Introduction

The value of the global market for packaging is estimated to be approximately US$ 417 billion per year and represented by 100,000 industries and 5 million employees (Lacroix 2015). Food packaging represents 65% of the market share, accounting for nearly US$ 100 billion in the United States of America, US$ 80 billion in Japan, US$ 29 billion in Germany and US$ 19 billion in France (Lacroix 2015). Despite the size of the packaging industry, it is estimated that nearly 50% of agricultural products are lost due to the absence of proper packaging, bad weather and physiological processes such as breathing, perspiration and the degradation of vitamins, pigments and carbohydrates, which could deteriorate the food quality and compromise its safety (Lacroix 2015). Improvements in the packaging technology of foodstuff could reduce food spoilage and shortages in certain regions of the world (Khan et al. 2012).

Historically, food packaging was created to provide a barrier to protect from microorganisms, moisture, gas and the migration of substances towards food. Other basic functions have been added, for instance, communication, convenience and containment (Silvestre et al. 2011). However, these

*Corresponding author: robin.zuluaga@upb.edu.co

functionalities are no longer sufficient. The Centers for Disease Control and Prevention (CDC) of the United States of America estimates that 48 million people are affected by foodborne diseases each year (Lacroix 2015). In addition, the World Health Organization (WHO) reports that annually one in every 10 people have exhibited health problems derived from the consumption of contaminated products; it is estimated that 420,000 people died as a result (Chaib 2015) of the same. Therefore, the prevention or minimization of the growth of pathogenic microorganisms in food represents a challenge for the food packaging industry (Lacroix 2015). However, the industry also has other issues to solve, such as increasing consumer demand for ready to eat foods, requests to decrease or remove additives from packaging and preservation materials, changes in retail and distribution practices associated with globalization, and stricter requirements regarding consumer health and safety (Lacroix 2015).

Currently, the most common packaging materials used to overcome all of these challenges are polyethylene or copolymer-based materials, which have been in use by the food industry for over 50 years (Khan et al. 2012). However, the increasing environmental degradation exerted by these materials has encouraged studies of different renewable resources for use as potential replacements for synthetic polymers (Lavoine et al. 2012). These materials must provide enough protection for the products to obtain a satisfactory shelf life at the same levels as those obtained with the synthetic polymers and be able to overcome the challenges mentioned above (Robinson and Morrison 2010). Among the desirable characteristics, they must have adequate mechanical properties and provide a barrier to oxygen, water vapour, light, bacteria and/or other contaminants to prevent food spoilage (Spence et al. 2010).

The use of products derived from renewable sources, such as materials based on annually renewable agricultural and biomass feedstock, to develop food packaging can provide a solution to international issues, for instance, petroleum shortage, global warming and waste production, among others (Isikgor and C. Remzi Becer 2015; Mohanty et al. 2002). Different starch (Avella et al. 2005; Montoya et al. 2014), lignocellulosic (Isikgor and C. Remzi Becer 2015), and unsaturated oil-based materials (Sharma and Kundu 2006) have been used. Lignocellulosic and starch-based materials are the most commonly used for food packaging (Lavoine et al. 2012).

Conventionally, paper and paperboard made of lignocellulosic or cellulosic fibres have been used in packaging for a wide range of food categories such as dried food products, frozen or liquid foods and beverages, even extending to fresh food (Spence et al. 2010). However, the use of lignocellulosic materials and products is generally motivated by its low cost (Wüstenberk 2014). To fully take advantage of the potentials of these materials, it is necessary to use them as nanostructured high-performance constituents in the form of nanocellulose (Henriksson et al. 2008). Nanocellulose is a nanomaterial that is isolated from plant or vegetable sources, for instance, natural fibres, wood or cotton. It was developed in 1977 by researchers at the ITT Rayonier Eastern Research Division Lab in Whippany, USA (Turbak et al. 1983). It has demonstrated considerable potential as a food ingredient (Gómez H. et al. 2016; Robson 2012; Ström et al. 2013; Winuprasith and Suphantharika 2013; Yano et al. 2012). In addition, it exhibits interesting mechanical, optical and oxygen barrier properties. Coupled with its renewable and biodegradable characteristics, these features make nanocellulose a desirable material for the packaging industry (Aulin et al. 2010; Lavoine et al. 2012; Li et al. 2015; Nair et al. 2014; Spence et al. 2010; Syverud and Stenius 2009; Zhao et al. 2013, 2015).

This work aims to review the impacts, benefits and challenges of the use of nanocellulose in the food packaging industry. Relevant publications on the potential of it in various food packaging applications were studied. The literature review indicated that nanocellulose films have barrier, optical and mechanical properties comparable with synthetic polymers; however, these properties depends on the relative humidity because of its hydrophilic nature. Therefore, nanocellulose can be used as a biobased food packaging material to develop nanocomposites and edible coating, as shown in Fig. 1. However, the process method also influences the properties of nanocellulose. For example, Fig. 1 shows two different vegetable nanocellulose films: an opaque film was produced by vacuum filtration followed by freeze-drying, while a translucent film was produced by oven vacuum filtration followed by oven drying.

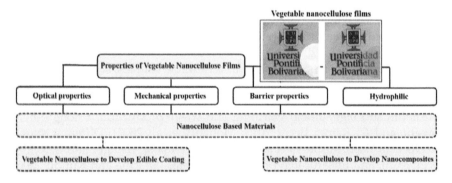

Fig. 1. Scheme of properties and applications of vegetable nanocellulose in food packaging.

Vegetable Nanocellulose

Cellulose is one of the most abundant natural polymers and the source of several sustainable materials on an industrial scale (Klemm et al. 2011). It has been used in the form of fibres, as an energy source (Mckendry 2002), and for building materials, clothing, and paper since the Egyptian papyri (Klemm et al. 2011). It has prompted the creation of novel types of materials.

For example, celluloid, the first thermoplastic polymer, was developed by causing cellulose to react with nitric acid and sulphuric acid to form cellulose nitrate and combining it with camphor under heat and pressure in 1868 by the Hyatt Manufacturing Company (Saunders and Taylor 1990). Another example is the fabrication of regenerated cellulose filaments by spinning a solution of cellulose in a mixture of copper hydroxide and aqueous ammonia. This development was followed by the process for producing rayon fibre and filaments (Klemm et al. 2011).

In recent decades, the production, characterization and development of applications for novel forms of cellulose, such as whiskers and nanofibrils, has promoted novel research (Andrade et al. 2015; Cunha et al. 2014; Dufresne 2013; Henriksson et al. 2008; Klemm et al. 2009, 2011; Klemm et al. 2005; Lähtinen et al. 2012; Li et al. 2015; Siró and Plackett 2010; Ström et al. 2013; Wüstenberk 2014). Whiskers, also known as cellulose nanocrystals (CNCs), are needle-like cellulose crystals with diameters between 10-20 nm. They are a high-purity and highly crystalline cellulose nanomaterial produced by strong acid hydrolysis to remove non-cellulosic components and most of the amorphous cellulose from the source materials (Azizi Samir et al. 2005).

Nanocellulose refers to cellulose fibrils composed of aligned β-D-(1→4) glucopyranose polysaccharide chains with diameters between 5-10 nm depending on the source and several micrometers in length. It is isolated from plants or vegetable sources using a multi-step sequence (Velásquez-Cock et al. 2016), starting with a series of chemical treatments to reduce the presence of other components such as hemicellulose and lignin (Zuluaga et al. 2009). Finally, the sequence concludes with mechanical treatments to deconstruct the hierarchical structure of cellulose fibres (Gañan et al. 2008; Velásquez-Cock et al. 2016). Waring blender (Uetani and Yano 2011; Zuluaga et al. 2009), grinding (Velásquez-Cock et al. 2016) and high-pressure homogenization (Turbak et al. 1983; Velásquez-Cock et al. 2016) are the most common techniques used as mechanical treatments.

However, these mechanical treatments are normally associated with high values of energy consumption for fibre delamination. For example, the homogenizer requires 78,800 kJ/kg, assuming a processing pressure of 55 MPa and 20 passes (Spence et al. 2011). Therefore, homogenization needs to be combined with a chemical pretreatment to decrease the energy consumption. Different pretreatments have been proposed, for instance, enzymatic pretreatment (Ankerfors 2015) and the introduction of charged groups through carboxymethylation of the hydroxyl groups (Aulin et al. 2010) or 2,2,6,6-tetramethylpiperidine-1-oxyl (TEMPO)-mediated oxidation (Saito and Isogai 2004).

The term nanocellulose refers to isolated cellulose fibrils with one dimension in the nanometer range (TAPPI 2011), produced using both top-down or bottom-up approaches. The top-down approach involves enzymatic, chemical and/or physical methodologies for their isolation from plant or vegetable sources such as lignocellulosic and agroindustrial wastes.

The bottom-up approach produces nanocellulose, also known as cellulose nanoribbons, from glucose using bacteria (Castro et al. 2011). In this work, the term nanocellulose refers only to nanoscaled fibrils isolated from renewable sources such as plants or vegetable sources.

According to the report published as a result of the EU concerted action project (Weber 2000), *"Biobased food packaging materials are materials derived from renewable sources. These materials can be used for food applications"*. Nanocellulose isolated from plants or vegetable sources can be used to produce films, aerogels/foams and nanocomposites with barrier, mechanical and optical properties that are useful for the development of food packaging materials. Therefore, it can be considered a biobased food packaging material.

The development of applications for vegetable nanocellulose as a biobased food packaging material depends, among other factors, on understanding the structure-property relationships of such hierarchical systems. In this sense, this work presents current advances in the properties of these materials as well as various applications of vegetable nanocellulose films, nanocomposite-based materials and edible coatings in food packaging.

Properties of Vegetable Nanocellulose Films

Nanocellulose films, sometimes called "nanopaper" (Ferrer et al. 2012a; Henriksson et al. 2008; Saunders and Taylor 1990), are produced from nanocellulose after a homogenization process (see Fig. 2a) by means of vacuum filtration, followed by dehydration techniques, such as oven drying (Rämänen et al. 2012), freeze drying (Missoum et al. 2012), spray drying (Peng et al. 2012), and supercritical fluids (Žepič et al. 2014) among others, as shown in Fig 2b. As the water is removed from the nanocellulose, a cellulose nanofibrils network is developed through interfibrillar hydrogen bonding (Lavoine et al. 2012), Fig. 2c. The physicochemical properties of the film depend on the film formation process and on the nature of the nanocellulose used (Spence et al. 2010). Depending on its properties, many applications such as transparent films for food packaging, electronic devices or reinforcement in nanocomposites have been proposed (Lavoine et al. 2012). Most of the publications about nanocellulose films address their optical (Aulin et al. 2010; Aulin and Ström 2013; Dufresne 2013; Labuschagne et al.2008; Lavoine et al. 2012; Liu et al. 2011; Rodionova et al.2011; Spence et al. 2010), mechanical (Henriksson et al. 2008; Nakagaito and Yano 2005) and barrier properties (Dufresne 2013).

Optical Properties of Vegetable Nanocellulose Films

The optical properties of nanocellulose films are studied by the determination of the light transmittance using a UV-visible spectrometer in the wavelength range of 200–1000 nm. Usually, the light transmittance at 600 nm is reported, as it is the middle of the visible spectrum (Dufresne 2013). Optically

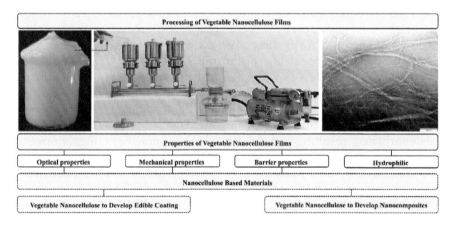

Fig. 2. Scheme of nanocellulose film processing. (a) Nanocellulose after homogenization process, (b) vacuum filtration of nanocellulose suspension, and (c) transmission electron microscopy image of vegetable nanocellulose from banana rachis isolated using the G30 method proposed by Velazquez-Cock et al. (2016).

transparent nanocellulose films are achieved using densely packed cellulose nanofibrils with interstices between the fibres that are small enough to avoid light scattering (Dufresne 2013). The optical properties depend on the film preparation procedure and the size distribution of the nanocellulose fibrils (Dufresne 2013). Films prepared by slow filtration, drying and compression are translucent; however, films prepared by slow filtration freeze-drying do not show optical transparency, as shown in Fig. 3.

Fig. 3. Photographs of vegetable nanocellulose films formed from banana rachis isolated using the G30 method (Velásquez-Cock et al. 2016). (a) Prepared by slow filtration and drying at 40 °C for four days and compression at 3 ton at 120 °C for 4 min, and (b) slow filtration and freeze-drying.

Other optical properties that are relevant to most packaging applications are brightness and opacity. These properties are measured using a spectrophotometer (TAPPI 1994, 1996, 1998). Samples that scatter less light (less opaque) appear less bright (Spence et al. 2010). Nanocellulose films decrease in brightness when the fibres are refined, but the brightness increases after homogenization. Fibre cutting, which occurs during refining, results in smaller fibre sizes that scatter less light, resulting in lower opacity. With homogenization, these fibre components are not reduced in length but are defibrillated, resulting in the ability to scatter more light (higher opacity) (Spence et al. 2010). The brightnesses and opacities of nanocellulose films are also influenced by their compositions. Nanocellulose films from hardwood present lower optical transparencies because of the presence of xylan, which could influence the dispersion degree of the nanofibrils in water (Dufresne 2013; Spence et al. 2010).

Mechanical Properties of Vegetable Nanocellulose Films

The mechanical properties of nanocellulose depend on the vegetable source and isolation process, for example, Ferrer et al. (2012b) produced different cellulose pulps from empty palm fruit bunch fibres (EPFBF) using three different treatments NaOH-AQ, FoOH and Milox. The tensile strengths obtained for these nanocellulose materials from EPFBF pulps were 137 ± 7 MPa, 106 ± 9 MPa and 124 ± 12 MPa, respectively (Ferrer et al. 2012a). These values were higher than the values corresponding to those obtained from birch pulps using similar manufacture procedures, ca. 100 MPa (Ferrer et al. 2012b).

Different strategies, both theoretical and experimental, have been used to determine the mechanical properties of nanocellulose. A broad range of values has been reported, as the results depend on the crystallinity, RH and polymerization degree (DP), among other variables (Dufresne 2013). Henriksson et al. (2008) reported that the tensile strength, toughness, and strain-to-failure correlate with the average molar mass and porosity for a film made of wood nanocellulose. The tensile strength varies from 129 ± 8.7 MPa to 214 ± 6.8 MPa for samples with DP between 410 and 1100, respectively. In the same study, it was determined that an increase in the porosity from 19% to 40% decreased the elastic modulus from 14.7 ± 0.5 GPa to 7.4 ± 0.6 GPa and the tensile strength from 205 ± 13 MPa to 95 ± 8 MPa, respectively. In the more porous networks, the interfibril bonds may be weakened by the reduced hydrogen-bonding density. This film was prepared using acetone, which is less hydrophilic than water, resulting in a film with a less porous network.

Barrier Properties of Vegetable Nanocellulose Films

Nanoscale pores in nanocellulose films have led researchers to anticipate their eventual use in barrier applications and, therefore, to investigate the influence of the preparation process on this property (Lavoine et al. 2012). Packaging materials should provide sufficient protection against water vapour and oxygen. Therefore, the water vapour transfer rate (WVTR) (volume of water vapour passing through a film per unit area and time under specific conditions), water vapour permeability (WVP), oxygen permeability and oxygen transmission rate (OTR) are the most commonly studied barrier properties in nanocellulose films.

The barrier properties of nanocellulose can be attributed to several causes. The dense network formed by nanofibrils (Belbekhouche et al. 2011) and the ability of nanocellulose to form hydrogen bonds, resulting in a strong network (see Fig. 4), make it an excellent material for barrier applications, as it is difficult for gas molecules to pass through the network (Fukuzumi 2012; Svagan et al. 2016). Figure 4 shows a transmission electronic microscopy (TEM) image of nanocellulose isolated from banana rachis following the G30 treatment (Velásquez-Cock et al., 2016). A strong network of nanocellulose fibrils with a spiral shape that was formed through the defibrillation process can be observed in the image. This shape is related to the hierarchical structure of banana rachis (Gañan et al. 2008).

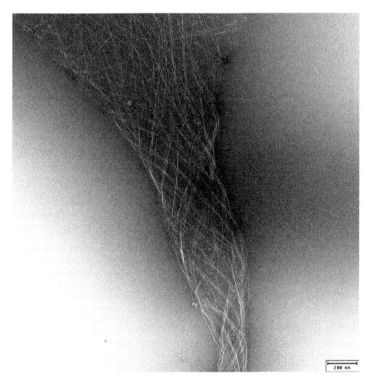

Fig. 4. Transmission electron microscopy image of vegetable nanocellulose from banana rachis isolated using the G30 method proposed for Velazquez-Cock et al. (2016).

There are studies indicating that hydrogen bonds and Van der Waals interactions lead to a more compact packing, higher cohesive energy density and lower free-volumes associated with lower gas permeability (Labuschagne et al. 2008). Finally, the crystalline regions of a polymer film are non-permeable, which increases the diffusion path length (Aulin et al. 2010). In semi-crystalline polymers, the crystalline regions are considered to be impermeable to gases (Aulin et al. 2010; Wüstenberk 2014). However, Belbekhouche and co-workers showed that parameters such as entanglement and nanoporosity exhibited a higher influence on water barrier than crystallinity (Belbekhouche et al. 2011).

According to previous investigations, the chemical compositions of wood pulps (Spence et al. 2010) and the microstructure of the nanocellulose play significant roles in the oxygen and water permeability properties of films (Aulin et al. 2010). Spence et al. (2010) determined that samples with increased lignin contents had higher WVTR because the increased content of hydrophobic components resulted in non-adsorbing pores through the film structure (Spence et al. 2010).

Film thickness also influences barrier properties. The oxygen permeability increases with film thickness, supporting the pore blocking theory, that is, fewer connected pores throughout the film (Aulin et al. 2010; Rodionova et al. 2011). The oxygen permeabilities of nanocellulose films with thicknesses of 21 ± 1 µm were 17 ± 1 ml m^{-2} day^{-1}. This value makes nanocellulose competitive with synthetic polymers such as ethylene vinyl alcohol (EVOH) (3–5 ml m^{-2} day^{-1}) and polyester-oriented coated polyvinylidene chloride (PVDC) films (9–15 ml m^{-2} day^{-1}) of roughly the same thickness (Nair et al. 2014).

However, the oxygen permeabilities of nanocellulose films also depend on the relative humidity (RH) (Aulin et al. 2010), as shown in Table 1. At low RH (0%), nanocellulose films show very low oxygen permeabilities compared with films prepared from plasticized starch and have oxygen permeability values in the same range as conventional synthetic films, such as polyamide. At higher RHs, the oxygen permeability increases exponentially because of the plasticizing and swelling of nanocellulose by water molecules (Aulin et al. 2010). According to the literature, the oxygen permeability of carboxylated nanocellulose films at 50% RH is lower than the oxygen permeability of glycerol-plasticized starch but comparable with the permeability value of ethylene vinyl alcohol (EVOH), which is commonly used in barrier applications.

Although the oxygen barrier properties of nanocellulose films are competitive with some of the currently available commercial films made from synthetic polymers, their water vapour barrier remains low mainly because of the strong hydrophilic nature of nanocellulose (Nair et al. 2014). Alternatively, chemical modifications to the hydroxyl groups on the surface have also been studied using silane reagents, fluorination and acetylation (Ashori et al. 2014). These chemical treatments create a higher hydrophobic film surface, which is indicated by the increase in the water contact angle.

Table 1. Oxygen permeabilities of nanocellulose films and literature values for some renewable and synthetic polymers commonly used in food packaging.

Material	Oxygen permeability $(cm^3 \, lm)/(m^2 \, day \, kPa)$	Conditions	Reference
Nanocellulose films	3.52–5.03	50% RH	Syverud and Stenius 2009
Nanocellulose (TEMPO-oxidized)	0.004	0% RH	Fukuzumi et al. 2009
Nanocellulose (carboxymethylated)	0.0006	0% RH	Aulin et al. 2010
Nanocellulose (carboxymethylated)	0.85	50% RH	Aulin et al. 2010
Amylose–glycerol (2.5:1)	7	50% RH	Rindlav-Westling et al. 1998
Amylopectin–glycerol (2.5:1)	14	50% RH	Rindlav-Westling et al. 1998
Polyvinylidene chloride (PVDC)	0.1–3	50% RH	Lange and Wyser 2003
Polyvinyl alcohol (PVOH)	0.20	0% RH	Lange and Wyser 2003
Polyamide (PA)	1–10	0% RH	Lange and Wyser 2003
Ethylene vinyl alcohol (EVOH)	0.01–0.1	0% RH	Lange and Wyser 2003

However, they are not accepted for the development of food contact materials by the European food safety authority or the Food and Drug Administrator in the United States of America (FDA) (Commissions of the European Communities 1990; European Commission 2011).

Another option to improve the oxygen barrier properties of nanocellulose films at high relative humidity (RH) levels is the production of hybrid clay-nanocellulose films. Liu et al. (2011) observed that the addition of montmorillonite to nanocellulose films improves the barrier properties considerably at higher relative humidities. The oxygen transmission rate (OTR) of dry clay-nanopaper (50:50) was less than 0.001 cm^3 mm m^{-2} day^{-1} atm^{-1}. When the RH was increased to 95%, the OTR of film increased to 3.5 cm^3 mm m^{-2} day^{-1} atm^{-1}, which is significantly lower than the OTR for the neat nanocellulose film, 17.8 cm^3 mm m^{-2} day^{-1} atm^{-1}. The authors suggested that hybrid clay-nanocellulose films could be further developed into barrier layers in packaging applications (Liu et al. 2011).

On the other hand, most works are focused on developing nano-composites. Composites are mixtures of traditional materials (polymers, metals or ceramics) with fillers, such as fibres, flakes, spheres or particulates, among others (Khan et al. 2012). Nanocomposites are composites containing fillers that have at least one dimension on the nanoscale (Azizi Samir et al. 2005). These materials represent an alternative to improve several physicochemical properties of conventional polymer composites. In addition, nanocellulose-based composites will have better barrier properties than vegetable nanocellulose films; for example, the use of moisture protective layers over nanocellulose films and coatings. The water vapour permeabilities of paperboard and papers decrease with a multilayer coating of nanocellulose and alkyd resins. Those nanocomposite coatings reached values considered as high barriers in packaging applications even at 50% RH (Aulin and Ström 2013).

Vegetable Nanocellulose to Develop Nanocomposites

Nanocellulose is used to reinforce matrices that can be processed at lower temperatures than its maximal degradation temperature of approximately 330 °C, as shown in the thermogram presented in Fig. 5 for nanocellulose isolated from banana rachis, following the G30 method (Velásquez-Cock et al. 2016). Therefore, petrochemical polymers that can be processed under 300 °C are the most commonly used matrices to produce cellulose nanocomposites.

According to Cho and Park (2011), the degradation temperature of nanocellulose-reinforced poly(vinyl alcohol) nanocomposites increased as the nanocellulose content increased, and the major degradation peak temperature shifted to a higher temperature. In the case of the 1 wt% and 3 wt% nanocellulose loadings, there was no significant influence on the thermal stability. However, the thermal stability of the nanocomposite

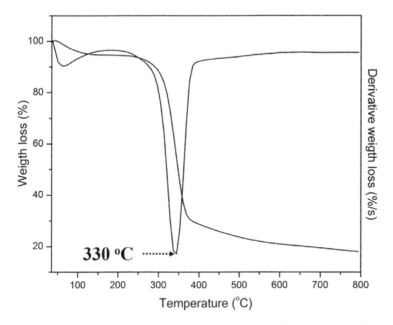

Fig. 5. Thermal analysis and derivative thermal analysis of vegetable
nanocellulose from banana rachis.

increased above these loading levels of nanocellulose, especially at the 5 wt%
loading (Cho and Park 2011).

Additionally, nanocellulose materials have been reported to have a
great effect in improving the moduli of polymer matrices (Avella et al. 2005;
Azeredo 2009; Isikgor and C. Remzi Becer 2015; Khan et al. 2012; Mohanty
et al. 2002; Montoya et al. 2014; Saunders and Taylor 1990; Sehaqui et al.
2011; Anna J. Svagan et al. 2008; Weber 2000). In addition, the ability of
nanocellulose to form a dense percolating network held together by strong
intermolecular bonds suggests their use as a barrier film nanocomposite
(Avella et al. 2005; Azeredo 2009; Isikgor and C. Remzi Becer 2015).

Effect of Vegetable Nanocellulose Addition on Polymer Matrices

The first studies on cellulose nanocomposites of a nanocellulose content up
to 7.4 wt.% were published by Nakagaito and Yano in 2005. The authors
developed composites using phenolic resin as a binder and achieved values
for the Young's modulus and tensile strength up to 19 GPa and 370 MPa,
respectively, with a density of 1.45 g/cm^2.

A porous network of nanocellulose from wood pulp was impregnated
by liquid low molar mass polyphenol formaldehyde (PF) precursors, and
the liquid was polymerized. The materials showed a high modulus and
tensile strength; however, the material was quite brittle, as the hydrophilic
nature of nanocellulose related to its OH-rich structure often results in

poor compatibility with non-polar polymer matrices, which leads to poor dispersion and low stress transfer efficiency between the matrix and the reinforcement (Nakagaito and Yano 2005).

The hydrophilic nature of nanocellulose can induce a dimensional instability and reduce the mechanical properties of the composites (Foresti et al. 2015). Therefore, it is necessary to partially or completely replace the hydroxyl groups on the surface with less hydrophilic groups (Azeredo 2009). This process can increase its dispersion within a wide range of matrix polymers (Foresti et al. 2015). However, the surface modification of nanocellulose uses catalysts and solvents that are not environmentally friendly and not allowed in the development of food contact materials. Recently, the surface modification of nanocellulose using naturally occurring α-hydroxycarboxylic acids as catalysts appeared as a promising alternative route for ecofriendly derivatization of nanocellulose (Foresti et al. 2015).

The previously mentioned challenges and the global environmental concern regarding the use of petroleum-based packaging materials are encouraging researchers and industries to search for packaging materials made from renewable resources. There is an increasing interest in nanocellulose biocomposites, as nanocomposite materials with high transparency and improved mechanical and barrier properties can be produced using renewable resources (Belbekhouche et al. 2011). However, as mentioned previously, the applications of biobased materials, for instance, nanocellulose, in the packaging industry are limited by their high water permeabilities and the influence of this water on the oxygen permeability.

Several researchers have investigated biobased composites reinforced with nanocellulose, which is commonly used to increase both mechanical and permeability properties (Aulin et al. 2010; Li et al. 2015; Nair et al. 2014; Svagan et al. 2007; Syverud and Stenius 2009). Fukuzumi et al. (2011) prepared high-oxygen-barrier nanocellulose films via TEMPO mediated oxidation. The 0.80 μm thick nanocellulose film, when coated on a 25 μm polylactic acid film, reduced the oxygen permeability from 7.4 ml m^{-2} day^{-1} kPa^{-1} for a pure polylactide acid (PLA) film to 9.9×10^{-4} ml m^{-2} day^{-1} Pa^{-1} for a PLA film with a nanocellulose layer.

In addition, starch-based materials have been extensively investigated as a choice product to improve the biodegradability of a variety of plastics. Starch is one of the most studied raw materials for the development of composites because it is readily available, cheap and biodegradable (Montoya et al. 2014). However, the brittleness of starch requires the use of plasticizers such as polyols (Azeredo 2009), which improve starch flexibility but decrease its thermomechanical properties. The addition of nanocellulose to starch systems enhances their thermal and mechanical properties (Montoya et al. 2014), reduces their water sensitivity and preserves their biodegradability properties (Azeredo 2009).

Dufresne et al. (2000) studied the mechanical properties at room temperature of nanocellulose and starch using a tensile test, and both glycerol and water to plasticize the starch matrix. The highest mechanical properties were obtained with an unplasticized starch matrix in a dry atmosphere. The reinforcing effect was more significant in plasticized starch due to the decrease in the Tg of the matrix down to temperatures lower than room temperature. In addition, the authors reported that water sensitivity linearly decreases with nanocellulose content (Alain et al. 2000).

Nanocellulose and starch can also be used to develop foams, as pure nanocellulose foams or as composite foams reinforced with nanocellulose. Svagan et al. (2008) used a freeze-drying technique to reinforce starch foams with nanocellulose. Ultrahigh porosity foams prepared from nanofibril suspensions of native cellulose were studied by Sehaqui et al. (2011). The compression resistance of the nanocomposite was improved by controlling the density and nanofibril interaction in the foams.

Nanocomposites of starch foam reinforced with nanocellulose are developed to substitute polystyrene-based foams (Klemm et al. 2011; Wüstenberk 2014). Pure nanocellulose foams are also produced using various freeze-drying and super critical CO_2 drying techniques (Klemm et al. 2011; Wüstenberk 2014).

Vegetable Nanocellulose for the Development of Edible Coatings

Edible films or coatings have provided an interesting means for controlling the quality and stability of several food products. Depending on the properties of the coating (e.g. organoleptic, mechanical, gas and solute barrier), it can be used to wrap various products, for example, protection of individual dried fruits, meat and fish and control of internal moisture transfer in cakes and pies, among others (Guilbert et al. 1995). Edible coatings act as barriers to moisture, undesirable microorganisms, gases and/or UV light. The coating should be essentially harmless to consumers, transparent in the visible region so the coated product is visible to the consumer, and it should impart no significant odour or taste to the foods (Zhao et al. 2013). Edible films and coatings from biopolymers have been widely used as lipid, water vapour, gas, and flavour barriers for fresh fruits and vegetables, frozen foods and meat products. Corn zein and sucrose fatty acid ester coatings have been applied successfully on fresh fruits and vegetables, such as apples, bananas and tomatoes, as oxygen and water vapour barriers to extend their shelf lives.

However, the edible coatings and films that have potential as packaging materials due to one or more functional properties often suffer from poor water resistance. In 2013, Zhao et al. patented a nanocoating made of nanocellulose, nanocalcium carbonate and/or other additives with improved water resistance. This coating protects fresh fruits and vegetables both pre- and post-harvest against moisture loss and UV damage. It also acts as a barrier coating for fresh and processed foods by reducing or preventing leaching of food substances, such as anthocyanins and other water-soluble compounds, as well as the loss and/or gain of moisture and gases such as O_2 and CO_2 during food processing and storage (Zhao et al. 2013).

A fundamental prerequisite for edible coatings and food contact materials in general is safety. Wood pulp and powdered cellulose are generally recognized as safe (GRAS) and can be used as raw materials for food contact materials and even as food additives (Lähtinen et al. 2012; Gómez et al. 2016). However, their biological impacts on ecosystems and humans cannot be evaluated merely on the chemical characteristics of the cellulose because the size, shape, aggregation properties and different unknown factors might still affect the interactions of nanocellulose with cells and other living organisms. Therefore, it is necessary to study the cytotoxicity and genotoxicity of nanocellulose.

Nanocellulose Safety Issues

Currently, few *in vitro* and *in vivo* studies looking at the toxic properties of nanocellulose isolated from different sources are available (Andrade et al. 2015; Hannukainen et al.2012; Pitkänen et al. 2010; Vartiainen et al. 2011). Table 2 summarizes recent reports on toxicology experiments and conclusions for vegetable nanocellulose. In recent decades, the nanotoxicology research on nanoparticles has built a comprehensive assessment system for metallic nanoparticles and carbon nanotubes. However, the toxicology study of nanocellulose and nanocellulose-based biocomposites is still restricted to a very preliminary stage (mainly on the level of cytotoxicity) (Lin and Dufresne 2014). Therefore, to receive more unambiguous information on the safety impacts of nanocellulose nanomaterials, internationally standardized methods are needed to ensure their safety (Lähtinen et al. 2012; Szakal et al. 2014).

According to the results from the tests of cytotoxicity and genotoxicity summarized in Table 2, vegetable nanocellulose materials have not shown any evidence for serious influence or damage at both the cellular and genetic levels as well as in *in vivo* organ and animal experiments. However, some authors have reported that the inhalation of nanocellulose may induce pulmonary inflammation due to the easy self-aggregation and non-degradation of nanocellulose in the bodies of animals (Lin and Dufresne 2014).

Conclusions

The nanocellulose properties summarized in this literature review indicate its great potential for packaging materials. The mechanical, optical and

Table 2. Summary of toxicological evaluations of vegetable nanocellulose.

Description of toxicological evaluation	Conclusions	References
Authors evaluated the worker exposures to particles in air during grinding and spray drying of birch nanocellulose. Mouse macrophages and human monocyte derived macrophages were exposed to nanocellulose, and the viabilities and cytokine profiles of the cells were studied thereafter.	No evidence of inflammatory effects or cytotoxicity on mouse and human macrophages was observed after 6 and 24 h exposure to the materials studied. Processing of nanocellulose with either a friction grinder or a spray dryer did not cause significant exposure to particles during normal operation.	Vartiainen et al. 2011
Nanocellulose was prepared by Masuko grinding using enzymatic pretreatment (NFC-1) and the other by microfluidization using pretreatment by tempo oxidation (NFC-2). Authors evaluated potential genotoxicity of two NFCs in human bronchial epithelial BEAS 2B cells. Cytotoxicity was analyzed after exposure for 4, 24 and 48 h at 9.5–950 µg/cm^2 by propidium iodine staining and luminometric assay. As cytotoxicity did not reach the 50% level at any of the doses tested, 950 µg/cm^2 was chosen as the highest dose for the genotoxicity assays.	NFC-1-induced DNA damage in BEAS 2B cells, but no induction of oxidative DNA damage was seen. In addition, NFC-2 was able to produce DNA damage in BEAS 2B cells, and there was a sporadic indication of oxidative DNA damage at 5 µg/cm^3.	Hannukainen et al. 2012
Nanocellulose from peach palm	An *in vivo* study in which male mice (*Rattus norvegicus albinus*) were used demonstrated that nanocellulose did not have harmful effects on the animal metabolism.	Andrade et al. 2015
Nanocellulose from Eucalyptus and Pinus radiata pulp fibres. The nanofibrillated materials were manufactured using a homogenizer without pretreatment and with 2,2,6,6-tetramethylpiperidine-1-oxy radical as a pretreatment.	Cytotoxicity tests were applied to the samples, which demonstrated that the nanofibres do not exert acute toxic phenomena on the tested fibroblast cells.	Alexandrescu et al. 2013

barrier properties of nanocellulose make it suitable for the development of food packaging materials. However, because of the hydrophilic character of nanocellulose, these properties strongly depend on the relative humidity. Therefore, nanocellulose is commonly used with polymer matrices. However, the hydrophilic nature of nanocellulose and the hydrophobic character of petrochemical polymers lead to poor dispersion and low stress transfer efficiency between the matrix and the reinforcement. The hydrophilicity of nanocellulose can be reduced by modifying the nanocellulose surface, but this process uses catalysts and solvents that are not allowed in the development of food contact materials. For this reason, there is increasing interest in nanocellulose biocomposites using biopolymers, such as PLA and starch, in the development of nanostructured films or nanostructured foams. The addition of nanocellulose to biopolymers improves the mechanical and barrier properties.

The development of edible coatings made of nanocellulose, nanocalcium carbonate and/or other additives is also presented. A fundamental prerequisite for the development and commercialization of edible coatings and for food contact materials in general is safety. Several *in vitro* and *in vivo* studies developed in recent years have suggested that nanocellulose is not cytotoxic or genotoxic. However, the inhalation of nanocellulose may induce pulmonary inflammation.

Acknowledgments

The authors acknowledge the National Financing Fund for Science, Technology and Innovation Francisco José de Caldas of the Colombian Government (COLCIENCIAS) and the Research Center for Investigation and Development (CIDI) from the Universidad Pontificia Bolivariana for their financial support through the program *Es tiempo de volver*. The authors also thank PhD J-L Putaux from the *Centre de Recherches sur les Macromolécules Végétales (CERMAV-CNRS)*, who is affiliated with Université Joseph for the transmission electronic microscopy images (TEM).

References

Alain, D., D. Danièle and R.V. Michel. 2000. Cellulose microfibrils from potato tuber cells: Processing and characterization of starch-cellulose microfibril composites. J. Appl. Polym. Sci. 76: 2080–2092.

Alexandrescu, L., K. Syverud, A. Gatti and G. Chinga-Carrasco. 2013. Cytotoxicity tests of cellulose nanofibril-based structures. Cellulose 20: 1765–1775.

Andrade, D.R.M., M.H. Mendonça, C.V. Helm, W.L.E. Magalhães, G.I.B. de Muniz and S.G. Kestur. 2015. Assessment of nano cellulose from peach palm residue as potential food additive: Part II: Preliminary studies. J. Food Sci. Tech. 52: 5641–5650.

Ankerfors, M. 2012. Microfibrillated cellulose: Energy-efficient preparation techniques and applications in paper. Licentiate Thesis, KTH Royal Institute of Technology, Stockholm, Sweden.

Ashori, A., M. Babaee, M. Jonoobi and Y. Hamzeh. 2014. Solvent-free acetylation of cellulose nanofibers for improving compatibility and dispersion. Carbohyd. Polym. 102: 369–375.

Aulin, C., M. Gällstedt and T. Lindström. 2010. Oxygen and oil barrier properties of microfibrillated cellulose films and coatings. Cellulose 17: 559–574.

Aulin, C. and G. Ström. 2013. Multilayered alkyd resin/nanocellulose coatings for use in renewable packaging solutions with a high level of moisture resistance. Ind. Eng. Chem. Res. 52: 2582–2589.

Avella, M., J.J. De Vlieger, M.E. Errico, S. Fischer, P. Vacca and M.G. Volpe. 2005. Biodegradable starch/clay nanocomposite films for food packaging applications. Food Chem. 93: 467–474.

Azeredo, H.M.C. 2009. Nanocomposites for food packaging applications. Food Res. Int. 42: 1240–1253.

Azizi Samir, M.A.S., F. Alloin and A. Dufresne. 2005. Review of recent research into cellulosic whiskers, their properties and their application in nanocomposite field. Biomacromolecules 6: 612–626.

Belbekhouche, S., J. Bras, G. Siqueira, C. Chappey, L. Lebrun, B. Khelifi, et al. 2011. Water sorption behavior and gas barrier properties of cellulose whiskers and microfibrils films. Carbohyd. Polym. 83: 1740–1748.

Castro, C., R. Zuluaga, J-L. Putaux, G. Caro, I. Mondragon and P. Gañán. 2011. Structural characterization of bacterial cellulose produced by Gluconacetobacter swingsii sp. from Colombian agroindustrial wastes. Carbohyd. Polym. 84: 96–102.

Chaib, F. 2015. World health organization. Retrieved March 27, 2016, from http://www.who.int/mediacentre/news/releases/2015/foodborne-disease-estimates/en/

Cho, M-J. and B.-D. Park. 2011. Tensile and thermal properties of nanocellulose-reinforced poly(vinyl alcohol) nanocomposites. J. Ind. and Eng. Chem. 17: 36–40.

Commissions of the European Communities. 1990. Relating to plastic materials and articles intended to come into contact with foodstuffs.

Cunha, A.G., J-B. Mougel, B. Cathala, L. Berglund and I. Capron. 2014. Preparation of double pickering emulsions stabilized by chemically tailored nanocelluloses. Langmuir 30: 9327–9335.

Dufresne, A. 2013. Nanocellulose: A new ageless bionanomaterial. Mater. Today 16: 220–227.

European Commission. 2011. Commission Regulation (EU) No 10/2011 of 14 January 2011 on Plastic materials and articles intended to come into contact with food. Official Journal of the European Union 15: 12–88.

Ferrer, A., I. Filpponen, A. Rodriguez, J. Laine and O.J. Rojas. 2012a. Valorization of residual Empty Palm Fruit Bunch Fibers (EPFBF) by microfluidization: Production of nanofibrillated cellulose and EPFBF nanopaper. Bioresour. Technol. 125: 249–255.

Ferrer, A., E. Quintana, I. Filpponen, I. Solala, T. Vidal, A. Rodríguez, et al. 2012b. Effect of residual lignin and heteropolysaccharides in nanofibrillar cellulose and nanopaper from wood fibers. Cellulose 19: 2179–2193.

Foresti, M.L., J.A. Avila Ramirez, C. Gómez Hoyos, S. Arroyo and P. Cerrutti. 2015. Naturally occurring α-Hydroxy Acids: Useful organocatalysts for the acetylation of cellulose nanofibres. Curr. Organocatal. 2: 1–1.

Fukuzumi, H., T. Saito, S. Iwamoto, Y. Kumamoto, T. Ohdaira, R. Suzuki, et al. 2011. Pore size determination of TEMPO-oxidized cellulose nanofibril films by positron annihilation lifetime spectroscopy. Biomacromolecules 12: 4057–4062.

Fukuzumi, H. 2012. Studies on structures and properties of TEMPO-oxidized cellulose nanofibril films. PhD Thesis, University of Tokyo.

Gañan, P., R. Zuluaga, J. Cruz, J.M. Velez, I. Mondragon and A. Retegi. 2008. Elucidation of the fibrous structure of Musaceae maturate. Cellulose 15: 131–139.

Gómez Hoyos, C., A. Serpa, J. Velásquez-Cock, P. Gañán, C. Castro, L.Vélez, et al. 2016. Vegetable nanocellulose in food science: A review. Food Hydrocolloid. 57: 178–186.

Guilbert, S., N. Gontard and B. Cuq. 1995. Technology and applications of edible protective films. Packag. Technol. Sci. 8: 339–346.

Hannukainen, K-S., S. Suhonen, K. Savolainen and H. Norppa. 2012. Genotoxicity of nanofibrillated cellulose *in vitro* as measured by enzyme comet assay. Toxicol. Lett. 211: S71.

Henriksson, M., L. Berglund, P. Isaksson, T. Lindström, and T. Nishino. 2008. Cellulose nanopaper structures of high toughness. Biomacromolecules 9: 1579–1585.

Isikgor, F.H. and C. Remzi Becer. 2015. Lignocellulosic biomass: A sustainable platform for production of bio-based chemicals and polymers. Polym. Chem. 6: 4497–4559.

Khan, A., T. Huq, R. Khan, B. Riedl and M. Lacroix. 2012. Nanocellulose based composites and bioactive agents for food Packaging. Crit. Rev. Food Sci. Nutr. 54: 163–174.

Klemm, D., B. Heublein, H-P. Fink and A. Bohn. 2005. Cellulose: Fascinating biopolymer and sustainable raw material. Angew. Chem. Int. Ed. 44: 3358–3393.

Klemm, D., D. Schumann, F. Kramer, N. Heßler, D. Koth and B. Sultanova. 2009. Nanocellulose materials – Different cellulose, different functionality. Macromol. Symp. 280: 60–71.

Klemm, D., F. Kramer, S. Moritz, T. Lindström, M. Ankerfors, D. Gray, et al. 2011. Nanocelluloses: A new family of nature-based materials. Angew. Chem. Int. Ed. 50: 5438–5466.

Labuschagne, P.W., W.A. Germishuizen, S.M. Sabine and F.S. Moolman. 2008. Improved oxygen barrier performance of poly(vinyl alcohol) films through hydrogen bond complex with poly(methyl vinyl ether-co-maleic acid). Eur. Polym. J. 44: 2146–2152.

Lacroix, M. 2015. New development of application of nanocellulose for food packaging application. *In*: International Conference and Exhibition on Biopolymers and Bioplastics. San Francisco, USA, 10–12 August.

Lähtinen, K., H. Valve, T. Jouttijärvi, P. Kautto, S. Koskela, P. Leskinen, et al. 2012. Piecing together research needs: Safety, environmental performance and regulatory issues of nanofibrillated cellulose (NFC). Helsinki. Retrieved from file:///C:/Users/User/Desktop/FINAL_nanosellu_130812.pdf

Lavoine, N., I. Desloges, A. Dufresne and J. Bras. 2012. Microfibrillated cellulose – Its barrier properties and applications in cellulosic materials: A review. Carbohydr. Polym. 90: 735–764.

Li, F., E. Mascheroni and L. Piergiovanni. 2015. The potential of nanocellulose in the packaging field: A review. Packag. Technol. Sci. 28: 475–508.

Lin, N. and A. Dufresne. 2014. Nanocellulose in biomedicine: Current status and future prospect. Eur. Polym. J. 59: 302–325.

Liu, A., A. Walther, O. Ikkala, L. Belova and L. Berglund. 2011. Clay nanopaper with tough cellulose nanofiber matrix for fire retardancy and gas barrier functions. Biomacromolecules 12: 633–641.

Mckendry, P. 2002. Energy production from biomass (part 1): Overview of biomass. Bioresour. Technol. 83: 37–46.

Missoum, K., J. Bras and M.N. Belgacem. 2012. Water redispersible dried nanofibrillated cellulose by adding sodium chloride. Biomacromolecules 13: 4118–4125.

Mohanty, A.K., M. Misra and L.T. Drzal. 2002. Sustainable bio-composites from renewable resources: Opportunities and challenges in the green materials world. J. Polym. Environ. 10: 19–26.

Montoya, U., R. Zuluaga, C. Castro, S. Goyanes and P. Gañán. 2014. Development of composite films based on thermoplastic starch and cellulose microfibrils from Colombian agroindustrial wastes. J. of Thermoplast. Compos. Mater. 27: 413–426.

Nair, S.S., J. Zhu, Y. Deng and A.J. Ragauskas. 2014. High performance green barriers based on nanocellulose. Sustain. Chem. Process. 2: 1–7.

Nakagaito, A.N. and H. Yano. 2005. Novel high-strength biocomposites based on microfibrillated cellulose having nano-order-unit web-like network structure. Appl. Phys. A: Mater. Sci. Process. 80: 155–159.

Peng, Y., D.J. Gardner and Y. Han. 2012. Drying cellulose nanofibrils: In search of a suitable method. Cellulose 19: 91–102.

Pitkänen, M., U. Honkalampi, A. Von Wright, A. Sneck, H.-P. Hentze, J. Sievänen et al. 2010. Nanofibrillar cellulose – in vitro study of cytotoxic and genotoxic properties. *In*: TAPPI – International Conference on Nanotechnology for the Forest Products Industry. Otaniemi, Espoo, Finland, 27–29 September.

Rämänen, P., P.A. Penttilä, K. Svedström, S.L. Maunu and R. Serimaa. 2012. The effect of drying method on the properties and nanoscale structure of cellulose whiskers. Cellulose 19: 901–912.

Rindlav-Westling, A., M. Stading, A.-M. Hermansson and P. Gatenholm. 1998. Structure, mechanical and barrier properties of amylose and amylopectin films. Carbohydrate Polymers 36(2–3): 217–224. http://doi.org/10.1016/S0144-8617(98)00025-3

Robinson, D.K.R. and M.J. Morrison. 2010. Nanotechnologies for food packaging. Reporting the science and technology research trends. Report for the ObservatoryNANO. Retrieved from www.observatorynano.eu

Robson, A. 2012. Tackling obesity: Can food processing be a solution rather than a problem? Agro Food Ind. Hi-Tech 23: 10–11.

Rodionova, G., M. Lenes, O. Eriksen and O. Gregersen. 2011. Surface chemical modification of microfibrillated cellulose: Improvement of barrier properties for packaging applications. Cellulose 18: 127–134.

Saito, T. and A. Isogai. 2004. TEMPO-mediated oxidation of native cellulose. The effect of oxidation conditions on chemical and crystal structures of the water-insoluble fractions. Biomacromolecules 5: 1983–1989.

Saunders, C.W. and L.T. Taylor. 1990. A review of the synthesis, chemistry and analysis of nitrocellulose. J. Energ. Mater. 8: 149.

Sehaqui, H., Q. Zhou and L. Berglund. 2011. High-porosity aerogels of high specific surface area prepared from nanofibrillated cellulose (NFC). Compos. Sci. Technol. 71: 1593–1599.

Sharma, V. and P.P. Kundu. 2006. Addition polymers from natural oils—A review. Prog. Polym. Sci. 31: 983–1008.

Silvestre, C., D. Duraccio and S. Cimmino. 2011. Food packaging based on polymer nanomaterials. Prog. Polym. Sci. 36: 1766–1782.

Siró, I. and D. Plackett. 2010. Microfibrillated cellulose and new nanocomposite materials: A review. Cellulose 17: 459–494.

Spence, K.L., R.A., Venditti, O.J. Rojas, Y. Habibi and J.J. Pawlak. 2010. The effect of chemical composition on microfibrillar cellulose films from wood pulps: Water interactions and physical properties for packaging applications. Cellulose 17: 835–848.

Spence, K.L., R.A. Venditti, O.J. Rojas, Y. Habibi and J.J. Pawlak. 2011. A comparative study of energy consumption and physical properties of microfibrillated cellulose produced by different processing methods. Cellulose 18: 1097–1111.

Ström, G., C. Öhgren and M. Ankerfors. 2013. Nanocellulose as an additive in foodstuff. Stockholm. Retrieved from http://www.innventia.com/Documents/Rapporter/Innventia report403.pdf

Svagan, A.J., M.A.S. Azizi Samir and L. Berglund. 2007. Biomimetic polysaccharide nanocomposites of high cellulose content and high toughness. Biomacromolecules 8: 2556–2563.

Svagan, A.J., M.A.S. Azisi Samir and L. Berglund. 2008. Biomimetic foams of high mechanical performance based on nanostructured cell walls reinforced by native cellulose nanofibrils. Adv. Mater. 20: 1263–1269.

Svagan, A.J., C. Bender Koch, M.S. Hedenqvist, F. Nilsson, G. Glasser, S. Baluschev, et al. 2016. Liquid-core nanocellulose-shell capsules with tunable oxygen permeability. Carbohyd. Polym. 136: 292–299.

Syverud, K. and P. Stenius. 2009. Strength and barrier properties of MFC films. Cellulose 16: 75–85.

Szakal, C., S.M. Roberts, P. Westerhoff, A. Bartholomaeus, N. Buck, I. Illuminato, et al. 2014. Measurement of nanomaterials in foods: Integrative consideration of challenges and future prospects. ACS Nano 8: 3128–3135.

TAPPI. 1994. Color of paper and paperboard. 2000–2001 TAPPI Test Methods.

TAPPI. 1996. Diffuse opacity of paper. 2000–2001 TAPPI Test Methods.

TAPPI. 1998. Brightness of pulp, paper, and paperboard (directional reflectance at 457 nm)

TAPPI. 2011. Proposed New TAPPI Standard: Standard Terms and Their Definition for Cellulose Nanomaterial.

Turbak, A.F., F.W. Snyder and K.R. Sandberg. 1983. Microfibrillated cellulose, a new cellulose product: Properties, uses, and commercial potential. J. Appl. Polym. Sci. Appl. Polym. Symp. 37: 815–827.

Uetani, K. and H. Yano. 2011. Nanofibrillation of wood pulp using a high-speed blender. Biomacromolecules 12: 348–353.

Vartiainen, J., T. Pöhler, K. Sirola, L. Pylkkänen, H. Alenius, J. Hokkinen, et al. 2011. Health and environmental safety aspects of friction grinding and spray drying of microfibrillated cellulose. Cellulose 18: 775–786.

Velásquez-Cock, J., P. Gañán, P. Posada, C. Castro, A. Serpa, C. Gómez Hoyos, et al. 2016. Influence of combined mechanical treatments on the morphology and structure of cellulose nanofibrils: Thermal and mechanical properties of the resulting films. Ind. Crops Prod. 85: 1–10.

Weber, C.J. 2000. Biobased packaging materials for the food industry: Status and perspectives: A European concerted action. KVL Department of Dairy and Food Science. Retrieved from https://books.google.com.co/books?id=W9OQAAAACAAJ

Winuprasith, T. and M. Suphantharika. 2013. Microfibrillated cellulose from mangosteen (*Garcinia mangostana L.*) rind: Preparation, characterization, and evaluation as an emulsion stabilizer. Food Hydrocolloid. 32: 383–394.

Wüstenberk, T. 2014. Cellulose and Cellulose Derivatives in the Food Industry. Fundamental and Applications. Wiley VCH, Weinheim.

Yano, H., K. Abe, Y. Kase, S. Kikkawa and Y. Onishi. 2012. Frozen dessert and frozen dessert raw material. EU Patent # 2 756 762 A1.

Žepič, V., E.Š. Fabjan, M. Kasunič, R.C. Korošec, A. Hančič, P. Oven, et al. 2014. Morphological, thermal, and structural aspects of dried and redispersed nanofibrillated cellulose (NFC). Holzforschung 68: 657–667.

Zhao, Y., J. Simonsen, G. Cavender, J. Jung and L.H. Fuchigami. 2013. Nanocellulose edible coatings and uses thereof. US Patent # WO2014153210A1.

Zhao, Y., J. Simonsen, G. Cavender, J. Jung and L.H. Fuchigami. 2015. Nanocellulose coatings to prevent damage in foodstuffs. US Patent # 20140272013 A1.

Zuluaga, R., J-L. Putaux, J. Cruz, J. Vélez, I. Mondragon and P. Gañán. 2009. Cellulose microfibrils from banana rachis: Effect of alkaline treatments on structural and morphological features. Carbohyd. Polym. 76: 51–59.

Xylan Polysaccharide Fabricated into Biopackaging Films

Xiaofeng Chen[1], Junli Ren[1,*], Chuanfu Liu[1], Feng Peng[2] and Runcang Sun[2]

[1] State Key Laboratory of Pulp and Paper Engineering, South China University of Technology, Guangzhou, 510640, Guangdong, China
[2] Beijing Key Laboratory of Lignocellulosic Chemistry, College of Materials Science and Technology, Beijing Forestry University, Beijing, 100083, China

Introduction

Due to the beautiful, cheap and convenient manufacturing process, plastic packaging materials are widely used in food and other industry. However, they are dominated by petroleum-derived polymers, which are not environmentally safe and degradable. They do not meet the need for the sustainable development in the future. With diminishing fossil fuel resources and the increasing impact of global climate change, more and more researches have been devoted to the exploration of biodegradable films from renewable resources mainly due to the promising alternative for non-biodegradable petroleum-based packaging films. Therefore, utilizing biodegradable polymers such as natural polymers to produce packaging films becomes an inevitable trend in food science and technology development. Recently, hemicelluloses, which rank second to cellulose in plant cell walls, have received an increasing interest especially due to their abundance, degradability and renewability in the last decade for the production of biodegradable films and coating (Hansen and Plackett 2008; Mikkonen and Tenkanen 2012).

In recent years, xylan, as one of the main hemicelluloses, has been widely applied for preparation of biodegradable packaging films as the promising

*Corresponding author: renjunli@scut.edu.cn

raw material. The chemical and thermal stability of hemicelluloses is generally lower than cellulose, due to the lack of crystallinity and low DP. It was found that xylan has the poor film-forming ability without any treatment or additives. Recently, more efforts such as modification of xylan, blending with other polymers or mixing with nanoparticles have been made to prepare xylan-based films with improved properties. This chapter covers the structure of xylan-type hemicelluloses, the influence of the composition and structure of xylan on film properties and approaches to improve the xylan's film-forming abilities and the films' properties.

The Structue of Xylan-type Hemicelluloses

Hemicelluloses are considered to be the second most abundant biopolymer in the plant kingdom. As plant constituents, hemicelluloses are not only present in wood but also in various other plants such as grasses, cereals and herbs (Ebringerova and Heinze 2000). Hemicelluloses constitute complex branched heteropolysaccharides composed by pentose (xylose and arabinose) and hexose (glucose, galactose and mannose). As heteropolymers, hemicelluloses mainly include xyloglucans, xylans, mannans, glucomannans and β-(1→3, 1→4)-glucans (Scheller and Ulvskov 2010). Different plant species have various hemicellulose contents such as wheat straw 32%, barley straw 32%, oat straw 27%, rye straw 31%, rice straw 25%, sunflower husk 23%, sugarcane rind 22% and corn cobs 37% (Fang et al. 2000; Peng et al. 2012). Due to the difference from cellulose with a high molecular weight, the average degree of polymerization (DP) of hemicelluloses is in the range of 80-200, which leads to low-molecular-weight polymers (Sun et al. 2000).

Xylan is the most common hemicelluloses which are present in hard wood and perennial plants such as grasses, cereals and herbs (Petzold-Welcke et al. 2014). However, the chemical structure of xylan varies among different plants, including the molecular architecture and the degree of substitution (DS). Xylan can be divided into the sub-groups of heteroxylans (HX) and homoxylans (X), and they have different chemical structures. HX is common in the cell wall of higher plants (Table 1). In hardwoods, glucuronoxylan (GX) consists of a backbone of (1→4)-linked β-D-xylopyranosyl (Xyl*p*) and is substituted with glucurononosyl and 4-*O*-methylglucuronosyl by α-(1→2) linkages. In addition, most of xylose residues contain an acetyl group at C-2 or C-3. In softwoods, the β-(1→4)-D-xylopranose backbone of arabinoglucuronoxylan (AGX) is substituted with single 4-*O*-methyl-D-glucuronic acid (MeGlcA) and α-L-arabinofuranosyl (α-L-Ara*f*) units at positions 2 and 3. Arabinoxylan (AX) in gramineae has the linear β-(1→4)-D-Xyl*p* as backbone and is substituted by α-L-Ara*f* units in the positions 2-*O* and/or 3-*O* (Ebringerova and Heinze 2000).

Table 1. Categories of xylan.

Plant type	Xylan type	Biological origin	Structure
Higher plants	Heteroxylan	Glucuronoxylan	Hardwoods	
		Arabinoglucuronoxylan	Softwoods	
		Arabinoxylan	Gramineae	
Others	Homoxylan			

The Influence of the Composition and Structure of Xylan on Film Properties

Different plants and various isolation and purity methods could yield the various chemical compositions of xylan; consequently the different structure of xylan could have an important influence on its properties. Some researchers have focused more attentions on the investigation of the relationship between the xylan structure and xylan-based films properties. Table 2 summarizes the films properties of xylan with different Ara/Xyl ratios.

Researchers found that the DS of xylan had a crucial effect on the solubility, film formation and material properties. Egues et al. (2013) confirmed that xylan with higher Ara/Xyl ratio had a better film forming property. Sternemalm et al. undertook the investigation of the effect of arabinose content on film properties (Sternemalm et al. 2008). A rye arabinoxylan (AX) with an Ara/Xyl ratio of 0.52 and an average number molecular weight of 305 kDa was hydrolyzed by oxalic acid, which could attempt the selective removal of the arabinose substituent on the xylan main chain. The debranching of the polymer was coupled with a decrease in molecular weight. It was found that the removal of arabinose substituent resulted in a gradual association of unsubstituted chains. A linear correlation was present between arabinose substitution and the moisture content of arabinoxylan at high RH, but it was different at low RH. This agreed with the observations by Mikkonen et al. (2012). A decrease in arabinose content resulted in the loss of a plasticizing effect, as determined by the dynamic mechanical analysis, which was correlated to the water binding capacity. Zhang et al. (2011) also studied the impact of composition and structure of wheat bran AX on physico-chemical properties when it was used for packaging films. The crystalline morphology, the beta transitions as well as the water sorption properties of AX films were well correlated to the substitution degree by arabinose groups. When the substitution degree was low, AX was less soluble in water and ethanol, less hydrophilic when there exist high relative humidity or crystalline morphology, and more surprisingly, the local macromolecular mobility increased. Thus, xylan-type hemicelluloses based materials showed physico-chemical properties strongly linked to the native Ara/Xyl ratios. AX with different Ara/Xyl ratios was also investigated for the properties of xylan films (Ying et al. 2013). The films based on xylan with the low Ara/Xyl ratio (0.33) showed different properties (diffusion of water and mechanical properties) from the AX films with higher Ara/Xyl ratios (0.53 and 0.7). The AX film (Ara/Xyl ratio of 0.33) exhibited higher monolayer and M2 values, which reflected the amount of water strongly adsorbed to specific sites and the decrease of the dipolar interaction strengths (Robert et al. 2011). Moreover, it was found that low-substituted films exhibited higher strength than high-substituted films. The increase in the slope of the plastic part of stress-strain curves for AX films (Ara/Xyl ratio of 0.33; Ara/Xyl ratio of 0.53) confirmed that the presence of arabinose residues could stop chains slippage.

Table 2. Optimum properties of xylan films with different Ara/Xyl ratios.

The optimum properties			Xylan type	Source	Ara/Xyl ratio	Arabinose Proportion	Reference
Mechanical properties	Tensile strength (MPa)	55	AGX	Norway spruce	–	6.9%	Escalante et al. 2012
		40	AX	Wheat flour	0.56	–	Heikkinen et al. 2013
		60	Acetylated AX	Rye	0.27	–	Stepan et al. 2012
		54.5	AX	Wheat flour	0.33	–	Ying et al. 2013
		50	AX	Barley husks	0.22	–	Höije et al. 2005
		53.56	Bleached AX	Corncob	0.13	–	Egues et al. 2014
	Elongation (%)	6.5	AX	Wheat flour	0.56	–	Heikkinen et al. 2013
		22	Acetylated AX	Rye	0.52	–	Stepan et al. 2012
		8.17	Bleached AX	Corncob	0.13	–	Egues et al. 2014
	Elastic modulus (MPa)	2,375	AGX	Norway spruce	–	6.9%	Escalante et al. 2012
		990	AX	Wheat flour	0.56	–	Heikkinen et al. 2013
		2,400	Acetylated AX	Rye	0.27	–	Stepan et al. 2012
		2,926	AX	Wheat flour	0.33	–	Ying et al. 2013
		2,930	AX	Barley husks	0.22	–	Höije et al. 2005
		1,662	Washed AX	Corncob	0.13	–	Egues et al. 2014

(Contd.)

Permeability properties	WVP	7.2 (g mm)/(kPa m² d)	AX	Wheat flour	0.34	--	Heikkinen et al. 2013
	OP	0.12 cm³ μm m² d⁻¹ kPa⁻¹	AGX	Norway spruce	--	6.9%	Escalante et al. 2012
		4.6 (cm³ m)/(m² d kPa)	AX	Wheat flour	0.34	--	Heikkinen et al. 2013
Hydrophobic property	Contact angle of water (°)	66	Acetylated AX	Rye	0.52	--	Stepan et al. 2012
		78.8	Bleached AX	Corncob	0.13	--	Egues et al. 2014
	Moisture content (%)	57	AX	Rye	0.36	--	Sternemalm et al. 2008

$OP = 0.12\ cm^3\ \mu m\ m^2\ d^{-1}\ kPa^{-1}$

Enzymes offer more controlled ways of making xylan modifications due to their high specificity. Two selective α-L-arabinofuranosidases (AXH-m and AXH-d3) were selected to modify wheat arabinoxylan (WAX) so as to understand the effect of fine structure on properties of WAX films (Heikkinen et al. 2013). The AXs with Ara/Xyl ratios of 0.29 (WAX-m) and 0.34 (WAX-d3) were obtained by removing about one-third of the α-l-arabinofuranosyl groups. High Ara/Xyl ratio had the beneficial influence on tensile strength of films. The film without enzyme addition (WAX-ref film), with the highest Ara/Xyl ratio, had a highest tensile strength of 40 MPa. This trend was similar to the conclusion obtained by Höije's group (Höije et al. 2008) and Mikkonen's group (Mikkonen et al. 2012), but was different from the previous report by Ying et al. (2013). The decreased tensile strength in WAX-m film was due to the inadequate interactions between the crystalline and amorphous parts and that in WAX-d3, the film was attributed to the differences in the molecular mobility of the xylan chains. The elongation at break (EAB) of the film from WAX-m was lower than that of WAX-ref film and WAX-d3 film. It could be explained by the fact that the reduced ability of WAX chains to slide across each other resulted from the precipitation of WAX-m and semicrystallization of WAX-m film. The lower Young's modulus of AXH-d3 film indicated that the removal of Ara*f* groups from disubstituted Xyl*p* units decreased the stiffness of the films. In addition, the AXH-d3 film had lower OP (oxygen permeability) value than that of WAX-ref film. The OP results indicated that the lower Ara/Xyl ratio led to a more dense packing of xylan chains, thereby decreasing the permeability of oxygen gas.

The properties of films such as mechanical properties and OP were examined to understand the effects of chemical composition on the xylan film's properties for food packaging application (Escalante et al. 2012). Arabinoglucuronoxylan (AGX) was extracted from Norway spruce by alkali extraction to prepare films. The chemical composition of AGX contained (w/w) 80.5% xylose, 10.0% 4-O-methylglucuronic acid, 6.9% arabinose and 2.3% other sugars (mannose, glucose and galactose). There was a right balance between glucuronic acid and arabinose pendant groups and molecular weight of AGX, which led to excellent films properties. Thus, transparent and homogeneous AGX films were formed, which were strong, with stress at break around 55 MPa, and quite flexible, with the EAB value of around 2.7%, and were stiff, with Young's modulus of 2,735 MPa. Moreover, the AGX film exhibited good oxygen barrier properties with the low average OP value ($0.12 \ cm^3 \ \mu m \ m^2 \ d^{-1} \ kPa^{-1}$), which was lower than that of the GX films prepared by Grondahl et al. (2004).

Pretreatment processes such as extraction and purification could also affect the chemical composition of xylan and further affect the properties of xylan films. AXs from corncob agricultural waste: crude extract (CCAX), purified by a washing step and bleaching CCAX (initial CCAX, purified and acetylated CCAXs) were used to prepare films (Egues et al. 2014). Mechanical properties and contact angles of films were evaluated to understand the influences of extraction, purification and acetylation on the AX films. It was found that untreated CCAX films were water soluble, yellowish in color and had poor mechanical properties. After the purification processes, the Young's modulus increased from ~293 MPa to ~1400-1600 MPa, and strength was improved from ~9 MPa to ~53 MPa, while the strain at break was kept at ~8% both in untreated and purified CCAX. The contact angles increased from ~21.3° to 67-74° after washing or bleaching CCAX. Acetylation of bleached CCAX showed the highest thermal resistance (325 °C), low T_g (125 °C) and a high contact angle (80°), and its films were stronger (strength ~67 MPa; Young's modulus ~2241 MPa) and more flexible (~13%). Therefore, crude extract CCAX film had poor mechanical properties due to the presence of CH_3COONa. Removing CH_3COONa, by washing or bleaching, could greatly improve properties of AX films. Furthermore, lignin could hinder the acetylation of washed CCAX, which resulted in bad film formation. The acetylation could improve the mechanical properties of films. Thus, Bleached-acetylated CCAX represented a suitable material for further material applications.

Some other physicochemical properties of xylan also had effects on the film-forming abilities and properties of films. As the polar polymers, xylan endows films with good intrinsic barrier properties against apolar migrants (oxygen or aromas) (Hansen and Plackett 2008; Hartman et al. 2006). However, xylan with poor film-forming abilities is ascribe it to their physicochemical properties, including the lower molar mass, rigidity and insufficient chain length of the polymer, high glass transition temperature and poor solubility (Gabrielii et al. 2000; Grondahl et al. 2004). Hydroxyl groups in the xylan molecular chain could form considerable intermolecular and intramolecular hydrogen bonds during the formation of films. Large numbers of active hydroxyl groups are the available sites for the reaction with other polymers. Nevertheless, the hydrophilic hydroxyl groups in the molecular chains make the xylan films sensitive to water (Mikkonen and Tenkanen 2012). Therefore, the hygroscopic nature which results in a low protection level against water is a disadvantage for xylan to prepare films.

Approaches to Improve the Xylan Film-forming Abilities and Xylan Films Properties

In order to improve the film-forming abilities of xylans and to obtain xylan films with good properties (mechanical properties, hydrophobic property, barrier property etc.), several approaches were employed, such as preparing plastified xylan films, reinforcing xylan films and modified xylan films (Bayati et al. 2014).

Plastified Xylan Films

Due to the lack of side chains or substitutes, the xylan chains are supposed to be rigid, which makes films that are less deformed (Höije et al. 2008). The addition of the plasticizers in polymeric materials could lead to modifications in the molecular three-dimensional organization and therefore in functional properties (Cuq et al. 1997). Thus, one effective method to enhance the flexibilities and extensibilities of xylan films is adding plasticizers into the xylan solution. Meanwhile, plasticizers could improve some other properties of the xylan films such as moisture and oxygen barrier properties (Grondahl et al. 2004). Table 3 reviews the influence of different plasticizer on properties of xylan films.

Polyalcohols. Polyalcohols such as glycerol, sorbitol, xylitol and propylene glycol were used as plasticizers. The hydroxyl groups in alcohols would tend to form hydrogen bonds with the hydroxyl groups of xylan, which could increase the interstitial volume of the material or the macromolecular mobility of xylan (Wang et al. 2014). The poly(ethylene glycol) methyl ether (mPEG) as plasticizer in AX/sepiolite composite films showed similar effects (Sarossy et al. 2012).

X extracted from corn hull with 4% dilute alkali, was employed as the raw material to prepare thin edible films. Smooth and transparent AX films were obtained because of the uniform deposition. Glycerol, propylene glycol and sorbitol, as plasticizers, were added to determine their effects on the functional properties of the films (Zhang and Whistler 2004). They found that plasticizers content had no effect on the mechanical properties but exhibit positive dependency for moisture content. Propylene glycol and sorbitol had less effect on elongation than glycerol (12.1%) due to the lower molecular weight and more hydroxyl groups of glycerol. Plasticizers also showed effects on barrier properties of films. All the xylan films with plasticizers had lower water vapour permeability (WVP) values than that of unplasticized ones. AX films had the lowest WVP value (0.23×10^{-10} g m^{-1} Pa^{-1} s^{-1}) when the sorbitol content reached 13%, because the bulky and poorly hygroscopic sorbitol had less effect on hydrogen bonding between AX chains. Moreover, grapes coated with AX-sorbitol films exhibited lower weight loss rate during storage, indicating AX films could potentially be applied in food preservation industry. The effects of glycerol, xylitol and sorbitol on plasticized oat spelt AX films were also investigated (Mikkonen et al. 2009). Oat spelt AX was not soluble because it contained less arabinose substituent. For preparing homogeneous and cohesive films, the insoluble part was removed and external polyols were added as plasticizers. Glycerol and sorbitol were selected due to the migration and/or crystallization of xylitol. It was consistent with the literature reported for Grondahl et al. (2004). Both glycerol and sorbitol had beneficial effects on improving the EAB of AX films. Similar effects of sorbitol on flexible oxygen barrier films were observed by Escalante et al. (2012). With a glycerol content of 40%,

Table 3. Optimum properties of plastified xylan films with different plasticizers.

The optimum properties		Xylan type	Source	Plasticizer	Other components	Reference
Mechanical properties	Tensile strength (MPa)	AX	Corn hull	0.163 propylene glycol urea	--	Zhang and Whistler 2004
		AX	Oat spelt	10% glycerol	--	Mikkonen et al. 2009
		Xylan	Beech wood	10% glycerol + 1% urea	PVA	Wang et al. 2013
		Xylan	Beech wood	10% citric acid	PVA	Wang et al. 2014
		Xylan	Beech wood	20% BTCA	PVA	Gao et al. 2014
	Elongation (%)	AX	Corn hull	0.200 glycerol	--	Zhang and Whistler 2004
		AX	Oat spelt	40% glycerol	--	Mikkonen et al. 2009
		Xylan	Beech wood	10% glycerol + 2% urea	PVA	Wang et al. 2013
		Xylan	Beech wood	50% citric acid	PVA	Wang et al. 2014
		Xylan	Beech wood	20% BTCA	PVA	Gao et al. 2014
	Elastic modulus (MPa)	AX	Oat spelt	10% glycerol	--	Mikkonen et al. 2009
Permeability properties	WVP	0.23×10^{-10} g·m^{-1}·Pa^{-1}·s^{-1}	AX	Corn hull	0.128 sorbitol	--	Zhang and Whistler 2004

(Contd.)

Table 3. (*Contd.*)

The optimum properties		Xylan type	Source	Plasticizer	Other components	Reference
	1.1 g· mm· m^{-2}·d^{-1}·kPa^{-1}	AX	Oat spelt	10% sorbitol	–	Mikkonen et al. 2009
	4.11 g/m·s·Pa	Xylan	Beech wood	1% urea	PVA	Wang et al. 2013
	0.015 g·μm· cm^{-2}·h^{-1}	Xylan	Beech wood	5% BTCA	PVA	Gao et al. 2014
OP	3.0 cm^3· μm/(m^2·d· kPa) (50 ~ 70 % RH)	AX	Oat spelt	10% glycerol	–	Mikkonen et al. 2009
	0.21 cm^3 μm m^2 d^{-1} kPa^{-1}	GX	Aspen wood	35% sorbitol	–	Grondahl et al. 2004
Hydrophobic property	Contact angle of water (°) 114.68	Xylan	Beech wood	1% urea	PVA	Wang et al. 2013

the EAB value reached about 11%. It was also found that sorbitol was more effective in its water vapour barrier properties than glycerol. The films containing 10% sorbitol had the lowest WVP (1.1 g mm m^{-2} d^{-1} kPa^{-1}). The oxygen permeability of films containing 10% glycerol or sorbitol was similar, and sorbitol became more effective when the content was 40%.

Other additives. Besides polyalcohols, the plasticized effects of some other additives (urea, citric acid, 1,2,3,4-butane tetracarboxylic acid etc.) on the xylan films were studied. Urea, glycerol, citric acid and 1,2,3,4-butane tetracarboxylic acid (BTCA) were regarded as plasticizers or cross-linking agents for the preparation of PVA/xylan films. The effects of these additives on properties of PVA/xylan composite films were investigated (Gao et al. 2014; Wang et al. 2013; Wang et al. 2014). The mechanical properties testing results indicated that both glycerol and urea mainly serve as important plasticizers in the composite films, as the EAB value decreased with the increasing plasticizers content. An opposite trend was noted when the amount of the citric acid increased, which meant that citric acid played a role as a cross-linking agent in the composite films. The citric acid could provide better intermolecular interactions between molecules, leading to higher tensile strength (Reddy and Yang 2010). Moreover, it was found that TS values increased with an increment of BTCA amount from 5% to 20%. With the 20% amount of BTCA, both TS and EAB reached the maximum values for 22.4 MPa and 92.4% respectively. As the additives with hydroxyl and carboxyl groups caused stronger and more flexible films compared to those with only hydroxyl groups, the hydroxyl-rich chemicals with carboxyl functional groups provided by BTCA strengthened the interactions of polymers (Yoon et al. 2006). In addition, the urea acted as a water-resistant agent at relatively low amounts (below 2%), which made films more hydrophobic with the increase of hydrophobic groups. The maximum value of contact angle (114.68°) was achieved with 1% urea amount. At higher amounts above 1%, the urea caused higher WVP similar to the effect of a plasticizer like glycerol, while the increase of citric acid content led to a decrease in the WVP values due to the existence of the esterification. The WVP value was observed to be the lowest for 0.015 g μm cm^{-2} h^{-1} with a BTCA content of 5%, which was attributed to the water-resistant nature of BTCA. But WVP increased with a further increment of BTCA amount due to the formation of intermolecular interactions of BTCA and polyhydroxylic compounds. The moisture barrier properties of films improved with the addition of BTCA, which resulted from the reducing water absorption (WA) values of blending films.

Emulsified Xylan Films

Another approach to reduce the moisture sensibility of xylan films was adding lipids, which resulted in the formation of emulsion-based films. Lipids offer high water barrier properties, thus they could make films be efficient against water transfer (Kokoszka et al. 2010; Peroval et al. 2003; Yang and Paulson 2000). Therefore, some studies have focused on the influences of lipids on xylan films. Table 4 shows the properties of emulsified xylan films with different lipids.

Maize bran AX was chosen to prepare emulsified xylan films by emulsifying a fat (palmitic acid, stearic acid, triolein or hydrogenated palm oil) with glycerol monostearate (GMS) as an emulsifier. The effects of lipid type on water vapour barrier properties, film structure and mechanical properties were investigated (Peroval et al. 2002a). Smooth, brown and translucent AX films were obtained before adding lipids because lipids made films more opaque. The pure AX film had a high WVP value (1.77 $\times$ 10^{-10} g/m s Pa) because of the hydrophily of natural AX polymers. The WVP of AX films with the addition of lipids confirmed that lipids could improve the water barrier properties of films efficiently, and the chain length of lipids was an important factor. The length of the hydrocarbon chain, the unsaturation degree and polarity of lipids were related to the water vapour resistance of films. A long lipid hydrocarbon chain had a positive influence on improving the barrier properties of films because the polar properties and mobility of short chains had less effect on organizing an interlocking network (Debeaufort and Voilley 1995).

Consequently, the WVP of AX-stearic acid (C18) film was lower than that of AX-palmitic acid (C16) film. The AX-triolein film showed the highest water vapour transmission rate (WVTR) ascribed to the polar property of triglyceride. For this reason, the AX-triolein film exhibited the lowest contact angle (39°). On the contrary, AX-OK35 film had the best hydrophobic surface with the contact angle of 94.4°. Most lipids had a negative influence on the elongation of AX films because lipids are unable to form a cohesive and continuous matrix. However, a significant increase in film elongation was observed when triolein was added, which had the highest elongation value (10.8%). The contrary result was obtained due to the plastication of unsaturated and lower molecular weight of triglycerides. However, the strength of all emulsified xylan films were lower than the AX films without lipids, whose tensile strength and elastic modulus were 26.5 and 72.4 MPa, respectively. It was consistent with the results that the addition of lipid was not conducive to the mechanical properties of films due to low interactions formed between fats and polysaccharides (Yang and Paulson 2000).

Lipid (HPKO) was added to the AX-glycerol solution and four sucroesters (SP10, SP30, SP40 and SP70) were tested in order to study their effects on the stability of film-forming emulsions (Peroval et al. 2002b). Optimal sucroester concentrations, which were SP70 (2.5%), SP40 (12.5%), SP30 (6.25%) and SP10 (2.5 to 12.5%), could be obtained according to the smallest globule size and highest stability of emulsion. AX films based on SP10 and SP70 at a 2.5% concentration had lower WVP values because SP10 increased the melting point and hydrophobicity of lipids. SP70 provided a small size of HPKO globules in the AX solution homogeneously. But the films without emulsifiers had the lowest WVP (9.31 $\times$ 10^{-11} g m^{-1} s^{-1} Pa^{-1}) and highest contact angle (79°),

Table 4. Optimum properties of emulsified xylan films.

The optimum properties			Xylan type	Source	Lipid	Emulsifier	Other components	Reference
Permeability properties	WVP	1.18×10^{-10} g/m·s·Pa	AX	Maize bran	27% Triolein	3% GMS	-	Peroval et al. 2002a
		9.51×10^{-11} g m^{-1} s^{-1} Pa^{-1} (DT = 30 °C)	AX	Maize bran	25% HPKO	2.5% SP10	15% glycerol	Peroval et al. 2002b
		7.72×10^{-11} g m^{-1} s^{-1} Pa^{-1} (DT = 80 °C)	AX	Maize bran	25% HPKO	2.5% SP10	15% glycerol	Phan et al. 2002
Hydrophobic property	Contact angle of water (°)	94.4	AX	Maize bran	27% OK 35	3% GMS	-	Peroval et al. 2002a
		79 (DT = 30 °C)	AX	Maize bran	25% HPKO	-	15% glycerol	Peroval et al. 2002b
		108 (DT = 40 °C)	AX	Maize bran	25% HPKO	-	15% glycerol	Phan et al. 2002

DT: Drying Temperature.

which confirmed that the films without emulsifiers exhibited better barrier property. Thus, two ways for improving film barrier ability were to enhance the stability of the emulsion by using a more efficient emulsifier, or increasing the emulsion destabilization during drying.

Reinforcing Xylan Films

In order to prepare xylan films possessing beneficial mechanical properties, many additives were employed into the mixture such as cellulose and clay. The additives may improve some other properties such as film formation ability, water vapour barrier property and homogeneity of xylan films. Table 5 displays the optimum properties of reinforcing xylan films with the different additives.

Cellulose based reinforcing agents. Cellulose whiskers (nanocrystalline cellulose, nanofibrillated cellulose, microfibrillated cellulose and bacterial cellulose) as cellulose-based reinforcing agents, were considered as effective additives for improving the strength properties of xylan films. It might be attributed to the strong mechanical strength and high specific ratio of the reinforcements, as well as the formation of a rigid hydrogen-bonded network. The lateral dimension of nanofibrillated cellulose (NFC) with a high aspect ratio (i.e. length to width ratio) was 5-40 nm (Paakko et al. 2007; Svagan et al. 2007). Microfibrillated cellulose (MFC) was mainly composed of liberated semicrystalline microfibrils with a width of 5-15 nm (Wagberg et al. 2008). MFC was also used to form films with high tensile strength and transparency (Sehaqui et al. 2010). In addition, using bacterial cellulose (BC) as a reinforcing agent lay in its high purity and fine fibrils (Dammstrom et al. 2005).

Rye arabinoxylan (rAX) and NFC were mixed to prepare composite films. Different proportions of NFC (0, 25%, 50%, 75% and 100%) were added into the solutions to examine the effects of NFC (Stevanic et al. 2012). The opaque, surface roughness and tensile properties of films, except the pure NFC film, increased with a higher amount of NFC added. Meanwhile, the structure of films became less homogeneous. The 100rAX film showed the lowest stiffness, strength and ductility, but the 75% NFC film showed the highest values, whose Young's modulus, stress at break and strain at break were 7.3 GPa, 143 MPa and 7.2%, respectively. According to the moisture sorption isotherms of films, NFC would be efficient to reduce the moisture adsorption of films, which would decrease the amount of available water sorption sites. It confirmed that strong permanent hydrogen-bonds and other interactions were created between xylan and NFC. The OP of the rAX film was low and composite films with a higher amount of the NFC even showed lower values. The 50% NFC film had the lowest OP value (0.79 cm^3 μm/m^2 day kPa). It was explained by the low permeability of NFC itself.

Modified NFCs were further added into the xylan films as reinforcing agents. The effects of sulfonated and hydrochloride NFC on properties

Table 5. Optimum properties of reinforcing xylan films.

The optimum properties		Xylan	Source	Reinforcing agent	Concn of reinforcing agent	Other components	Reference
Mechanical properties	Tensile strength (MPa)	0.25 g xylan	Oat spelt	Sulfonated cellulose whiskers	10%	50% sorbitol	Saxena et al. 2009
	6						
	143	AX	Rye	NFC	75%	--	Stevanic et al. 2012
	95	85% AX	Rye	MFC	15%	--	Mikkonen et al. 2012
	51	Acetylated xylan	Corn cob	Acetylated cellulose	5%	--	Gordobil et al. 2014
	68	AX	Rye	BC	5%	--	Stevanic et al. 2011
	73.6	AX	Rye flour	Sepiolite	2.5%	--	Sarossy et al. 2012
	Elongation (%)	0.25 g xylan	Oat spelt	Sulfonated cellulose whiskers	7%	50% sorbitol	Saxena et al. 2009
	30						
	7.2	AX	Rye	NFC	75%	--	Stevanic et al. 2012
	8.1	AX	Rye	BC	5%	--	Stevanic et al. 2011
	44.4	AX	Rye flour	Sepiolite	2.5%	30% mPEG	Sarossy et al. 2012

(Contd.)

Table 5. (*Contd.*)

The optimum properties		Xylan	Source	Reinforcing agent	Concn of reinforcing agent	Other components	Reference
Elastic modulus (MPa)	7300	AX	Rye	NFC	75%	–	Stevanic et al. 2012
	2,650	Enzyme modified AX	Rye	MFC	15%	–	Mikkonen et al. 2012
	3,200	Acetylated xylan	Corn cob	Acetylated cellulose	5%	–	Gordobil et al. 2014
	3,700	AX	Rye	BC	15%	–	Stevanic et al. 2011
	4,300	AX	Rye flour	Sepiolite	10%	–	Sarossy et al. 2012
Permeability properties WVTR	174 g mil/h·m²	0.25 g xylan	Oat spelt	Sulfuric nanocrystalline cellulose	10%	0.25g sorbitol	Saxena et al. 2011
OP	0.79 cm³·µm/m²·day·kPa (50% RH)	AX	Rye	NFC	50%	–	Stevanic et al. 2012
	0.16~3.2 cm³·µm/(m²·d·kPa) (50% RH)	AX	Rye flour	Sepiolite	5%	–	Sarossy et al. 2012
Hydrophobic property Contact angle of water (°)	72	Acetylated xylan	Corn cob	Acetylated cellulose	5%	–	Gordobil et al. 2014

of xylan films were investigated (Saxena et al. 2009; Saxena et al. 2011). Mechanical properties and water transmission of films were evaluated respectively. Addition of sulfonated whiskers resulted in a gradual improvement in tensile energy absorption (TEA) and strength properties of xylan films. An inverse variation tendency was observed in a strain of HCl whisker reinforcing xylan films. The film with 7% addition of sulfonated whisker showed a highest strain value and 5% of HCl whisker caused highest TEA and tensile strength values. Generally, an improvement in the mechanical properties for the xylan films could be achieved, which was attributed to the formation of a rigid hydrogen-bonded network of cellulose whiskers that was governed by a percolation mechanism (Samir et al. 2005). Furthermore, their impacts on water transmission of films were examined. Xylan films reinforced with 10% sulfuric nanocrystalline cellulose and 10% softwood fibre exhibited the lowest (174 g mil/h m²) and highest (807 g mil/h m²) water transmission rate respectively. However, 10% acacia fibre and 10% hydrochloric nanocrystalline cellulose had no significant influence on changing the water transmission rate of xylan films. The results explained that the uneven structure and agglomeration of xylan caused a higher water transmission rate of the pure xylan film. After the addition of the reinforcing agents, pulp fibres could not form an integrated matrix, which was different from that of cellulose whiskers. Thus cellulose whiskers led to lower WVTR values.

A set of rye rAXs with different DS and DP was modified by a controlled enzymatic. MFC was added to examine its effects on properties of self-standing films (Mikkonen et al. 2012). The rAX without enzyme and modified by Shearzyme were named rAX-H and rAX-L, and the partial debranching ones were called rDAX-L and rDAX-H accordingly. It was found that Shearzyme treatment decreased the weight average molar mass and increased the poly-dispersity of rAXs. Even though all films were transparent, the transparency of films decreased slightly when MFC was added. Moreover, the addition of MFC increased the haze of films, which indicated that rAX-H and MFC interacted and formed structures. Meanwhile, MFC slightly decreased the moisture uptake of the rAX-MFC composite films. The mechanical properties of films were evaluated by the storage modulus and tensile testing. The films with the addition of MFC showed higher storage modulus values because of the decreasing moisture content of films. Nevertheless, the addition of MFC did not affect the elongation at the break of films.

The rAX and enzymatically debranched rye arabinoxylan (rDAX) were combined with BC to prepare films with the AX/BC ratio of 100/0, 95/5 and 85/15 (Stevanic et al. 2011). Two series of cohesive films were formed without any plasticizer. The result indicated that the stiffness of the films could be improved by a relatively low amount of BC. But for the rAX films with a content of 15% BC, the stiffness decreased due to an agglomeration of BC. In general, stronger and stiffer films were produced when BC was added to xylan. The rDAX:15BC film exhibited the best elasticity with the maximum Young's modulus value for 3,700 MPa. The rAX:5BC film exhibited the maximum stress at break and strain at break values for 68 MPa and 8.1%, respectively. BC had less effect on the moisture uptake of films than the removal of arabinose units due to its relatively high crystallinity (Yoshinaga et al. 1997). Dammstrom et al. (2005) also found the difference in moisture uptake between the GX/BC composite film and the pure bacterial cellulose was small.

Clay reinforcing agents. Some clay particles were also used as a reinforcing agent for materials, not only because of their reactive surfaces (Manjaiah et al. 2010), but also because they would make the materials contain heat resistance, strength and low gas permeability after preparing clay-polymer nanocomposites (Wang et al. 2004).

AX from rye flour and a fibrous clay sepiolite were used to prepare sepiolite-hemicellulose composite films (Sarossy et al. 2012). Sepiolite at loadings between 2.5 and 10 wt% were regarded as the reinforcing agent to enhance the properties of films. Even though the light transmittance decreased slightly with an increase in the clay content, nonetheless, cohesive and semitransparent films were obtained. The mechanical and barrier properties of films were mainly studied to observe the reinforcing effects of sepiolite. The results suggested that the sepiolite had a positive impact on improving the stiffness and strength of films. The film with the 10 wt% addition of sepiolite showed the highest Young's modulus value (4.3 GPa). The highest tensile strength value (73.6 MPa) was observed from the film with the 2.5 wt% addition of sepiolite. The improvement of mechanical properties was not only explained by the interaction through hydrogen bonding between the sepiolite fibres and the rye AX chains, but also by the network formed between sepiolite fibres themselves. The interaction was similar to that of xylan/montmorillonite biocomposites produced by Unlu et al. (2009). However, the addition of sepiolite had no significant influence on WVP due to the hydrophily of sepiolite fibres.

Others. Goksu et al. (2007) used cotton stalk xylan to prepare films. Meanwhile lignin was used as an additive to enhance film formation. The lignin content in the mixture depended on controlling the ethanol washing solution volume during xylan extraction from cotton stalk. The pure xylan showed a poor film-forming capacity when lignin was almost removed from xylan. Compared to pure xylan, the xylan-lignin mixture resulted in continuous and self supporting film formation. The number of cracks after evaporation was observed to determine the optimum lignin concentration for continuous film formation. They found that 1% lignin was enough for film formation.

Modified Xylan-based Films

The poor water vapour barrier properties of xylan films may be owing to the hydrophilic hydroxyl groups in the xylan polymer. Therefore, some modifications were done for the hydroxyl groups on xylan chains, which were carried out for weakening the hydrophilies of xylans. The problem of xylan films with a poor water-resisting property could be solved by chemically modification of the xylan polymer. The relatively high molar mass and highly branched chemical structure of AX offers possibilities for modifications (Sternemalm et al. 2008). Graft polymerization is one method to modify the structure of xylan films. As hydrophobic long carbon chains could effectively prevent xylan chains from aggregating by disrupting the intramolecular and intermolecular hydrogen bonds, some long-chains polymers were chosen to graft on the xylan backbone (Peng et al. 2015; Peroval et al. 2004; Peroval et al. 2003; Zhong et al. 2013). Table 6 summarizes the xylan films with long-chain polymers modification.

A treatment: oxygen plasma pre-activation, immersion in a solution of acrylates and electron beam irradiation at 40 kGy; B treatment: emulsified AX-based films treated by direct irradiation; and C treatment: emulsified AX-based films treated by double irradiation.

Omega-3 fatty acids were grafted on the AX polymeric chains by oxygen plasma associated with EB irradiation to study their effects on WVP, structure and hydrophobicity of AX films (Peroval et al. 2003). Three fish oils (A, B, C) and one plant oil (D) were selected to be grafted on the AX polymeric chains. Grafting level of the modified AX films were 5.7, 6.4, 5.1, 6.1 for A, B, C, D oils respectively, which was achieved by breaking of double bond in AX. The A and C films had better surface hydrophobicity, which showed higher contact angles (91.2° and 115.1°) than the untreated one (66.8°). But B film had the contrary result because of the polar character of the B oil, which enhanced water affinity of the polymer (Peroval et al. 2002a). Compared to the present results, the AX films grafting of stearyl acrylate (SA) and stearyl methacrylate (SM) showed higher contact angles for 113.8° and 122.8° respectively (Peroval et al. 2004). But the contact angles of films treated in pure vinyl acetate and vinyl stearate were lower for 87° and 73° though they were higher than that of the untreated blank AX films (57°). It was respectively due to the covalently bound acetate esters and the possible deposition of stearate esters on the films surfaces (Stepan et al. 2013). For the water vapour barrier properties, A, B and C films showed similar WVP values. Because the A and C films had a high grafting level and surface hydrophobicity, and B film contained a greater amount of immobilized molecule onto the AX network. The lowest WVP value (1.09×10^{-10} g m^{-1} s^{-1} Pa^{-1}) was obtained from D film. It was explained by the composition of the D oil that it was mainly constituted with R-linolenic acid, which would be grafted onto the AX polymeric chains more easily. However, SA-emulsified film treated by double irradiation had lower WVP value for 0.68×10^{-10} g m^{-1} s^{-1} Pa^{-1}, which was the most efficient barrier against water vapour (Peroval et al. 2004).

Xylans modified by long-chain anhydride were used to prepare the film, which was an efficient strategy to improve some properties (mechanical strength, less moisture–sensitive properties) of xylan films (Zhong et al.

Table 6. Properties of xylan films with long-chain polymers modification.

Xylan type	Source	Modification methods	Grafting level (%)	Other components	Contact angle of water (°)	WVP	Barrier efficiency (%)	Reference
AX	Maize bran	Grafting of Omega-3 fatty acids	5.1	20% glycerol	115.1	1.45×10^{-10} g m⁻¹ s⁻¹ Pa⁻¹ (22% RH)	--	Peroval et al. 2003
AX	Maize bran	Grafting of Omega-3 fatty acids	6.1	20% glycerol	71.8	1.09×10^{-10} g m⁻¹ s⁻¹ Pa⁻¹ (22% RH)	--	Peroval et al. 2003
AX	Maize bran	Grafting of stearyl acrylate (A treatment)	--	15% glycerol	113.8	1.34×10^{-10} g m⁻¹ s⁻¹ Pa⁻¹ (22% RH)	27.1	Peroval et al. 2004
AX	Maize bran	Grafting of stearyl methacrylate (A treatment)	--	15% glycerol	122.8	1.32×10^{-10} g m⁻¹ s⁻¹ Pa⁻¹ (22% RH)	9.4	Peroval et al. 2004
AX	Maize bran	Grafting of stearyl acrylate (C treatment)	--	15% glycerol	--	0.68×10^{-10} g m⁻¹ s⁻¹ Pa⁻¹ (22% RH)	50	Peroval et al. 2004
AX	Maize bran	Grafting of stearyl methacrylate (B treatment)	--	15% glycerol	--	1.06×10^{-10} g m⁻¹ s⁻¹ Pa⁻¹ (22% RH)	0	Peroval et al. 2004
Xylan	Bamboo	Grafting of 2–OSA	0.29	--	71.9	--	--	Zhong et al. 2013
CMX	Bamboo	Grafting of Poly(propylene oxide)	--	60% chitosan + 1% glycerol	111.2	--	--	Peng et al. 2015
CMX	Bamboo	Grafting of Poly(propylene oxide)	--	40% chitosan + 1% glycerol	--	0.65×10^{-6} g/m/h/Pa (30% RH)	--	Peng et al. 2015

2013). 2-Octenylsuccinic anhydride (2-OSA) and 2-dodecenyl-1-succinic anhydride (2-DSA) were selected to modify xylan in order to prepare 2-OSA-xylan and 2-DSA-xylan films. The DS of 2-DSA-xylan increased from 0.081 to 0.27 as the reactant ratio increased from 1:8 to 1:1, and the film samples were named 2-OSA-0.067, 2-OSA-0.098, 2-OSA-0.17, 2-OSA-0.22 and 2-OSA-0.29 accordingly. However, the poor film-forming performance of 2-DSA-xylan and the rough surfaces of 2-DSA-xylan films indicated that 2-DSA had no significant improvement on xylan films. Thus, most properties measurements were carried out for the 2-OSA-xylan films. A sorbitol plasticized xylan film (2-OSA-0) was employed as the control sample. It was stiffer than the 2-OSA-xylan films, especially the ones with higher DS. The 2-OSA-xylan films were flexible, resulted in significant tensile strain increase with the increase of DS. The highest tensile strain value (35.1%) was observed from the 2-OSA-0.29 xylan film. The strength of the 2-OSA-xylan films, exhibited the highest tensile strength value (44.0 MPa) with a DS of 0.064, which were also higher than the sorbitol plasticized xylan film. However, the 2-OSA-xylan films showed low tensile strength with relatively high substitutions (DS=0.098-0.29). It could be explained that the further increasing distance of xylan moieties decreased the hydrogen bonds of xylan chains, producing xylan films with low tensile stress and high strain (Klebert et al. 2009). Moreover, the equilibrium moisture content and the contact angle of xylan films indicated that long-chain anhydride modification could produce internally plasticized film with hydrophobic performance. The 2-OSA-xylan films showed low sensitivity toward moisture, which had lower moisture contents (5.90-11.67 wt%) and higher contact angles (39.5-71.9°) than the sorbitol plasticized xylan film, whose moisture content and contact angle were 13.92 wt% and 31°. The effect of 2-OSA on improving the hydrophobic performance of films was similar to the previous report (Peroval et al. 2003).

Poly(propylene oxide) chains were grafted onto xylan from bamboo using the $Al(Oi\text{-}Pr)_3$ initiated ring-opening polymerization of propylene oxides. Then it was further carboxymethylated with sodium chloroacetate under microwave irradiation to obtain carboxymethyl xylan-g-poly(propylene oxide) (CMX-g-PPO) (Peng et al. 2015). A series of composite films with different CMX-g-PPO/chitosan weight ratio (20/80, 40/60, 60/40 and 80/20) were prepared. Xylan/chitosan film and plasticized xylan/chitosan film were employed as control samples. The water contact angles of CMX-g-PPO/chitosan films ranged from 100.3° to 111.2°, indicating that CMX-g-PPO/chitosan films had excellent hydrophobic properties. It was due to the hydrophobic alkyl groups from the grafted PPO side chains onto xylan chains, which was similar to the effect of other long-chain polymers (Peroval et al. 2004). The lowest WVP value for 0.65×10^{-6} g/m/h/Pa was observed at RH of 30% with 60% CMX-g-PPO, because excessive CMX-g-PPO with negative charges could decrease the intermolecular forces and

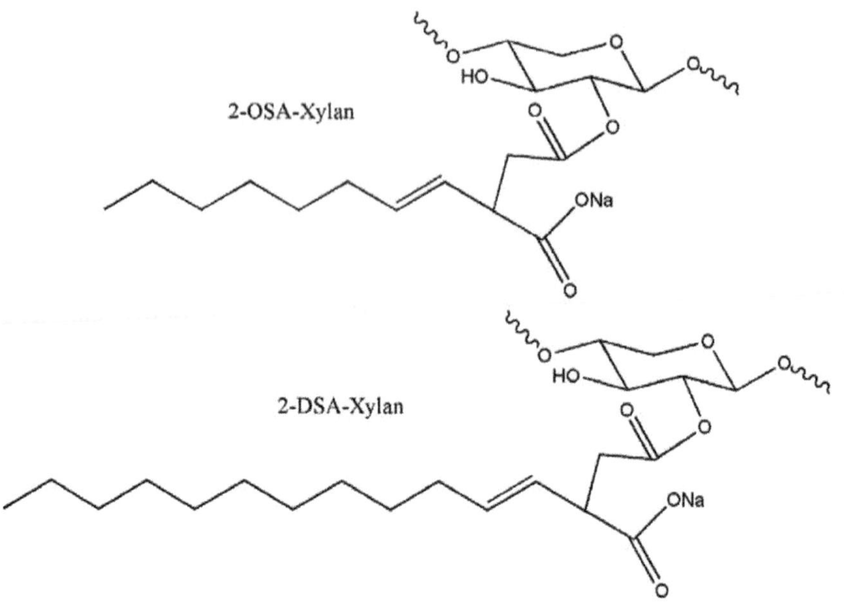

Fig. 1. Structures of 2-OSA-xylan and 2-DSA-xylan.

loosen the inner structure of CMX-g-PPO and the chitosan matrix. However, the CMX-g-PPO/chitosan films had a better moisture barrier property than the plasticized xylan/chitosan film, but worse than the xylan/chitosan film. With the increasing content of CMX-g-PPO from 20% to 80% in the films, the tensile strength, tensile strain at break, and Young's modulus significantly decreased to 3.2 MPa, 2.0% and 270 MPa, respectively. The decrease of mechanical properties was attributed to the lack of chitosan. The CMX-g-PPO/chitosan (20/80) composite film showed relatively better mechanical properties with tensile strength, tensile strain at break and Young's modulus for 27.8 MPa, 9.0% and 843 MPa, respectively.

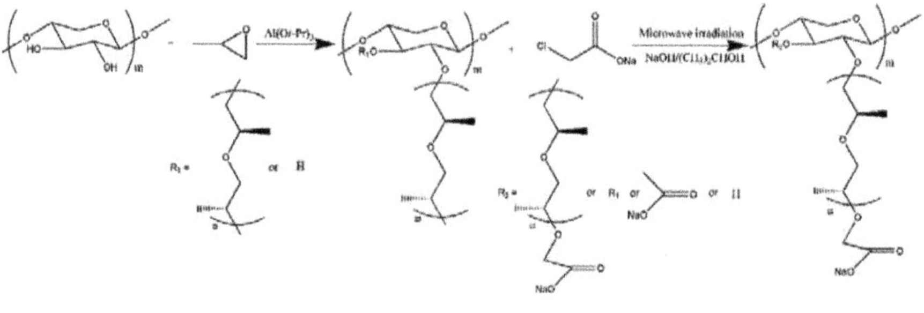

Fig. 2. Synthetic routes of CMX-g-PPO.

The AX extracted from barley husks was used to prepare films. Fluorination in the gas phase was used to improve the hydrophobicity of the films with trifluoroacetic anhydride (TFAA) (Grondahl et al. 2006). The reaction that occurred between TFAA and the hydroxyl groups of the AX formed an ester, which reduced the amount of hydrophilic hydroxyl groups (Hutt and Leggett 1997). AX films with different fluorine content (1%-7%) at the surfaces were obtained by a surface modification. The results showed that the contact angle and equilibrium moisture content had the opposite variation tendency when the fluorine content increased. The increasing contact angle from 30° to 70° confirmed that the modification made the films less hydrophilic. The decreasing equilibrium moisture content (18%-12%) at 50% RH illustrated that the modification also had a positive influence on improving the hydrophobicity of AX films. Even so, the AX films were hydrophilic as the contact angles were lower than 90°.

Fig. 3. Reaction of AX with TFAA.

Simkovic et al. have developed modified xylan films through preparing cationic, anionic and amphoteric xylans (Simkovic et al. 2011a, b; Simkovic et al. 2014a; Simkovic et al. 2014b). Xylans extracted from beech holocellulose including water soluble sample (WSX) and water insoluble residue (WIX), were modified by hydroxypropylsulfonation, quaternization and hydroxypropylsulfonation/quaternization (Simkovic et al. 2011a). The modified xylans obtained with molar masses between M_n of 3.8 and 948.9 kg/mol with polydispersity index D from 1.7 to 26.8 were employed for preparing xylan films. A series of films were obtained with both these water soluble/insoluble xylan and xylan from different sources. They found that the molar mass of modified samples was not the predominant factor for increasing tensile strength of films, but it had a positive impact on increasing the Young's modulus of xylan films. This conclusion was confirmed by comparing the TS and Young's modulus values of quaternized and sulfonated samples.

Simkovic et al. (2011b) also prepared xylan sulphate films with xylans extracted from beech sawdust holocellulose. Xylans were sulphated in different solvents, which achieved sulphate derivatives with DS up to 1.40 at 90% yield. The result showed that xylan sulphate films with DS 0.20 had weaker mechanical properties (Young's modulus and tensile strength) than the xylan films (Simkovic et al. 2011a), which was attributed to less intermolecular hydrogen bonds. But the films with the insoluble component of xylan exhibited higher Young's modulus (5,390 MPa) and tensile strength (140 MPa) which was owing to their higher molar mass than the soluble part. Therefore, the sulphated xylan with high DS had a negative impact on the mechanical properties of films. But the adhesion of insoluble parts with solubilized xylan sulphate had opposite effects.

Quaternized xylan (QX), fully sulfated xylan (XS) and partially sulfated xylan (SX) were prepared (Simkovic et al. 2014b). Then QX and SX were further sulfated and quaternized to obtain fully substituted QXS and partially substituted SXQ. In comparison with xylan (X), the decrease of modulus values of all modified samples indicated that modifications bring radical change both in the supramolecular structure of xylan and interactions between macromolecules. In addition, the decreasing elongation of QXS implied that modification procedure would make the films become brittle.

Carboxymethyl, 2-hydroxypropylsulfonate and trimethylammonium-2-hydroxypropyl groups were introduced into xylan to prepare carboxymethyl/trimethylammonium-2- hydroxypropyl xylan (CQX), carboxymethyl/2-hydroxypropylsulfonate xylan (CSX) and carboxymethyl/2-hydroxypropylsulfonate/trimethylammonium-2-hydroxypropyl xylan (CSQX) (Simkovic et al. 2014a). The obtained xylan derivatives were used to prepare films (Simkovic et al. 2011a, b; Simkovic et al. 2014b). The testing results showed the mechanical properties values of different modified xylan films were relatively good and balanced, as all the modulus values, tensile strength values and elongation values of them were higher than 2,805 MPa, 44.4 MPa and 0.8% respectively.

Carboxymethylxylan (CMX) with different degrees of substitution (DSs) for 0.36, 0.58, and 1.13 were applied to prepare xylan films (Alekhina et al. 2014). All films showed increase of moisture uptake with increasing RH as the carboxymethyl groups were naturally water soluble. The moisture uptake of CMX films increased with decreasing DS at RH below 40%. In contrast, opposite results were obtained at higher RH from 50% to 90%. It could be explained that the carboxymethyl groups showed similar behaviour to bind water as external polyol plasticizers, occupying the water sorption sites of films at low RH and be hydrophilic at high RH. It proved that water acted as a plasticizer of polysaccharides films (Mikkonen and Tenkanen 2012; Talja et al. 2007). Owing to the hydrophilicity and softness of CMX 1.13 film, the examination of mechanical and barrier properties were only carried out for CMX 0.36 and CMX 0.58 films. CMX 0.36 film with a WVP for 19 g mm/(m^2 d kPa) was lower than that of CMX 0.58 film. Nevertheless, CMX 0.58 film with an OP value for 6 cm^3 µm/(m^2 d kPa) was lower than that of CMX 0.36 film. These results indicated that the hydrophilicity and better water solubility of carboxymethyl would lead to a more densely organized structure, further increasing the WVP and reducing the OP of CMX films.

Others

Xylans can also be utilized as additives to other film-forming biopolymers such as gluten, chitosan, cellulose, PVA and so on. Xylan extracted from grass and corncob was used to prepare composite films with wheat gluten

(Kayserilioglu 2003). The tensile strength of the films was not affected by adding birchwood xylan. The stretchability of films would be improved when the films were mixed with increasing amounts of xylan, dried at 80°C/35% RH and produced at pH 11. The xylan type also affected the mechanical properties of the films. Films containing birchwood xylan gave higher tensile strength but the ones containing corncob xylan gave lower values. Opposite effects were observed on the stretchability of films. Solubility of films decreased with increasing xylan content, which was due to the formation of the hydrogen bond. Both xylan content and xylan type had little influence on the WVTR of composite films. The results indicated it was possible to use xylan as an additive for films preparation.

Low content of xylan was immersed in chitosan film to prepare chitosan/xylan composite films (Luo et al. 2014). Chitosan films were soaked in three NaOH solutions with 1, 2 and 3% xylan to get the composite films designated as CSXY-1, CSXY-2 and CSXY-3. Compact composite films were obtained because of the formation of strong hydrogen bonds and electrostatic interactions between -OH, -NH$_2$ groups in chitosan molecules and -OH groups in xylan molecules. The addition of xylan had little effect on the contact angles of chitosan/xylan composite films because of the hydrophilicity of xylan chains, and chitosan film had a high contact angle due to the hydrophobic backbone of chitosan chains (Almeida et al. 2010). The contact angle of CSXY-2 film (92.60°) showed a little higher than that of pure chitosan film (92.53°) due to the increasing surface roughness. The increase of xylan concentration led to a decrease of OP values of the films, which confirmed the good intrinsic barrier properties of xylan films (Hansen and Plackett 2008; Hartman et al. 2006). CSXY-3 film exhibited the best oxygen barrier property with an OP value for 7.30 cm^3 µm m^{-2} day^{-1} kPa^{-1}. The WVTR of chitosan/xylan composite films depended on whether xylan nodules connected tightly with chitosan molecules. The composite film with 3% xylan formed a dense network structure that led to a lower WVTR value for 2.66 × 10^{-5} g/mm^2 h. Nevertheless, the films that contained xylan show poorer moisture barrier property than that of pure chitosan film, which was ascribed to the strong water-absorption property of xylan. The dense internal structure formed between xylan and chitosan also had favourable influence on the strength and flexibility of films, which validated the former argument of Peng et al. (2015). The CSXY-3 film had the highest tensile strength and elongation at the break values for 71.31 MPa and 7.69% respectively.

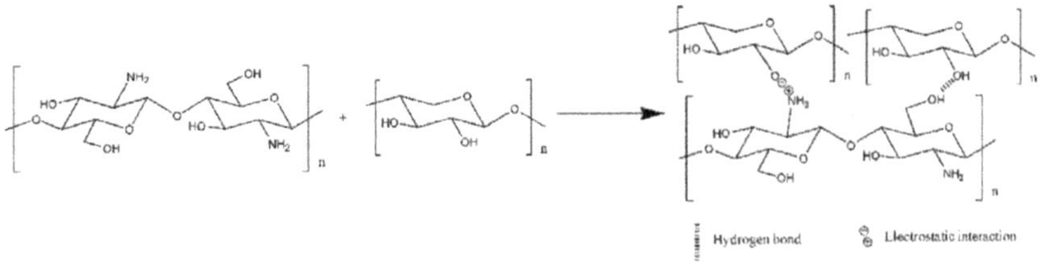

Fig. 4. Interaction between chitosan and xylan.

Different amounts of AGX (0 %, 5 %, 10 %, 20 % and 33 %) were dissolved in 1-ethyl-3-methylimidazolium (EmimAc) with cellulose and regenerated with ethanol into transparent cellulose/xylan blend films (Sundberg et al. 2015). All the regenerated films were amorphous because the AGX content did not affect the crystallinity of the regenerated films. The AGX content did not affect the measured storage modulus of the regenerated films either. Similar tensile strength and deformation of different films indicated that a cellulose network could form independent of the AGX content. But AGX could increase the Young's modulus of the regenerated films at humidity below a critical plasticization level. The structure of the films was disrupted after leaching of AGX and it further removed the strengthening effect.

The effects of cotton stalk xylan concentration and plasticizer on properties of xylan/lignin films were determined as well (Goksu et al. 2007). Cotton stalk xylan concentration was adopted in the range of 8% to 14%. Even though an increasing xylan concentration was beneficial for improving mechanical properties of films, the films exhibited low tensile strength (1.08-1.39 MPa) and low elastic modulus (0.11-0.49 MPa), which meant poor heavy-duty use potential and intrinsic stiffness. But an efficient increasing for the strain at break values was obtained with the increasing xylan concentration. The strain at break values between approximately 45% and 57% indicated a good flexibility for the reinforcing xylan films, which might be due to the increase of the entrapped water amount. The xylan concentration had no influence on the solubility of films. Nevertheless, it reduced the WVTR values of the films by enhancing the packing of the xylan molecules.

The effects of xylan content on the properties of PVA/xylan composite films were investigated as well (Wang et al. 2014). With a same content of citric acid, different weight ratio of PVA and xylan (1, 1.5, 2, 3, 4) were selected to prepare composite films. When the weight ratio of PVA and xylan increased (1-4), the TS of composite films was enhanced (7.6-15.2 MPa), while the EAB decreased. The maximum EAB value for 249.5% was observed when the weight ratio of PVA and xylan was 3. The solubility of composite films increased with an increase of xylan content, as more ester bonds formed between carboxyl groups in xylan and citric acid. However, little change occurred on the WVP values of composite films which indicated that xylan content had little effect on the permeability of films.

Conclusion

Considering the abundance, low cost and biodegradability, xylan as one of the important polysaccharides has attracted more and more interest for producing xylan-based films for packaging. Xylans have promising application for packaging due to their good intrinsic barrier properties against apolar migrants. The chemical compositions and the structure of xylan, especially Arx/Xyl ratio, have significant impacts on the properties of xylan-based films. Furthermore, to improve the xylan film-forming ability, different efficient approaches such as the addition of plasticizer, the modification and blending with other biopolymers and clay have been developed for achieving the desirable properties in the different application. Thus, xylan films could replace the oil-products to be potentially applied in agriculture and food packaging.

Acknowledgements

This work was supported by grants from National Natural Science Foundation of China (No. 21406080), the Programme of Introducing Talents of Discipline to Universities (Project 111) and Science and Technology Planning Project of Guangdong Province (No. 2015B020241001) and the Fundamental Research Funds for the Central Universities (2015PT011), SCUT.

References

Alekhina, M., K.S. Mikkonen, R. Alen, M. Tenkanen and H. Sixta. 2014. Carboxymethylation of alkali extracted xylan for preparation of bio-based packaging films. Carbohydr. Polym. 100: 89–96.

Almeida, E.V.R., E. Frollini, A. Castellan and V. Coma. 2010. Chitosan, sisal cellulose, and biocomposite chitosan/sisal cellulose films prepared from thiourea/NaOH aqueous solution. Carbohydr Polym. 80: 655–664.

Bayati, F., Y. Boluk and P. Choi. 2014. Diffusion behavior of water at infinite dilution in hydroxypropyl xylan films with sorbitol and cellulose nanocrystals. ACS Sustain. Chem. Eng. 2: 1305–1311.

Cuq, B., N. Gontard and S. Guilbert. 1997. Thermal properties of fish myofibrillar protein-based films as affected by moisture content. Polymer. 38: 2399–2405.

Dammstrom, S., L. Salmen and P. Gatenholm. 2005. The effect of moisture on the dynamical mechanical properties of bacterial cellulose/glucuronoxylan nanocomposites. Polymer. 46: 10364–10371.

Debeaufort, F. and A. Voilley. 1995. Effect of surfactants and drying rate on barrier properties of emulsified edible films. Int. J. Food Sci. Technol. 30: 183–190.

Ebringerova, A. and T. Heinze. 2000. Xylan and xylan derivatives – biopolymers with valuable properties, 1. Naturally occurring xylans structures, isolation procedures and properties. Macromol. Rapid Commun. 21: 542–556.

Egues, I., A. Eceiza and J. Labidi. 2013. Effect of different hemicelluloses characteristics on film forming properties. Ind. Crops Prod. 47: 331–338.

Egues, I., A.M. Stepan, A. Eceiza, G. Toriz, P. Gatenholm and J. Labidi. 2014. Corncob arabinoxylan for new materials. Carbohydr. Polym. 102: 12–20.

Escalante, A., A. Goncalves, A. Bodin, A. Stepan, C. Sandstrom, G. Toriz, et al. 2012. Flexible oxygen barrier films from spruce xylan. Carbohydr. Polym. 87: 2381–2387.

Fang, J.M., R.C. Sun, J. Tomkinson and P. Fowler. 2000. Acetylation of wheat straw hemicellulose B in a new non-aqueous swelling system. Carbohydr. Polym. 41: 379–387.

Gabrielii, I., P. Gatenholm, W.G. Glasser, R.K. Jain and L. Kenne. 2000. Separation, characterization and hydrogel-formation of hemicellulose from aspen wood. Carbohydr. Polym. 43: 367–374.

Gao, C.D., J.L. Ren, S.Y. Wang, R.C. Sun and L.H. Zhao. 2014. Preparation of polyvinyl alcohol/xylan blending films with 1,2,3,4-Butane tetracarboxylic acid as a new plasticizer. J. Nanomater. 2014: 1–8.

Goksu, E.I., M. Karamanlioglu, U. Bakir, L. Yilmaz and U. Yilmazer. 2007. Production and characterization of films from cotton stalk xylan. J. Agric. Food Chem. 55: 10685–10691.

Gordobil, O., I. Egues, I. Urruzola and J. Labidi. 2014. Xylan-cellulose films: Improvement of hydrophobicity, thermal and mechanical properties. Carbohydr. Polym. 112: 56–62.

Grondahl, M., L. Eriksson and P. Gatenholm 2004. Material properties of plasticized hardwood xylans for potential application as oxygen barrier films. Biomacromolecules 5: 1528–1535.

Grondahl, M., A. Gustafsson and P. Gatenholm. 2006. Gas-phase surface fluorination of arabinoxylan films. Macromolecules 39: 2718–2721.

Hansen, N.M.L. and D. Plackett. 2008. Sustainable films and coatings from hemicelluloses: A review. Biomacromolecules 9: 1493–1505.

Hartman, J., A.C. Albertsson, M.S. Lindblad and J. Sjoberg. 2006. Oxygen barrier materials from renewable sources: Material properties of softwood hemicellulose-based films. J. Appl. Polym. Sci. 100: 2985–2991.

Heikkinen, S.L., K.S. Mikkonen, K. Pirkkalainen, R. Serirnaa, C. Joly and M. Tenkanen. 2013. Specific enzymatic tailoring of wheat arabinoxylan reveals the role of substitution on xylan film properties. Carbohydr. Polym. 92: 733–740.

Höije, A., M. Gröndahl, K. Tømmeraas and P. Gatenholm. 2005. Isolation and characterization of physicochemical and material properties of arabinoxylans from barley husks. Carbohydr. Polym. 61: 266–275.

Höije, A., E. Sternemalm, S. Heikkinen, M. Tenkanen and P. Gatenholm 2008. Material properties of films from enzymatically tailored arabinoxylans. Biomacromolecules 9: 2042–2047.

Hutt, D.A. and G.J. Leggett. 1997. Functionalization of hydroxyl and carboxylic acid terminated self-assembled monolayers. Langmuir 13: 2740–2748.

Kayserilioglu, B.S., U. Bakir, L. Yilmaz and N. Akkas. 2003. Use of xylan, an agricultural by-product, in wheat gluten based biodegradable films: Mechanical, solubility and water vapor transfer rate properties. Bioresour. Technol. 87: 239–246.

Klebert, S., L. Nagy, A. Domjan and B. Pukanszky. 2009. Modification of cellulose acetate with oligomeric polycaprolactone by reactive processing: Efficiency, compatibility, and properties. J. Appl. Polym. Sci. 113: 3255–3263.

Kokoszka, S., F. Debeaufort, A. Lenart and A. Voilley. 2010. Liquid and vapour water transfer through whey protein/lipid emulsion films. J. Sci. Food Agric. 90: 1673–1680.

Luo, Y., X. Pan, Y. Ling, X. Wang and R. Sun. 2014. Facile fabrication of chitosan active film with xylan via direct immersion. Cellulose 21: 1873–1883.

Manjaiah, K.M., S. Kumar, M.S. Sachdev, P. Sachdev and S.C. Datta. 2010. Study of clay-organic complexes. Curr. Sci. 98: 915–921.

Mikkonen, K.S., S. Heikkinen, A. Soovre, M. Peura, R. Serimaa, R.A. Talja, et al. 2009. Films from oat spelt arabinoxylan plasticized with glycerol and sorbitol. J. Appl. Polym. Sci. 114: 457–466.

Mikkonen, K.S. and M. Tenkanen. 2012. Sustainable food-packaging materials based on future biorefinery products: Xylans and mannans. Trends Food Sci. Technol. 28: 90–102.

Mikkonen, K.S., L. Pitkanen, V. Liljestrom, E.M. Bergstrom, R. Serimaa, L. Salmen, et al. 2012. Arabinoxylan structure affects the reinforcement of films by microfibrillated cellulose. Cellulose. 19: 467–480.

Paakko, M., M. Ankerfors, H. Kosonen, A. Nykanen, S. Ahola, M. Osterberg, et al. 2007. Enzymatic hydrolysis combined with mechanical shearing and high-pressure homogenization for nanoscale cellulose fibrils and strong gels. Biomacromolecules 8: 1934–1941.

Peng, F., P. Peng, F. Xu and R.C. Sun. 2012. Fractional purification and bioconversion of hemicelluloses. Biotechnol. Adv. 30: 879–903.

Peng, P., M.Z. Zhai, D. She and Y.F. Gao. 2015. Synthesis and characterization of carboxymethyl xylan-g-poly(propylene oxide) and its application in films. Carbohydr. Polym. 133: 117–125.

Peroval, C., F. Debeaufort, D. Despre and A. Voilley. 2002a. Edible arabinoxylan-based films. 1. Effects of lipid type on water vapor permeability, film structure, and other physical characteristics. J. Agric. Food Chem. 50: 3977–3983.

Peroval, C., F. Debeaufort, D. Despre, J.L. Courthaudon and A. Voilley. 2002b. Arabinoxylan-lipids-based edible films and coatings. 2. Influence of sucroester nature on the emulsion structure and film properties. J. Agric. Food Chem. 50: 266–272.

Peroval, C., F. Debeaufort, A.M. Seuvre, B. Chevet, D. Despre and A. Voilley. 2003. Modified arabinoxylan-based films. Part B. Grafting of omega-3 fatty acids by oxygen plasma and electron beam irradiation. J. Agric. Food Chem. 51: 3120–3126.

Peroval, C., F. Debeaufort, A.M. Seuvre, P. Cayot, B. Chevet, D. Despre, et al. 2004. Modified arabinoxylan-based films grafting of functional acrylates by oxygen plasma and electron beam irradiation. J. Memb. Sci. 233: 129–139.

Petzold-Welcke, K., K. Schwikal, S. Daus, and T. Heinze. 2014. Xylan derivatives and their application potential – Mini-review of own results. Carbohydr. Polym. 100: 80–88.

Phan, D., F. Debeaufort, C. Peroval, D. Despre, J.L. Courthaudon and A. Voilley. 2002. Arabinoxylan-lipid-based edible films and coatings. 3. Influence of drying temperature on film structure and functional properties. J. Agric. Food Chem. 50: 2423–2428.

Reddy, N. and Y.Q. Yang. 2010. Citric acid cross-linking of starch films. Food Chem. 118: 702–711.

Robert, P., F. Jamme, C. Barron, B. Bouchet, L. Saulnier, P. Dumas, et al. 2011. Change in wall composition of transfer and aleurone cells during wheat grain development. Planta 233: 393–406.

Samir, M.A.S.A., F. Alloin and A. Dufresne. 2005. Review of recent research into cellulosic whiskers, their properties and their application in nanocomposite field. Biomacromolecules 6: 612–626.

Sarossy, Z., T.O.J. Blomfeldt, M.S. Hedenqvist, C.B. Koch, S.S. Ray and D. Plackett. 2012. Composite films of arabinoxylan and fibrous sepiolite: Morphological, mechanical, and baffler properties. ACS Appl. Mater. Interfaces 4: 3378–3386.

Saxena, A., T.J. Elder, S. Pan and A.J. Ragauskas. 2009. Novel nanocellulosic xylan composite film. Compos. Part B: Eng. 40: 727–730.

Saxena, A., T.J. Elder and A.J. Ragauskas. 2011. Moisture barrier properties of xylan composite films. Carbohydr. Polym. 84: 1371–1377.

Scheller, H.V. and P. Ulvskov. 2010. Hemicelluloses. Annu. Rev. Plant Biol. 61: 263–289.

Sehaqui, H., A.D. Liu, Q. Zhou and L.A. Berglund. 2010. Fast preparation procedure for large, flat cellulose and cellulose/inorganic nanopaper structures. Biomacromolecules 11: 2195–2198.

Simkovic, I., O. Gedeon, I. Uhliarikova, R. Mendichi and S. Kirschnerova. 2011a. Positively and negatively charged xylan films. Carbohydr. Polym. 83: 769–775.

Simkovic, I., O. Gedeon, I. Uhliarikova, R. Mendichi and S. Kirschnerova. 2011b. Xylan sulphate films. Carbohydr. Polym. 86: 214–218.

Simkovic, I., I. Kelnar, I. Uhliarikova, R. Mendichi, A. Mandalika and T. Elder. 2014a. Carboxymethylated-, hydroxypropylsulfonated- and quaternized xylan derivative films. Carbohydr. Polym. 110: 464–471.

Simkovic, I., A. Tracz, I. Kelnar, I. Uhliarikova and R. Mendichi. 2014b. Quaternized and sulfated xylan derivative films. Carbohydr. Polym. 99: 356–364.

Stepan, A.M., A. Höije, H.A. Schols, P. de Waard and P. Gatenholm. 2012. Arabinose content of arabinoxylans contributes to flexibility of acetylated arabinoxylan films. J. Appl. Polym. Sci. 125: 2348–2355.

Stepan, A.M., G.E. Anasontzis, T. Matama, A. Cavaco-Paulo, L. Olsson and P. Gatenholm. 2013. Lipases efficiently stearate and cutinases acetylate the surface of arabinoxylan films. J. Biotechnol. 167: 16–23.

Sternemalm, E., A. Höije and P. Gatenholm. 2008. Effect of arabinose substitution on the material properties of arabinoxylan films. Carbohydr. Res. 343: 753–757.

Stevanic, J.S., C. Joly, K.S. Mikkonen, K. Pirkkalainen, R. Serimaa, C. Remond, et al. 2011. Bacterial nanocellulose-reinforced arabinoxylan films. J. Appl. Polym. Sci. 122: 1030–1039.

Stevanic, J.S., E.M. Bergstrom, P. Gatenholm, L. Berglund and L. Salmen. 2012. Arabinoxylan/nanofibrillated cellulose composite films. J. Mater. Sci. 47: 6724–6732.

Sun, R.C., J.M. Fang and J. Tomkinson. 2000. Characterization and esterification of hemicelluloses from rye straw. J. Agric. Food Chem. 48: 1247–1252.

Sundberg, J., G. Toriz and P. Gatenholm. 2015. Effect of xylan content on mechanical properties in regenerated cellulose/xylan blend films from ionic liquid. Cellulose 22: 1943–1953.

Svagan, A.J., M.A.S.A. Samir and L.A. Berglund. 2007. Biomimetic polysaccharide nanocomposites of high cellulose content and high toughness. Biomacromolecules 8: 2556–2563.

Talja, R.A., H. Helen, Y.H. Roos and K. Jouppila. 2007. Effect of various polyols and polyol contents on physical and mechanical properties of potato starch-based films. Carbohydr. Polym. 67: 288–295.

Unlu, C.H., E. Gunister and O. Atici. 2009. Synthesis and characterization of NaMt biocomposites with corn cob xylan in aqueous media. Carbohydr. Polym. 76: 585–592.

Wagberg, L., G. Decher, M. Norgren, T. Lindstrom, M. Ankerfors and K. Axnas. 2008. The build-up of polyelectrolyte multilayers of microfibrillated cellulose and cationic polyelectrolytes. Langmuir 24: 784–795.

Wang, S., J. Ren, W. Kong, C. Gao, C., Liu, C., Peng, F., et al. 2013. Influence of urea and glycerol on functional properties of biodegradable PVA/xylan composite films. Cellulose 21: 495–505.

Wang, S., J. Ren, W. Li, R. Sun and S. Liu. 2014. Properties of polyvinyl alcohol/xylan composite films with citric acid. Carbohydr. Polym. 103: 94–99.

Wang, Y.C., S.C. Fan, K.R. Lee, C.L. Li, S.H. Huang, H.A. Tsai, et al. 2004. Polyamide/ SDS-clay hybrid nanocomposite membrane application to water-ethanol mixture pervaporation separation. J. Memb. Sci. 239: 219–226.

Yang, L. and A.T. Paulson. 2000. Effects of lipids on mechanical and moisture barrier properties of edible gellan film. Food Res. Int. 33: 571–578.

Ying, R.F., C. Rondeau-Mouro, C. Barron, F. Mabille, A. Perronnet and L. Saulnier. 2013. Hydration and mechanical properties of arabinoxylans and beta-D-glucans films. Carbohydr. Polym. 96: 31–38.

Yoon, S.D., S.H. Chough and H.R. Park. 2006. Effects of additives with different functional groups on the physical properties of starch/PVA blend film. J. Appl. Polym. Sci. 100: 3733–3740.

Yoshinaga, F., N. Tonouchi and K. Watanabe. 1997. Research progress in production of bacterial cellulose by aeration and agitation culture and its application as a new industrial material. Biosci. Biotechnol. Biochem. 61: 219–224.

Zhang, P.Y. and R.L. Whistler. 2004. Mechanical properties and water vapor permeability of thin film from corn hull arabinoxylan. J. Appl. Polym. Sci. 93: 2896–2902.

Zhang, Y., L. Pitkanen, J. Douglade, M. Tenkanen, C. Remond and C. Joly. 2011. Wheat bran arabinoxylans: Chemical structure and film properties of three isolated fractions. Carbohydr. Polym. 86: 852–859.

Zhong, L.X., X.W. Peng, D. Yang, X.F. Cao and R.C. Sun. 2013. Long-chain anhydride modification: A new strategy for preparing xylan films. J. Agric. Food Chem. 61: 655–661.

9

Reactive Extrusion for the Production of Starch-based Biopackaging

Tomy J. Gutiérrez[1, 2, 3]*, M. Paula Guarás[3] and Vera A. Alvarez[3]

[1] Department of Analytical Chemistry, Faculty of Pharmacy, Central University of Venezuela, PO Box 40109, Caracas 1040-A, Venezuela

[2] Institute of Food Science and Technology, Faculty of Sciences, Central University of Venezuela, PO Box 47097, Caracas 1041-A, Venezuela

[3] Thermoplastic Composite Materials (CoMP) Group, Institute of Research in Materials Science and Technology (INTEMA), Faculty of Engineering, National University of Mar del Plata (UNMdP) and National Council of Scientific and Technical Research (CONICET), Colón 10850, Mar del Plata 7600, Buenos Aires, Argentina

Introduction

Reactive extrusion (REx) combines mass and heat transport operations, which together with simultaneous chemical reactions taking place inside the extruder, modify the properties of existing polymers or generate new ones. This process is increasingly becoming recognized as a powerful technique to develop and manufacture a variety of novel polymeric materials in a highly efficient and flexible way. The combination of chemical reactions and transport phenomena in an extruder provides an invaluable window of opportunity for the compatibilization of synthetic resins and biopolymers such as starch, or the compatibilization of synthetic resins and natural fillers.

REx was developed in the 1980´s primarily for the modification of synthetic polymers. Since then, however, the technology has been rapidly developed and applied in various areas such as viscosity-breaking, polymerization, grafting, cross-linking and coupling reactions, amongst others.

The use of extruders as chemical reactors enables high viscosity polymers to be processed in the absence of solvents giving large operational flexibility

*Corresponding author: tomy.gutierrez@ciens.ucv.ve; tomy_gutierrez@yahoo.es

due to the broad range of processing conditions available, for example, pressure (0-500 atm) and temperature (70-500 °C), the viability of multiple feed systems, and the capacity to control both residence time (distribution) and the degree of mixing (van Duin et al. 2001).

Examples of this process in the literature emphasize the fact that the use of extruders has changed in recent decades from the preparation and modification of polymers for the plastics industry, to the creation of thermoplastic materials, new and unusual resins and other synthetic polymers for industry using REx (Xie et al. 2006).

The most widely used reactive extruders are still single-screw and twin-screw extruders. Single-screw extruders generally have three basic zones: a primary feed zone, a melt zone and a pump zone (Fig. 1). The most common types of twin-screw extruders are co-rotating ones with intermeshing screws. These are suitable for many reactive processes as they give excellent control of residence times, superior mixing intensity, good heat transfer and are self-wiping (White et al. 1987).

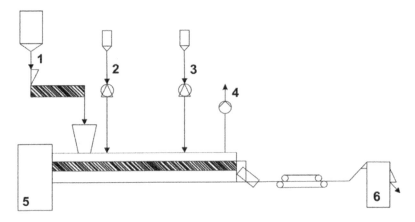

Fig. 1. Schematic representation of a reactive extrusion system used for starch modification: (1) weight starch feeder, (2) and (3) metering pumps, (4) vacuum pump, (5) twin-screw extruder and (6) forced strand pelletizer with "dry" cut.

Starch

Starch is a polysaccharide produced by mostly higher order plants as a means of storing energy. Starch is stored intracellularly in the form of spherical granules between 2 and 100 μm wide. Most commercially available starches are isolated from grains such as corn, rice and wheat, or from tubers such as potato and cassava. Starch is composed of two isomers, linear amylase and highly branched amylopectine (Mohanty et al. 2000).

As a natural polymer, starch has received growing interest recently as a possible alternative to petroleum-based polymers as it is renewable, biodegradable, readily available and economical, amongst other beneficial properties. However, because starch is synthesized by plants to meet their natural requirements, its microstructure is far more complex than that of conventional polymers. Thus, in practice native starches are simply not suitable for any specific application. This has driven the development of several modified starches for non-food applications.

Thermoplastic starch (TPS) has been shown to have poor mechanical properties and a high susceptibility to moisture, characteristics that are improved by the incorporation of fillers such as cellulose, clays, inorganic nanoparticles, etc. Another of the alternatives frequently employed to improve the properties of starch-based films is the incorporation of bioactive substances such as antioxidant compounds that are "generally recognized as safe" (GRAS). More recently, however, REx technology has awakened great interest as it enables the production of improved polymeric materials at an industrial level (Shogren 1993; Narayan 1994; Forssell et al. 1996; Van Soest et al. 1996; Forssell et al. 1997; Nayak 1999; de Graaf et al. 2003).

Reactive Extrusion of Starch

Extrusion technology occurs at a high temperature, shear and pressure over a short period of time. It has the advantage of being extremely versatile and does not produce effluents (Murúa-Pagola et al. 2009). In general, the REx from starch is a process whereby starch gelatinization occurs under typical extrusion conditions. It is well known that starch gelatinization under these conditions is controlled by water diffusion. Since the shear forces destroy the starch granules, water transfer is faster inside the biopolymer (Burros et al. 1987). The starch may thus be plasticized with very small amounts of water (16-20%) reducing subsequent drying requirements.

The application of extrusion for starch was initially studied for food processing. Food extrusion has been used over the past 50 years with early developments in the preparation of ready-to-eat cereals (Harper and Clark 1979). The use of twin-screw extruders for food processing began around 1970. Initial studies were based on the preparation of modified starches using the extruder as a continuous reactor (Fanta 1973; Donovan 1979; Carr and Bagby 1981; Eise et al. 1981; Gomez and Aguilera 1984; Wurzburg 1986; White et al. 1987; Wang et al. 1990; Harper et al. 1991; Valle et al. 1991; van Duin et al. 2001). Nonetheless, REx nowadays is a technique that combines conventional chemical processes (polymer synthesis and modification) with traditional extrusion techniques (melting, blending, structuring, devolatilization and shaping) into a single process carried out in the extruder (Kalambur and Rizvi 2006).

Specifically, the processing of starch-based polymers by extrusion involves multiple chemical and physical reactions, for example water diffusion, granule expansion, gelatinization, decomposition, melting and crystallization (Liu et al. 2009).

It must be pointed out, however, that starch extrusion presents some challenges, including:

- Several physical and chemical reactions may occur together during processing.
- The rheological behavior of TPS, especially when mixed with other polymers and additives, is much more complex than that of conventional polymers.
- TPS viscosity is much higher under regular extrusion conditions (and an increase in temperature will not lower the viscosity in many cases).
- The evaporation of water contained in starch can produce bubble formation.

In addition, one of the specific objectives of the REx of starch is to melt and mix the polymer. During this process, however, shear forces are applied that make the fragmentation and degradation of the starch inevitable. It is worth noting that the main causes of the fragmentation of starch under these conditions are screw speed, temperature, moisture content and even the type of starch.

Murúa-Pagola et al. (2009) prepared acetylated, n-octenylsuccinylated and phosphorylated starches from waxy corn starch by REx in a single-screw extruder. They found that the viscosity as well as the water solubility index of all the modified starches was reduced. Likewise, the water absorption index increased after the extrusion process. They also showed that the modified starches produced by REx exhibited good characteristics as shell materials for encapsulation by spray drying.

Moad (2011) recently reviewed different processes for the chemical transformation of starch by extrusion. This review focused on grafting monomers from starch using techniques such as: ring-opening of epoxides; esterification (with lactones, anhydrides, acids, halides or vinyl esters); phosphorylation and silylation; graft polymerization from starch by radical-induced grafting or the ring-opening polymerization (ROP) of lactones; reactive compatibilization of starch with polyesters and polyolefins by grafting; cross-linking of starch with epichlorohydrin or by phosphation; and the degradation of starch either thermally or catalyzed by acids or enzymes.

Reactive Extrusion for Obtaining Modified Starches

The chemical modification of starches has been carried out with different objectives such as (1) to increase the thermal resistance of starch; (2) prevent starch retrogradation; (3) decrease or increase the viscosity of the gels obtained; and (4) prevent or increase the swelling capacity of starch. However, chemical modification reactions in wet conditions take hours to complete. In contrast, modifying starches under REx conditions has the great advantage that the reactions only take a few seconds, besides being a continuous process. The reason that starches can be modified resides in the nucleophilic character of the hydroxyl groups contained within them. These modified starches in the presence of a plasticizer and under REx conditions enable the production of TPS materials for biopackaging. Some of the ways starch is modified under REx conditions are discussed below.

Phosphatation

Cationic starches with amino, ammonium, sulfonium, phosphonium and other groups have been studied for the past 30 years. For example, Salay and Ciacco (1990) demonstrated that it is possible to prepare phosphated starch (starch cross-linked with phosphate groups) with a low degree of substitution (DS) using the REx process. These authors were also able to obtain starches with a higher DS by changing the experimental extrusion conditions such as temperature, concentration of sodium tripolyphosphate and pH.

Chang and Lii (1992) compared conventional processes with REx for the phosphatation of cassava and corn starches. They found that REx required lower amounts of additives compared to conventional methods for the preparation of phosphated starch with a similar DS, thus reducing both water contamination and production costs.

Seker et al. (2003, 2004) and Seker and Hanna (2005, 2006) prepared cross-linked starch with sodium trimetaphosphate in single- and twin-screw extruders. All these authors were able to demonstrate that the DS was the same for the starches obtained independent of the type of extruder used.

O'Brien and Wang (2009) and O'Brien et al. (2009) used starches containing different amylose contents from various botanical sources to prepare phosphated starch by REx, with the aim of producing hydrogels for the controlled release of metoprolol tartrate. They were able to obtain extruded starches that formed hydrogels capable of sustaining drug release. Additionally, these authors found that REx modified the structural characteristics of the hydrogel by the phosphatation reaction, resulting in changes in the kinetics of drug release from the matrices. In this case, the modified extruded products obtained were weaker gels as demonstrated by their rheological properties, leading to a higher drug release rate. The reaction efficiency by phosphation through REx and the subsequent release of drugs can thus be affected by the shear rate of the extruder and the pH of the different starch types.

Manoi and Rizvi (2010) evaluated starches phosphated by REx under supercritical fluid conditions (supercritical carbon dioxide) finding that these starches were partially water soluble and partially monophosphated. These authors also concluded that the introduction of phosphate groups enables the production of crosslinked starches. This leads to an increase in gelatinization temperature caused by the limited molecular mobility.

Following from this, several studies have indicated that the use of two different crosslinking agents such as sodium trimetaphosphate and sodium tripolyphosphate (Fig. 2) influences the degree of phosphation of the starches obtained by REx.

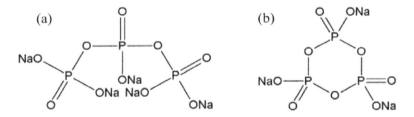

Fig. 2. Structures of (a) sodium tripolyphosphate and (b) sodium trimetaphosphate.

Esterification

Esterification as a starch modification reaction consists essentially in introducing an acetate group into the starch structure, preferably at an alkaline pH (~ 10) to allow the hydroxyl groups in starch become good nucleophilic groups, thus enabling them to react with the esters. The properties of the esterified starch will consequently depend on the esterifying agent used. In the particular case of acetylation, this reaction consists in the introduction of an acetyl group (CH_3CO-) frequently using acetic anhydride as the acetylating agent. Other esterifying agents have, however, also been employed such as propionic, ethanoic, palmitic, phthalic, dodecyl succinic, succinic and maleic anhydrides (MA) (Fig. 3), vinyl acetate and acrylic acid, among others (Ruggeri et al. 1983; Bratawidjaja et al. 1989; Oliphant et al. 1995; Wang et al. 1997; Moad 1999; Kelar and Jurkowski 2000; Lutfor et al. 2001; Martin et al. 2001; Pesetskii et al. 2001; Shi et al. 2001; Taniguchi et al. 2006; Xie et al. 2006).

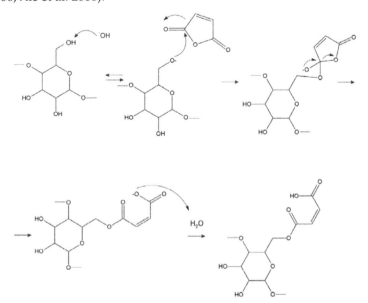

Fig. 3. Esterification reaction mechanism of starch with maleic anhydride.

Esterification reactions performed under REx conditions have been proposed in order to improve the cohesive properties between the starch chains. This helps to reduce the hydrophilicity of TPS, as well as improving the mechanical properties of these materials for their use in biopackaging (Gonzalez and Pérez 2002; Shogren 1996; Averous et al. 2000; Fang and Hanna 2000; Martin et al. 2001; Willett and Shogren 2002; Xu and Hanna 2005). Furthermore, these reactions produce grafted starch which is highly reactive and, in the presence of catalysts, facilitates cross-linking between biopolymers and/or cross-linking between biopolymers-synthetic polymers by REx.

An advantage of starch modification, especially modification by esterification through REx, is that it increases diffusion between polymer blends within the extruder. As is well known, the high molecular weights of starches (ranging from 100,000 to 500,000 g/mol, and more than millions for amylose and amylopectin) generate a highly viscose melted mass which hinders homogeneous blending. With modified starch, however, the gelatinization temperature can be increased whilst decreasing the viscosity of the melted mass, resulting in a more isotropic material, i.e. one with the same properties in any direction within the polymer.

The DS of starch can be affected by several mechanisms during the REx processes (Fig. 4). Despite this, the physicochemical and functional properties of TPS from acetylated starch are always improved, even those with a low DS (Xu et al. 2004).

Highly acetylated starches with a DS from 2 to 3 were thoroughly investigated from 1950 to 1980. Acetylated starches with a low DS (from 0.01 to 0.2) still show commercial promise as their use is permitted by the U.S. Food and Drug Administration (FDA). In addition, these starches show good film-forming, binding, adhesive, thickening, stabilizing and texture properties.

In this regard, Miladinov and Hanna (1999) prepared acetylated starches with different DS and processed them using REx with either water or ethanol in a single-screw extruder. They found that samples extruded with water showed lower spring indices and water absorption, and a higher density of solids than acetylated starch extruded with ethanol. They also found that samples with a lower DS required a lower specific mechanical energy and had greater density indices and absorption than starch with a higher DS. This demonstrates that both the two independent variables used influenced the density and water solubility of the starches.

Miladinov and Hanna (2000) obtained esterified starches from different fatty acid anhydrides (acetic, propionic, heptanoic and palmitic) at various concentrations and under REx conditions using sodium hydroxide (NaOH) as the catalyst in a single-screw extruder. The authors concluded that an increase in the DS was related to the anhydride content employed. Furthermore, the molecular weight of the anhydrides affected the molecular weight of the starches produced, such that low molecular weight anhydrides

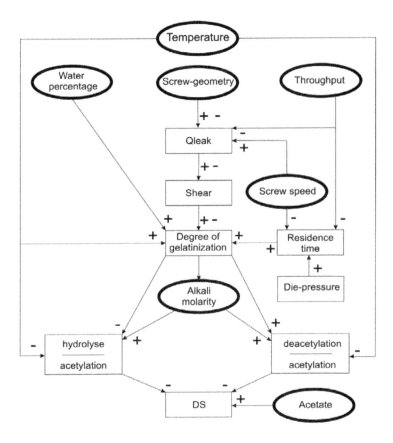

Fig. 4. Steady state interaction diagram for starch esterification reactions.

produced starches with lower molecular weights than those modified using anhydrides with higher molecular weights.

As previously indicated, modification by esterification enables grafts to be introduced onto the starch structure thus increasing its reactivity. Raquez et al. (2008a) prepared TPS under REx conditions using MA as the esterifying agent, glycerol as the plasticizer and mixed all this with polyester. These authors concluded that glycerol is fully incorporated into the matrix and esterification occurs preferentially on the starch backbone at carbon 6. In addition, they indicated that other reactions such as hydrolysis and glycosylation occurred simultaneously, reducing the intrinsic viscosity of the TPS. According to the authors, these results suggest an improvement in the processability of the material.

Oxidation

In recent decades, the REx of starch has been successfully used to obtain products (cationics, alkyl glucosides, carboxylates, etc.) with improved reaction efficiency and solubility (Marsman et al. 1990; Meuser et al. 1990; Chinnaswamy and Hanna 1991; Valle et al. 1991; Chang and Lii 1992; Carr 1994; Tomasik et al. 1995; Esan et al. 1996).

The thermo-chemical oxidation of starch using sodium hypochlorite is the oldest and most frequently used oxidant. Other oxidants such as permanganate, hydrogen peroxide, persulfate, periodate and dichromates have, however, also been employed to produce water-soluble starches containing high numbers of carboxyl and carbonyl groups (Fig. 5). Commercially, starches are oxidized in batches at room temperature and in the presence of low concentrations of an oxidant (near to 3 wt. %). Their production at an industrial level is thus fairly easy and practical (Wurzburg 1986; Wing and Willett 1997).

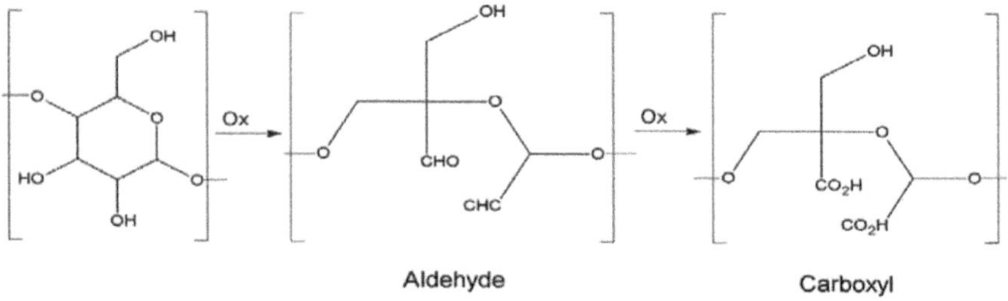

Fig. 5. Mechanism for starch oxidation.

While it is true that the production of oxidized starch is highly efficient and economical, modification by oxidation in the presence of a plasticizer and under REx conditions enables the production of TPS. The reactive TPS is then easily cross-linked with other biopolymers and/or synthetic polymers in the presence of a catalyst. Additionally, the oxidized starch interacts more strongly with the plasticizer, thus enabling the production of thermoformable materials for biopackaging.

Parovuori et al. (1995) oxidized potato starch using hydrogen peroxide under alkaline and acidic reaction conditions with copper, iron and tungstate catalysts in order to introduce carboxyl and carbonyl groups into the starch molecule. The oxidation reactions mainly resulted in the formation of carbonyl groups, and higher reaction yields were obtained at an acid pH. In addition, the molecular weights of the oxidized starches decreased markedly with the DS, and were dependent on the catalyst used. The authors also demonstrated that the oxidized starches presented reduced rheological properties, yielding starches with higher thermal resistance.

Wing and Willett (1997) investigated the oxidation of starches with different amylose contents with the aim of increasing the number of carboxyl groups contained within them and their water solubility. To do this they used three types of corn starch containing up to 70% amylose which were oxidized using hydrogen peroxide and a cupric-ferrous sulfate catalyst for the drying process by REx. An increase in the peroxide concentration produced an increase in the levels of oxidation and solubility of the starches. However, starches with a high amylose content showed lower solubility, but had more carboxyl groups, than starches with a lower amylose content.

Starches Modified by Reactive Extrusion and Their Compatibility with other Polymers

Biodegradable starch-based composites can be completely prepared by REx using biodegradable polymers, with the aim of reducing the high susceptibility to moisture of the starch. Usually, starch is blended with aliphatic polyesters, polyvinyl alcohol (PVA) and other biopolymers. The most widely used polyesters are poly (β-hydroxyalkanoates) (PHB) obtained by microbial synthesis, and poly (ε-caprolactone) (PCL) and polylactide (PLA) derived from chemical polymerization. Mixing biodegradable polyesters with starch enhances the competitiveness of these materials, whilst maintaining high performance levels (Ratto et al. 1999; Mani and Bhattacharya 2001). Composite starch-based blends are usually made from modified starch as this improves the intermolecular interactions between polymers. As previously discussed, the grafting of reactive substrates onto the starches results in an improvement in the compatibility of the composite materials derived from polymeric blends under REx conditions. Among the modifying agents that improve the compatibility of starch with other biopolymers peroxides (Sun et al. 1997), anhydrides (Cho et al. 1997) and oxazoline (Tan et al. 1996) all produce simple and inexpensive reactive mixtures. The mechanical properties and enhanced compatibility of these reaction mixtures are due to the cross-linking reactions or coupling between components as well as van de Waals intermolecular interactions (hydrogen bonds) (Jun 2000). For example, Kalambur and Rizvi (2005, 2006) developed oxidized starch-PCL nanocomposite blends under REx conditions (using Fenton's reagent to oxidize the starch and initiate cross-linking between the oxidized starch and PCL). The blends showed enhanced properties and improved interfacial adhesion between the starch and PCL (Fig. 6). These biodegradable nanocomposites contained up to 40% starch, resulting in tough materials with elongational properties comparable to those of 100% polyester.

Compared with conventional chemical mixing processes, REx has the advantage that several chemical operations are combined within an extruder to produce large quantities of composite materials easily and efficiently. In the next section, we describe some of the composite materials derived from starch blended with other biopolymers under REx conditions that are widely used for obtaining thermoplastic biopackaging materials.

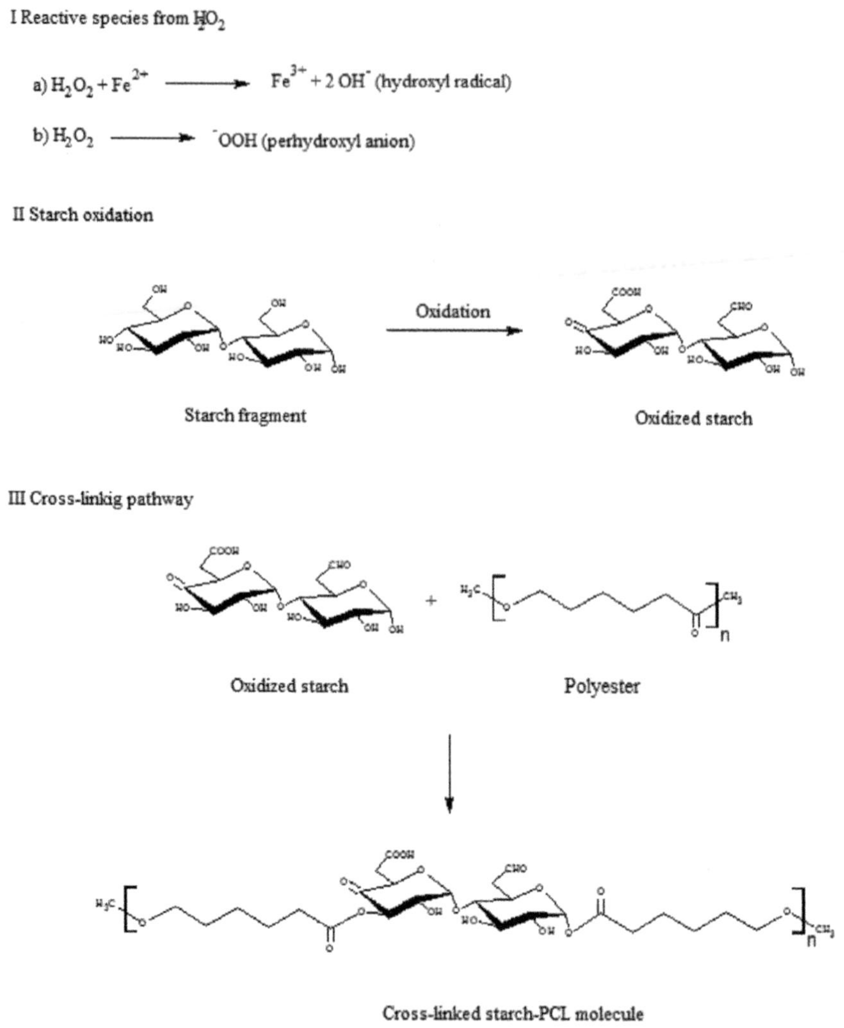

I Reactive species from H_2O_2

a) $H_2O_2 + Fe^{2+} \longrightarrow Fe^{3+} + 2\,OH^-$ (hydroxyl radical)

b) $H_2O_2 \longrightarrow {}^-OOH$ (perhydroxyl anion)

II Starch oxidation

Starch fragment   Oxidized starch

III Cross-linkig pathway

Oxidized starch   Polyester

Cross-linked starch-PCL molecule

Fig. 6. Oxidation and cross-linking steps used during reactive extrusion to produce oxidized starch/polyester composite blends.

Polylactic Acid (PLA)/Starch Blend

PLA is a polymer derived from renewable resources that has sparked interest as an alternative to conventional hydrocarbon polymers. The cyclic dimer of lactic acid (LA) is recovered after the fermentation of corn or sugar beets followed by a free-solvent polymerization/depolymerization process (Fig. 7). PLA has a relatively low crystallinity as it is largely transparent and rigid; a characteristic which makes it a promising candidate for processing biaxially oriented films, thermoformed containers and stretch-blown bottles.

PLA also has characteristics, such as good aroma barrier properties and low permeability to water vapour, oxygen and carbon dioxide, which make it suitable for use in high-volume packaging. It is also safe for use with articles requiring food contact, since the residual monomers have a natural origin and are non-toxic. PLA also has good biocompatibility and processability as well as high strength and Young's modulus values. Nevertheless, it is brittle under tension and bend loads, and develops physical aging during storage. Moreover, PLA is far more expensive than conventional polymers (Jun 2000).

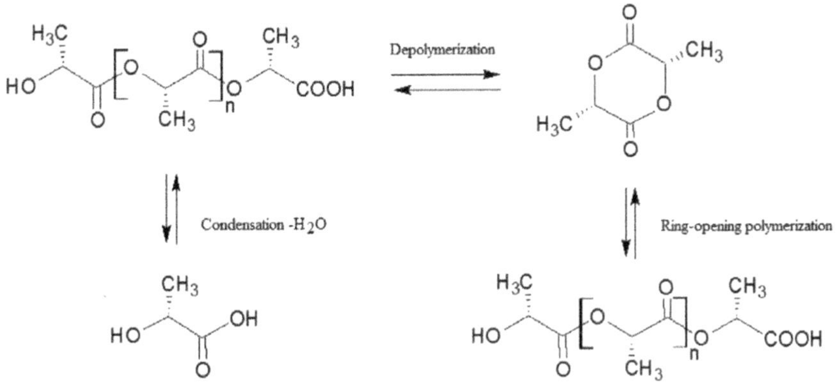

Fig. 7. Production of PLA *via* prepolymer and lactide.

The main problem encountered when mixing starch with PLA is the poor interfacial adhesion between them, resulting from the hydrophilic nature of the starch and the hydrophobicity of the PLA. This incompatibility results in mixtures with poor mechanical properties (Wang et al. 2008a). In order to obtain improved thermoplastic composite materials from starch/ PLA blends, glycerol, formamide and water have been used alone, or in combination, as plasticizers, a technique which improves interfacial affinity and dispersion of the blends (Wang et al. 2008b; Lu et al. 2009).

Wang et al. (2007) studied single stage TPS/PLA blending by REx using glycerol as a plasticizer and MA to promote the esterification reaction. The catalytic initiator used was dicumyl peroxide with the purpose of achieving cross-linking between the biopolymers. MA drastically improved the physical properties and compatibility between the TPS and PLA. The authors concluded that MA can work as a starch plasticizer providing greater fluidity to the extruded mixture (Wang et al. 2007).

Dubois and Narayan (2003) investigated the reaction mixtures of starch/ PLA obtained by extrusion and compatibilized with MA using a free radical process, with 2,5-dimethyl-2,5-di-(tert-butylperoxy) hexane as the free radical initiator. The processed un-modified PLA suffered considerable thermal degradation, demonstrated by a decrease in the molecular weight, solution viscosity and melt viscosity.

Huneault and Li (2007) prepared TPS/PLA blends in a twin-screw extruder with concentrations of TPS between 30-60%. Firstly, a mixture of starch/glycerol/water was placed into the main feeder. The water was then removed from the mid-extruder by vacuum ventilation. Unmodified PLA was fed through a single-screw feeder. Another sample of PLA was modified with MA by REx, by placing the compounds in the main hopper and degassing in the middle of the extruder in order to remove any unreacted agent. The modified PLA was dried and pelleted and then mixed with TPS in the same manner as described above. The use of modified starch resulted in mixtures with high ductility and considerable elongation at break values (100-200%) compared to unmodified blends, due to the improved interfacial adhesion.

Polycaprolactone (PCL)/Starch Blend

PCL is a semicrystalline, linear, hydrophobic polyester that can be consumed by microorganisms. Its physical properties, processability and availability make it an attractive candidate to replace non-biodegradable polymers. PCL has a melting temperature (T_m) of 60 °C and glass transition temperature –60 °C (Avella et al. 2000). Its high cost and low stiffness, however, put it at a disadvantage compared to other polymers commonly used for food packaging such as polyethylene (PE) and polypropylene (PP).

Mixing PCL directly with starch poses some problems as it has a hydrophobic nature and is thermodynamically immiscible with hydrophilic starch. This results in an incompatibility between phases and poor mechanical properties. Ideally, the starch and PCL could be linked by either using existing functional groups or introducing new ones. Simple mixing does not result in phase separation if the composition of starch is below a certain level, and below this critical level the deterioration of properties is negligible (Kalambur and Rizvi 2006).

Yu et al. (2007) studied the effect of a compatibilizer and its distribution in the microstructure on the mechanical properties of several mixtures made with corn starch and PCL. The mixtures were processed through a twin-screw extruder followed by injection moulding. Methylenediphenyl diisocyanate (MDI) was used as a starch modifying agent to improve the interface between the starch and the polyesters. An improvement in the mechanical and interfacial properties was observed, possibly due to the formation of a compatibilizing block copolymer formed between the isocyanate and hydroxyl groups of the starch, and the hydroxyl end groups of the polyester.

Ikeo et al. (2006) developed mixtures of PCL/plasticizer/starch reinforced with nanoclay and processed using REx. PCL grafted with MA (MA-g-PCL) and extruded in advance was added as compatibilizer resulting in greater compatibility between the TPS and PCL. However, a more significant improvement was achieved when montmorillonite was added to the mixtures, obtaining an increase in both Young's modulus and yield stress.

Polyvinyl Alcohol (PVA)/Starch Blend

PVA is a synthetic polymer produced from renewable resources. It is water-soluble and biodegradable, has excellent mechanical properties and is highly compatible with starch. It has thus gained much attention as a replacement for conventional polymers as packaging materials. Its application is limited, however, by its high cost (Chiellini et al. 2003). It is also difficult to melt-process PVA as its thermo-degradation temperature is slightly higher than its melting temperature.

Furthermore, TPS/PVA blends have proved extremely challenging to process by twin-screw extrusion due to their rheology. Most of these composite materials are thus prepared by casting (Park et al. 2005). Zou et al. (2008) extruded TPS/PVA blends with citric acid and borax as additives, which are economical and environmentally friendly, in order to obtain a material with a good water resistance. Analysis of the water content by the orthogonal matrix method showed that it increased with increasing starch content. The citric acid and borax both crosslinked to starch which contributed to a decrease in water absorption and an improvement in the mechanical properties of the modified starch. Liu et al. (1999) aimed to develop an economical and environmental friendly processing technology using conventional processing techniques. The mixtures were prepared for a single-screw extruder using glycerol and water as plasticizers. These authors found that glycerol was more effective than water as a plasticizer, but at the expense of a decrease in the yield stress and water barrier properties. This was somewhat mitigated by using a 50/50 (by weight) water/glycerol mixture as a plasticizer. The incorporation of PVA into a TPS matrix improved the mechanical properties but also increased the melt viscosity.

Microcrystalline Cellulose/Starch Blend

Cellulose is the principal structural component of plants, and the most abundant source of carbohydrate in the world (Engelhardt 1995). Microcrystalline cellulose (MCC) is a natural polymer whose chemical name is beta 1,4-glucan. Recently, MCC has attracted much interest for its role in improving the functional properties of food (Ang et al. 1991). Methylcellulose (MC) is a cellulose ether formed by the alkali treatment of cellulose, followed by reacting with methyl chloride. It exhibits thermal gelation and has excellent film forming properties thus making it an excellent material for biopackaging. It is also used in the preparation of edible films (Peressini et al. 2003).

Extruding corn starch/microcrystalline cellulose mixtures in the presence or absence of plasticizers (polyols) has been done by REx conditions to produce edible films. Increasing the cellulose content results in an increase in the breaking strength, but a decrease in the elongation at break and water vapour permeability of the films (Psomiadou et al. 1996). Furthermore, the films may be thermodynamically compatible with water-soluble carboxymethylcellulose (CMC) films when the starch content is below 25% by mass under REx conditions. They are also biodegradable in the presence of microorganisms (Suvorova et al. 2000).

Finally, starch/cellulose blends can be crosslinked using REx. This results in the production of innovative materials derived from natural resources with improved properties that are one hundred percent biodegradable.

The Use of Catalysts to Improve the Compatibility of Starch with other Polymers under Conditions of Reactive Extrusion

Several catalysts have been used in conjunction with REx to promote crosslinking reactions in a polymer blend (especially TPS). The temperatures and pressures used to process starch-based biodegradable composites activate the catalyst, thereby increasing the reaction rate, i.e. the kinetics of the crosslinking reaction are favoured. In this way, the compatibility of starch-based polymer blends has been improved *via* a process that can be scaled up to an industrial level. Among the different catalysts reported in the literature are: tin(II) bis(2-ethylhexanoate) ($Sn(Oct)_2$), lithium octanoate ($Y(Oct)_3$), aluminum sec-butoxide ($Al(O^{sec}Bu)_3$), aluminum triisopropoxide ($Al(O^iPr)_3$), sodium hydride (NaH), tributyltin methoxide (nBu_3SnOMe), tetrabutyl titanate ($Ti(OBu)_4$), triphenylphosphine ($P(C_6H_5)_3$), N-acetyl caprolactam, potassium persulfate ($K_2S_2O_8$), 4-dimethylaminopyridine (DMAP), 2,5-dimethyl-2,5-di-(tert-butylperoxy)hexane (Lupersol 101), tetraethyl orthosilicate (TEOS), dicumyl peroxide (DCP), di-tert-butyl peroxide, benzoyl peroxide, ceric ammonium nitrate (CAN), glyoxal, titaniumn-butoxide, aluminum mono- and trialkoxides (Dubois et al. 1989; Jacobs et al. 1991; Barakat et al. 1993; Du et al. 1995; Kricheldorf et al. 1995; Gaylord 1996; Carlson et al. 1999; Mani et al. 1999; Mecerreyes and Jérôme 1999; Tang et al. 1999; de Graaf and Janssen 2000; Penczek et al. 2000; Sailaja and Chanda 2000; Lutfor et al. 2001; Kim and White 2002, 2003, 2004, 2005; Wang et al. 2005; Zhu et al. 2005; Raquez et al. 2006; Wang et al. 2007; Frost et al. 2011; Song et al. 2011). The last two of these have been shown to be very effective initiators. These commercial catalysts can also be handled easily (i.e. they do not require high vacuum equipment) and are relatively easy to purify (at least down to ca. 2 mol% of proton-containing impurities) by distillation for semi-quantitative synthetic work (Penczek et al. 2000).

In particular $Sn(Oct)_2$ is widely used (Kowalski et al. 1998) and has been approved as a food additive by the FDA. The most probable mechanism involves the direct catalytic action of $Sn(Oct)_2$ (Du et al. 1995; Kricheldorf et al. 1995). Nevertheless, according to Raquez et al. (2008b), $Sn(Oct)_2$ first activates the monomer forming a donor-acceptor complex, which then participates directly in the first propagation step. $Sn(Oct)_2$ is thereafter liberated in every ensuing propagation step. This means that Sn^{2+} atoms are not covalently bound to the polymer at any stage of polymerization. Another likely mechanism, the "active chain-end" mechanism, has been recently proposed by Penczek et al. (2000). This involves the *in-situ* formation of Sn-alkoxide bonds at chain-ends as observed by MALDI-TOF and fully confirmed by kinetic studies (Kowalski et al. 1998). Thus, through a rapid exchange equilibrium, Sn(Oct)$_2$ and probably any other covalent metal carboxylate, is first converted by reacting with protic compounds (ROH) into tin (or other metal) alkoxides which act as active centers for polymerization (Fig. 8). Polymerization itself involves a "coordination insertion" mechanism similar to the previously discussed mechanism for covalent metal alkoxides and dialkylaluminum alkoxides (Fig. 9).

It is worth noting that, as already mentioned, the use of catalysts enables the cross-linking of polymers by REx. This results in an increase in the molecular weight, density, melting temperature and interfacial adhesion forces of composite materials (Carlson et al. 1999; de Graaf and Janssen 2000; Finkenstadt and Willett 2005). This last has a favorable effect on the tensile strength and Young's modulus of the obtained materials, due to the improved compatibility between the compounds (Mani and Bhattacharya 1998; Wang et al. 2005; Raquez et al. 2006; Wang et al. 2007; Nayak 2010; Frost et al. 2011; Shin et al. 2011; Zeng et al. 2011).

Nevertheless, increasing the screw speed and reducing the flow can lead to a severe reduction in molecular weight. Likewise, an increase in the polymer/catalyst ratio can cause changes to the molecular weight (Xanthos 1992; Gimenez et al. 2000; White et al. 2001; Gaspar-Cunha et al. 2002; Choulak et al. 2004; Zhu et al. 2005; Cassagnau et al. 2006; Vergnes and Berzin 2006; Cassagnau et al. 2007). The feed rate within the extruder enables the adjustment of the length of the grafted chains on the polymers and the degree of crosslinking between them (Goodman and Vachon 1984; Kim and White 2003; Kim and White 2005).

Other factors to take into consideration are the physical characteristics of the catalyst. For example, although $Al(O_iPr)_3$ has been shown to be an

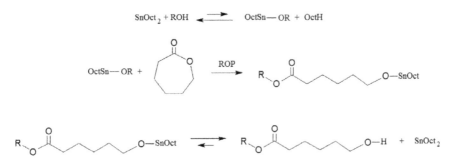

Fig. 8. Proposed activation mechanism for the ring-opening polymerization of ε-caprolactone with Sn(Oct)$_2$ as the catalyst.

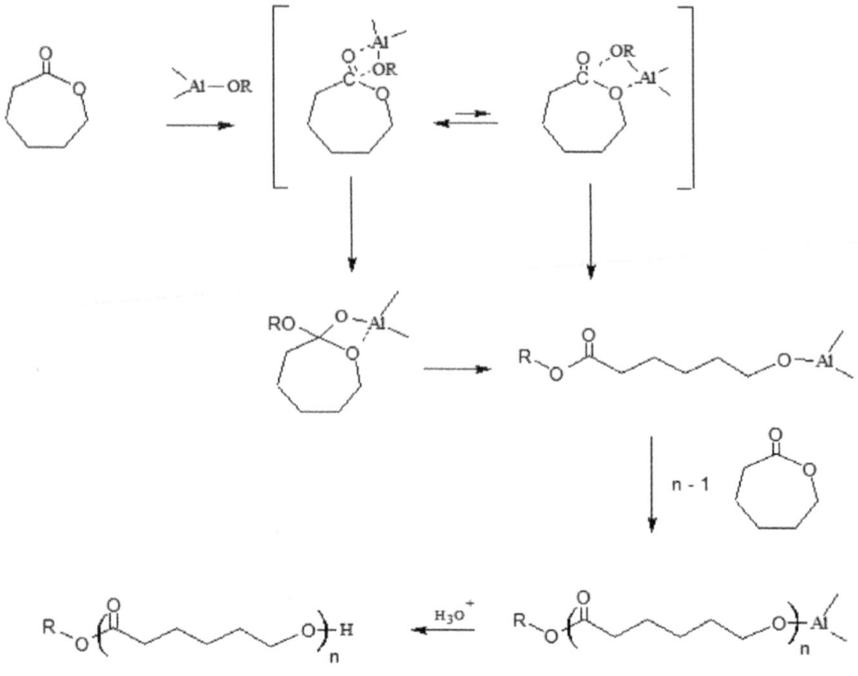

Fig. 9. "Coordination-insertion" mechanism of the ring-opening
polymerization of ε-caprolactone.

efficient catalyst for the ROP of PCL, it must first be sublimed and then dissolved in an organic solvent such as toluene in order to control the ROP process. In contrast, $Al(O^{sec}Bu)_3$ is commercially available as a pure liquid and therefore does not require any previous purification step before use. In addition, $Al(O^{sec}Bu)_3$ has proved to be an efficient initiator for the bulk ROP of PCL carried out in small reactors (Mecerreyes et al. 1999; von Schenck et al. 2002).

Finally, it is also worth mentioning that grafts can be included onto synthetic biopolymers in order to initiate reactions with native or modified starches in the presence of a catalyst.

Biodegradable Polymeric Nanocomposites Produced through Reactive Extrusion

By definition, a composite material is formed by the combination of different phases all of which have different structural and chemical compositions. This leads to a synergy of physical, chemical and/or mechanical properties that are distinct from each component taken separately. Composite materials may contain reinforcing materials such as fibres, particles, clays, minerals and so on. In addition, the matrix can include various types of materials: organic, mineral and metallic. The classification of composites is based on either the type of filler (fibres or particles) or the matrix itself.

In recent years the use of natural/bio-fibre reinforced composites has rapidly expanded due to the availability of natural/bio-fibre derived from annually renewable resources. These materials are versatile as they can be used as reinforcing fibres in both thermoplastic and thermosetting matrix composites, and are also environmentally friendly (Mohanty et al. 2002; Kulpinski 2007). Most of the research currently being undertaken focuses on the potential of natural fibres for reinforcing polymers as they have low density and acceptable specific strength properties, as well as being easy to prepare and biodegradable.

Fully green biocomposites were prepared by White et al. (2001) from PLA and recycled cellulose fibres. These authors concluded that the tensile strength and Young's modulus of the composites were significantly higher than materials without fibres due to the presence of the high-modulus cellulose fibres. However, a significant decrease in tensile strength with a high fibre content was observed due to a lack of interaction between the cellulose and PLA. Cellulose has a strong hydrophilic character due to the three hydroxyl groups per monomeric unit in contrast to the rather hydrophobic PLA.

Wood fibre has also been used as a reinforcing material in the preparation of corn gluten meal-based composites by REx (Wu et al. 2003). These authors found that the melt-viscosity of the medium increased with both an increase in wood fibre content, and a decrease in water content, leading to a decrease in melt-mobility. From flexural testing, it has been demonstrated that the flexural strengths of these biocomposites increased with the incorporation of 10-30 wt.% wood fibre, but then decreased when 40-50 wt.% wood fibre was added.

Liu et al. (2005) developed soy-based biocomposites reinforced with Indian grass fibre. These natural fibres were chemically pre-treated in order to enhance the interfacial adhesion between the natural fibres and polymer (soy protein), and hence also improve the properties of the resulting composites. For example, alkaline treatment eliminated the lignin fraction contained in the fibre thus reducing its size. The authors concluded that reinforcement with raw Indian grass fibre improved both the tensile and the flexural properties as well as the heat deflection temperature of the soy-based bioplastics, but did not affect their impact strength. Interestingly, Indian grass fibre treated with the alkali solution significantly increased the tensile strength, impact strength and flexural strength of the bioplastics due to a better dispersion of the fibre into the matrix, as well as enhancing the aspect ratio of the fibre.

Other authors have preferred to melt-blend biodegradable polymers and natural fibres by adding maleic anhydride-modified polymers as compatibilizers. Nitz et al. (2001) reported the melt-compounding of PCL with wood flour and lignin in a twin-screw extruder. They found that when low quantities of the MA-grafted PCL (less than 2.5 wt.% of overall product) were added, attractive properties could be obtained in these PCL-based composites. For example, PCL composites containing 2.5 wt.% in MA-g-PCL and 40 wt.% wood flour, showed an increase in both Young's modulus (450%) and the tensile strength (115%) compared to the properties of neat PCL. The mechanical properties of the wood flour composites were superior to those of the lignin-based compounds. However, more than 70 wt.% lignin could be added without the mechanical properties being impaired, while compositions containing 40 wt.% lignin showed an elongation at break exceeding 500%. Interestingly, biodegradation studies revealed that the addition of lignin enhanced the biostability of PCL compounds, which could prolong the lifetime of PCL-based compounds in outdoor applications.

On the other hand, inorganic fillers such as talc have also been utilized in the preparation of biodegradable polymeric composites (Whaling et al. 2006). Talc is a common filler used for the improvement of polymer properties such as strength, rigidity, durability and hardness. Melt-blending a polymer with talc results in moderate to significant improvements in the tensile, flexural and storage modulus. This can be explained by the poor filler dispersion and filler-matrix adhesion as revealed by scanning electron microscopy in the talc-filled polymeric composites.

Raquez et al. (2008c) prepared new biodegradable hybrid materials with high-performance talc by REx for blown film applications. In the first step, the polyester backbone was reactively modified through the free-radical grafting of MA in order to improve the interfacial adhesion between polymer and talc. The resulting MA-g-polymer was then reactively melt-blended with talc through esterification reactions of MA moieties grafted onto the polyester chains with the silanol present on the edge surface of the talc (Farmer 1974; Temuujin et al. 2002). $Sn(Oct)_2$ and 4-dimethylaminopyridine (DMAP) were used as the esterification catalysts (Fig. 10). The interfacial adhesion between both partners was substantially enhanced as demonstrated by SEM and the selective extraction of the polyester component. As a result, the biaxial tensile properties measured for the blown films prepared from these compatibilized composites were considerably improved compared to those of the conventional polymer-talc melt-blends. Extrapolation to a one-step REx process was successfully achieved by preparing *in situ* chemically modified polymer-talc composites containing up to 60 wt.% talc. Interestingly, the highest tensile properties were obtained by melt-blending 50 wt.% of native polymer with 50 wt.% of a prepared chemically modified polymer/talc hybrid, thereafter used as a masterbatch. This approach resulted in a reduction in the degradation of native polyester chains through undesirable reactions such as β-scission and transesterification reactions promoted by the MA free-radical treatment and esterification catalysts, respectively, used during the REx process. Finally, X-ray photoelectron spectroscopy measurements carried out on the reactively modified polymer-talc compositions, for example, containing 60 wt.% talc, confirmed the formation of covalent ester bonds between the silanol functions available on the edge surfaces of the talc particles, and the maleic anhydride moieties grafted onto the polyester backbones.

Similarly, nanoclays have been incorporated into starch matrices in REx processes. In this regard, several authors have indicated that the exfoliation of these fillers enables the production of improved materials (Messersmith and Stupp 1992; Okada and Usuki 1995; Giannelis 1996; Schöllhorn 1996; Wang et al. 2000; Bharadwaj 2001; Lange and Wyser 2003; Chaiko and Leyva, 2005). For example, Ikeo et al. (2006) developed nanoclays to reinforce PCL/starch composites blended by REx. As before, the modification of the nanoclay results in improved interleaving within the polymeric matrix and an increase in the biodegradability of these systems (Okamoto et al. 2003; Sinha Ray et al. 2003; Ray et al. 2005; Raquez et al. 2008b). In addition, Huang and Yu (2006) reported that the incorporation of nanoclays avoided the phenomenon of starch retrogradation by REx of activated montmorillonite and TPS.

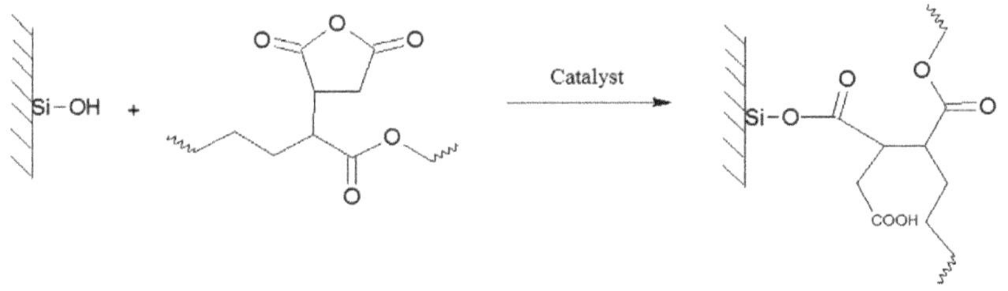

Fig. 10. Surface grafting mechanism of MA moieties onto the silanol functions on the edges of talc particles.

Furthermore, modified cellulose nanocomposites crosslinked by REx with MA-modified starch and Luperox 101 as a catalyst were found to exhibit extraordinary properties (Park et al. 2004a, b).

Conclusions and Outlooks

Current and future trends are directed towards the production of polymeric materials that are environmentally friendly. Reactive extrusion (REx) is currently the most efficient and economical way for producing eco-friendly plastics from starch, as it combines the versatility of today's industrial processes with polymer science and technology. In addition, the chemical reactions such as polymerization, grafting, branching and functionalization that occur during REx do not need solvents for their processing. This gives REx the advantage over other processes since energy consumption required for the evaporation of solvents is eliminated. In addition, this technology enables the compatibilization of synthetic polymers from the petrochemical industry with many naturally incompatible biopolymers, resulting in the production of improved TPS materials with higher rates of biodegradability.

Several catalysts have been used for the compatibilization of these biopolymers with synthetic polymers, with tin (II) bis (2-ethylhexanoate) $(Sn(Oct)_2)$ being the most promising as it is approved by the FDA as a food additive. It is worth remembering that most packaging materials are used to package food, making the development of edible and biodegradable packaging extremely advantageous. Finally, due to the large number of variables that affect these processes, extensive studies should be conducted in order to obtain the ideal conditions for the production of materials with the most desirable characteristics.

Acknowledgements

The authors would like to thank the Consejo Nacional de Investigaciones Científicas y Técnicas (CONICET) (Postdoctoral fellowship internal PDTS-Resolution 2417), Universidad Nacional de Mar del Plata (UNMdP) for the financial support, and Dr. Mirian Carmona-Rodríguez.

References

Ang, J.F. and W.B. Miller. 1991. Multiple functions of powdered cellulose as a food ingredient. Cereal Food. World 36(7): 558–564.

Avella, M., M.E. Errico, P. Laurienzo, E. Martuscelli, M. Raimo and R. Rimedio. 2000. Preparation and characterisation of compatibilised polycaprolactone/starch composites. Polymer 41(10): 3875–3881.

Averous, L., L. Moro, P. Dole and C. Fringant. 2000. Properties of thermoplastic blends: Starch-polycaprolactone. Polymer 41(11): 4157–4167.

Barakat, I., P. Dubois, R. Jerome and P. Teyssié. 1993. Macromolecular engineering of polylactones and polylactides. X. Selective end-functionalization of poly (D, L)-lactide. J. Polym. Sci., Part A: Polym. Chem. 31(2): 505–514.

Bharadwaj, R.K. 2001. Modeling the barrier properties of polymer-layered silicate nanocomposites. Macromolecules 34(26): 9189–9192.

Bratawidjaja, A.S., I. Gitopadmoyo, Y. Watanabe and T. Hatakeyama. 1989. Adhesive property of polypropylene modified with maleic anhydride by extrusion molding. J. Appl. Polym. Sci. 37(4): 1141–1145.

Burros, B.C., L.A. Young and P.A. Carroad. 1987. Kinetics of corn meal gelatinization at high temperature and low moisture. J. Food Sci. 52(5): 1372–1376.

Carlson, D., L. Nie, R. Narayan and P. Dubois. 1999. Maleation of polylactide (PLA) by reactive extrusion. J. Appl. Polym. Sci. 72(4): 477–485.

Carr, M.E. and M.O. Bagby. 1981. Preparation of cationic starch ether: A reaction efficiency study. Starch-Stärke 33(9): 310–312.

Carr, M.E. 1994. Preparation of cationic starch containing quaternary ammonium substituents by reactive twin-screw extrusion processing. J. Appl. Polym. Sci. 54(12): 1855–1861.

Cassagnau, P., J. Gimenez, V. Bounor-Legaré and A. Michel. 2006. New rheological developments for reactive processing of poly (ε-caprolactone). Chimie 9(11): 1351–1362.

Cassagnau, P., V. Bounor-Legaré and F. Fenouillot. 2007. Reactive processing of thermoplastic polymers: A review of the fundamental aspects. Internat. Polym. Process. 22(3): 218–258.

Chaiko, D.J. and A.A. Leyva. 2005. Thermal transitions and barrier properties of olefinic nanocomposites. Chem. Mater. 17(1): 13–19.

Chang, Y.H. and C.Y. Lii. 1992. Preparation of starch phosphates by extrusion. J. Food Sci. 57(1): 203–205.

Chiellini, E., A. Corti, S. D'Antone and R. Solaro. 2003. Biodegradation of poly (vinyl alcohol) based materials. Prog. Polym. Sci. 28(6): 963–1014.

Chinnaswamy, R. and M.A. Hanna. 1991. Extrusion-grafting starch onto vinylic polymers. Starch-Stärke 43(10): 396–402.

Cho, K., K.H. Seo and T.O. Ahn. 1997. Morphology and rheological behavior of amorphous polyamide/(styrene-acrylonitrile/styrene maleic anhydride) reactive blends. Polym. J. 29(12): 987–991.

Choulak, S., F. Couenne, Y. Le Gorrec, C. Jallut, P. Cassagnau and A. Michel. 2004. Generic dynamic model for simulation and control of reactive extrusion. Ind. Eng. Chem. Res. 43(23): 7373–7382.

de Graaf, R.A. and L.P.B.M. Janssen. 2000. The production of a new partially biodegradable starch plastic by reactive extrusion. Polym. Eng. Sci. 40(9): 2086–2094.

de Graaf, R.A., A.P. Karman and L.P. Janssen. 2003. Material properties and glass transition temperatures of different thermoplastic starches after extrusion processing. Starch-Stärke 55(2): 80–86.

Donovan, J.W. 1979. Phase transitions of the starch-water system. Biopolymers 18(2): 263–275.

Du, Y.J., P.J. Lemstra, A.J. Nijenhuis, H.A. Van Aert and C. Bastiaansen. 1995. ABA type copolymers of lactide with poly (ethylene glycol). Kinetic, mechanistic, and model studies. Macromolecules 28(7): 2124–2132.

Dubois, P., R. Jerome and P. Teyssie. 1989. Macromolecular engineering of polylactones and polylactides. Polym. Bull. 22(5-6): 475–482.

Dubois, P. and R. Narayan. 2003. Biodegradable compositions by reactive processing of aliphatic polyester/polysaccharide blends. In: Macromolecular Symposia, Vol. 198, No. 1, pp. 233–244. Wiley-VCH Verlag.

Eise, K., H. Herrmann, S. Jakopin, U. Burkhardt and H. Werner. 1981. An analysis of twin-screw extruder mechanisms. Adv. Polym. Tech. 1(2): 18–39.

Engelhardt, J. 1995. Sources, industrial derivatives, and commercial applications of cellulose. Carbohydr. Eur. 12: 5–14.

Esan, M., T.M. Bruemmer and F. Meuser 1996. Chemical and Processing Aspects of the Production of Cationic Starch using Extrusion Cooking. Starch-Staerke, Germany.

Fang, Q. and M.A. Hanna. 2000. Functional properties of polylactic acid starch-based loose-fill packaging foams 1. Cereal Chem. 77(6): 779–783.

Fanta, G.F. 1973. Synthesis of graft and block copolymers of starch. Block and graft copolymerization 1: 11.

Farmer, V.C. 1974. The Infrared Spectra of Minerals. Mineralogical Society, London.

Finkenstadt, V.L. and J.L. Willett. 2005. Reactive extrusion of starch-polyacrylamide graft copolymers: Effects of monomer/starch ratio and moisture content. Macromol. Chem. Phys. 206(16): 1648–1652.

Forssell, P., J. Mikkilä, T. Suortti, J. Seppälä and K.J.M.S. Poutanen. 1996. Plasticization of barley starch with glycerol and water. J. Macromol. Sci. A 33(5): 703–715.

Forssell, P.M., J.M. Mikkilä, G.K. Moates and R. Parker. 1997. Phase and glass transition behaviour of concentrated barley starch-glycerol-water mixtures, a model for thermoplastic starch. Carbohyd. Polym. 34(4): 275–282.

Frost, K., J. Barthes, D. Kaminski, E. Lascaris, J. Niere and R. Shanks. 2011. Thermoplastic starch-silica-polyvinyl alcohol composites by reactive extrusion. Carbohyd. Polym. 84: 343–350.

Gaspar-Cunha, A., A. Poulesquen, B. Vergnes and J.A. Covas. 2002. Optimization of processing conditions for polymer twin-screw extrusion. Int. Polym. Process. 17(3): 201–213.

Gaylord, N. 1996. Maleic Anhydride Grafting. CRC Press, Inc., New York.

Giannelis, E.P. 1996. Polymer layered silicate nanocomposites. Adv. Mater. 8(1): 29–35.

Gimenez, J., M. Boudris, P. Cassagnau and A. Michel. 2000. Control of bulk ε-caprolactone polymerization in a twin screw extruder. Polym. React. Eng. 8(2): 135–157.

Gomez, M.H. and J.M. Aguilera. 1984. A physicochemical model for extrusion of corn starch. J. Food Sci. 49(1): 40–43.

Gonzalez, Z. and E. Pérez. 2002. Effect of acetylation on some properties of rice starch. Starch-Stärke 54(3–4): 148–154.

Goodman, I. and R.N. Vachon. 1984. Copolyesteramides—II. Anionic copolymers of ε-caprolactam with ε-caprolactone. Preparation and general properties. Eur. Polym. J. 20(6): 529–537.

Harper, G.R., M.C. Davies, S.S. Davis, T.F. Tadros, D.C. Taylor, M.P. Irving, et al. 1991. Steric stabilization of microspheres with grafted polyethylene oxide reduces phagocytosis by rat Kupffer cells in vitro. Biomaterials 12(7): 695–700.

Harper, J.M. and J.P. Clark. 1979. Food extrusion. Crit. Rev. Food Sci. 11(2): 155–215.

Huang, M. and J. Yu. 2006. Structure and properties of thermoplastic corn starch/montmorillonite biodegradable composites. J. Appl. Polym. Sci. 99(1): 170–176.

Huneault, M.A. and H. Li. 2007. Morphology and properties of compatibilized polylactide/thermoplastic starch blends. Polymer 48(1): 270–280.

Ikeo, Y., K. Aoki, H. Kishi, S. Matsuda and A. Murakami. 2006. Nano clay reinforced biodegradable plastics of PCL starch blends. Polym. Adv. Technol. 17(11–12): 940–944.

Jacobs, C., P. Dubois, R. Jérôme and P. Teyssié. 1991. Macromolecular engineering of polylactones and polylactides. 5. Synthesis and characterization of diblock copolymers based on poly-ε-caprolactone and poly (L, L or D, L) lactide by aluminum alkoxides. Macromolecules 24(11): 3027–3034.

Jun, C.L. 2000. Reactive blending of biodegradable polymers: PLA and starch. J. Polym. Environ. 8(1): 33–37.

Kalambur, S. and S.S.H. Rizvi. 2005. Biodegradable and functionally superior starch-polyester nanocomposites from reactive extrusion. J. Appl. Polym. Sci. 96(4): 1072–1082.

Kalambur, S. and S.S. Rizvi. 2006. An overview of starch-based plastic blends from reactive extrusion. J. Plast. Film Sheet 22(1): 39–58.

Kelar, K. and B. Jurkowski. 2000. Preparation of functionalised low-density polyethylene by reactive extrusion and its blend with polyamide 6. Polymer 41(3): 1055–1062.

Kim, B.J. and J.L. White. 2002. Bulk polymerization of ε-Caprolactone in an internal mixer and in a twin screw extruder. Int. Polym. Process. 17(1): 33–43.

Kim, B.J. and J.L. White. 2003. Continuous polymerization of lactam–lactone block copolymers in a twin-screw extruder. J. Appl. Polym. Sci. 88(6): 1429–1437.

Kim, B.J. and J.L. White. 2004. Engineering analysis of the reactive extrusion of ε-caprolactone: The influence of processing on molecular degradation during reactive extrusion. J. Appl. Polym. Sci. 94(3): 1007–1017.

Kim, I. and J.L. White. 2005. Reactive copolymerization of various monomers based on lactams and lactones in a twin-screw extruder. J. Appl. Polym. Sci. 96(5): 1875–1887.

Kowalski, A., A. Duda and S. Penczek. 1998. Kinetics and mechanism of cyclic esters polymerization initiated with tin (II) octoate, 1. Polymerization of ε-caprolactone. Macromol. Rapid Comm. 19(11): 567-572.

Kricheldorf, H.R., I. Kreiser-Saunders and C. Boettcher. 1995. Polylactones: 31. Sn (II) octoate-initiated polymerization of L-lactide: A mechanistic study. Polymer 36(6): 1253–1259.

Kulpinski, P. 2007. Cellulose fibers modified by hydrophobic-type polymer. J. Appl. Polym. Sci. 104(1): 398–409.

Lange, J. and Y. Wyser. 2003. Recent innovations in barrier technologies for plastic packaging—a review. Packag. Technol. Sci. 16(4): 149-158.

Liu, H., F. Xie, L. Yu, L. Chen and L. Li. 2009. Thermal processing of starch-based polymers. Prog. Polym. Sci. 34(12): 1348–1368.

Liu, W., A.K. Mohanty, L.T. Drzal and M. Misra. 2005. Novel biocomposites from native grass and soy based bioplastic: Processing and properties evaluation. Ind. Eng. Chem. Res. 44(18): 7105–7112.

Liu, Z., Y. Feng and X.S. Yi. 1999. Thermoplastic starch/PVAl compounds: Preparation, processing, and properties. J. Appl. Polym. Sci. 74(11): 2667–2673.

Lu, D.R., C.M. Xiao and S.J. Xu. 2009. Starch-based completely biodegradable polymer materials. Express Polym. Lett. 3(6): 366–375.

Lutfor, M.R., M.Z.A. Rahman, S. Sidik, A. Mansor, J. Haron and W. Yunus. 2001. Kinetics of graft copolymerization of acrylonitrile onto sago starch using free radicals initiated by ceric ammonium nitrate. Des. Monomers Polym. 4: 253–260.

Mani, R. and M. Bhattacharya. 1998. Properties of injection moulded starch synthetic polymer blends—III. Effect of amylopectin to amylose ratio in starch. Eur. Polym. J. 34: 1467–1475.

Mani, R., M. Battacharya and J. Tang. 1999. Functionalization of polyesters with maleic anhydride by reactive extrusion. J. Polym. Sci., Part A: Polym. Chem. 37(11): 1693–1702.

Mani, R. and M. Bhattacharya. 2001. Properties of injection moulded blends of starch and modified biodegradable polyesters. Eur. Polym. J. 37(3): 515–526.

Manoi, K. and S.S. Rizvi. 2010. Physicochemical characteristics of phosphorylated cross-linked starch produced by reactive supercritical fluid extrusion. Carbohyd. Polym. 81(3): 687–694.

Marsman, J.H., R.T. Pieters, L.P.B.M. Janssen and A.A.C.M. Beenackers. 1990. Determination of degree of substitution of extruded benzylated starch by H-NMR and UV spectrometry. Starch-Stärke 42(5): 191–196.

Martin, O., E. Schwach and Y. Couturier. 2001. Properties of biodegradable multilayer films based on plasticized wheat starch. Starch-Stärke 53(8): 372–380.

Mecerreyes, D. and R. Jérôme. 1999. From living to controlled aluminium alkoxide mediated ring-opening polymerization of (di) lactones, a powerful tool for the macromolecular engineering of aliphatic polyesters. Macromol. Chem. Phys. 200(12): 2581–2590.

Mecerreyes, D., R. Jérôme and P. Dubois. 1999. Novel macromolecular architectures based on aliphatic polyesters: Relevance of the "coordination-insertion" ring-opening polymerization. Adv. Polym. Sci. 147: 1–59.

Messersmith, P.B. and S.I. Stupp. 1992. Synthesis of nanocomposites: Organoceramics. J. Mater. Res. 7(9): 2599–2611.

Meuser, F., N. Gimmler and J. Oeding. 1990. System Analytical Consideration of Derivatization of Starch with a Cooking Extruder as Reactor. Starch-Staerke, Germany, FR.

Miladinov, V.D. and M.A. Hanna. 1999. Physical and molecular properties of starch acetates extruded with water and ethanol. Ind. Eng. Chem. 38(10): 3892–3897.

Miladinov, V.D. and M.A. Hanna. 2000. Starch esterification by reactive extrusion. Ind. Crop. Prod. 11(1): 51–57.

Moad, G. 1999. The synthesis of polyolefin graft copolymers by reactive extrusion. Prog. Polym. Sci. 24(1): 81–142.

Moad, G. 2011. Chemical modification of starch by reactive extrusion. Prog. Polym. Sci. 36(2): 218–237.

Mohanty, A.K., M. Misra and G. Hinrichsen. 2000. Biofibres, biodegradable polymers and biocomposites: An overview. Macromol. Mater. Eng. 276(1): 1–24.

Mohanty, A.K., M. Misra and L.T. Drzal. 2002. Sustainable bio-composites from renewable resources: Opportunities and challenges in the green materials world. J. Polym. Env. 10(1-2): 19–26.

Murúa-Pagola, B., C.I. Beristain-Guevara and F. Martínez-Bustos. 2009. Preparation of starch derivatives using reactive extrusion and evaluation of modified starches as shell materials for encapsulation of flavoring agents by spray drying. J. Food Eng. 91(3): 380–386.

Narayan, R. 1994. Polymeric materials from agricultural feedstocks. ACS Symposium Series. American Chemical Society. Washington, DC 575: 2–28.

Nayak, P.L. 1999. Biodegradable polymers: Opportunities and challenges. J. Macromol. Sci.-Pol. R. 39(3): 481–505.

Nayak, S.K. 2010. Biodegradable PBAT/starch nanocomposites. Polym. Plast. Technol. Eng. 49: 1406–1418.

Nitz, H., H. Semke, R. Landers and R. Mülhaupt. 2001. Reactive extrusion of polycaprolactone compounds containing wood flour and lignin. J. Appl. Polym. Sci. 81(8): 1972–1984.

O'Brien, S. and Y.J. Wang. 2009. Effects of shear and pH on starch phosphates prepared by reactive extrusion as a sustained release agent. Carbohyd. Polym. 77(3): 464–471.

O'Brien, S., Y.J. Wang, C. Vervaet and J.P. Remon. 2009. Starch phosphates prepared by reactive extrusion as a sustained release agent. Carbohyd. Polym. 76(4): 557–566.

Okada, A. and A. Usuki. 1995. The chemistry of polymer-clay hybrids. Mater. Sci. Eng. C. 3(2): 109–115.

Okamoto, K., S. Sinha Ray and M. Okamoto. 2003. New poly (butylene succinate)/layered silicate nanocomposites. II. Effect of organically modified layered silicates on structure, properties, melt rheology, and biodegradability. J. Polym. Sci., Part B: Polym. Phys. 41(24): 3160–3172.

Oliphant, K.E., K.E. Russell and W.E. Baker. 1995. Melt grafting of a basic monomer on to polyethylene in a twin-screw extruder: Reaction kinetics. Polymer 36(8): 1597–1603.

Park, H.M., M. Misra, L.T. Drzal and A.K. Mohanty. 2004a. "Green" nanocomposites from cellulose acetate bioplastic and clay: Effect of eco-friendly triethyl citrate plasticizer. Biomacromolecules 5(6): 2281–2288.

Park, H.M., X. Liang, A.K. Mohanty, M. Misra and L.T. Drzal. 2004b. Effect of compatibilizer on nanostructure of the biodegradable cellulose acetate/ organoclay nanocomposites. Macromolecules 37(24): 9076–9082.

Park, H.R., S.H. Chough, Y.H. Yun and S.D. Yoon. 2005. Properties of starch/PVA blend films containing citric acid as additive. J. Polym. Environ. 13(4): 375–382.

Parovuori, P., A. Hamunen, P. Forssell, K. Autio and K. Poutanen. 1995. Oxidation of potato starch by hydrogen peroxide. Starch-Stärke 47(1): 19–23.

Penczek, S., A. Duda, A. Kowalski, J. Libiszowski, K. Majerska and T. Biela. 2000. On the mechanism of polymerization of cyclic esters induced by tin (II) octoate. Macromol. Symp. 157(1): 61–70.

Peressini, D., B. Bravin, R. Lapasin, C. Rizzotti and A. Sensidoni. 2003. Starch-methylcellulose based edible films: Rheological properties of film-forming dispersions. J. Food Eng. 59(1): 25–32.

Pesetskii, S.S., B. Jurkowski, Y.M. Krivoguz and K. Kelar. 2001. Free-radical grafting of itaconic acid onto LDPE by reactive extrusion: I. Effect of initiator solubility. Polymer 42(2): 469–475.

Psomiadou, E., I. Arvanitoyannis and N. Yamamoto. 1996. Edible films made from natural resources; microcrystalline cellulose (MCC), methylcellulose (MC) and corn starch and polyols-Part 2. Carbohyd. Polym. 31(4): 193–204.

Raquez, J.M., P. Degée, Y. Nabar, R. Narayan and P. Dubois. 2006. Biodegradable materials by reactive extrusion: From catalyzed polymerization to functionalization and blend compatibilization. Chimie 9(11): 1370–1379.

Raquez, J.M., Y. Nabar, M. Srinivasan, B.Y. Shin, R. Narayan and P. Dubois. 2008a. Maleated thermoplastic starch by reactive extrusion. Carbohyd. Polym. 74(2): 159–169.

Raquez, J.M., R. Narayan and P. Dubois. 2008b. Recent advances in reactive extrusion processing of biodegradable polymer-based compositions. Macromol. Mater. Eng. 293(6): 447-470.

Raquez, J.M., Y. Nabar, R. Narayan and P. Dubois. 2008c. Novel high-performance talc/poly [(butylene adipate)-co-terephthalate] hybrid materials. Macromol. Mater. Eng. 293(4): 310–320.

Ratto, J.A., P.J. Stenhouse, M. Auerbach, J. Mitchell and R. Farrell. 1999. Processing, performance and biodegradability of a thermoplastic aliphatic polyester/starch system. Polymer 40(24): 6777–6788.

Ray, S.S., M. Bousmina and K. Okamoto. 2005. Structure and properties of nanocomposites based on poly (butylene succinate-co-adipate) and organically modified montmorillonite. Macromol. Mater. Eng. 290(8): 759–768.

Ruggeri, G., M. Aglietto, A. Petragnani and F. Ciardelli. 1983. Some aspects of polypropylene functionalization by free radical reactions. Eur. Polym. J. 19(10–11): 863–866.

Sailaja, R.R.N. and M. Chanda. 2000. Use of maleic anhydride-grafted polyethylene as compatibilizer for polyethylene-starch blends: Effects on mechanical properties. J. Polym. Mater. 17: 165–176.

Salay, E. and C.F. Ciacco. 1990. Production and properties of starch phosphates produced by the extrusion process. Starch-Stärke 42(1): 15–17.

Schöllhorn, R. 1996. Intercalation systems as nanostructured functional materials. Chem. Mater. 8(8): 1747–1757.

Seker, M., H. Sadikoglu, M. Ozdemir and M.A. Hanna. 2003. Phosphorus binding to starch during extrusion in both single-and twin-screw extruders with and without a mixing element. J. Food Eng. 59(4): 355–360.

Seker, M., H. Saddikoglu and M.A. Hanna. 2004. Properties of cross-linked starch produced in a single screw extruder with and without a mixing element. J. Food Process Eng. 27(1): 47–63.

Seker, M. and M.A. Hanna. 2005. Cross-linking starch at various moisture contents by phosphate substitution in an extruder. Carbohyd. Polym. 59(4): 541–544.

Seker, M. and M.A. Hanna. 2006. Sodium hydroxide and trimetaphosphate levels affect properties of starch extrudates. Ind. Crop. Prod. 23(3): 249–255.

Shi, D., J. Yang, Z. Yao, Y. Wang, H. Huang, W. Jing, et al. 2001. Functionalization of isotactic polypropylene with maleic anhydride by reactive extrusion: Mechanism of melt grafting. Polymer 42(13): 5549–5557.

Shin, B.Y., S.H. Jang and B.S. Kim. 2011. Thermal, morphological, and mechanical properties of biobased and biodegradable blends of poly(lactic acid) and chemically modified thermoplastic starch. Polym. Eng. Sci. 51: 826–834.

Shogren, R.L., G.F. Fanta and W.M. Doane. 1993. Development of starch based plastics—A reexamination of selected polymer systems in historical perspective. Starch-Stärke 45(8): 276–280.

Shogren, R.L. 1996. Preparation, thermal properties, and extrusion of high-amylose starch acetates. Carbohyd. Polym. 29(1): 57–62.

Sinha Ray, S., K. Okamoto and M. Okamoto. 2003. Structure-property relationship in biodegradable poly (butylene succinate)/layered silicate nanocomposites. Macromolecules 36(7): 2355–2367.

Song, D., Y.S. Thio and Y. Deng. 2011. Starch nanoparticle formation via reactive extrusión and related mechanism study. Carbohyd. Polym. 85: 208–214.

Sun, Y.J., V. Flars and W.E. Baker. 1997. Evaluation and characterization of vector fluids and peroxides in a process of *in situ* compatibilization of polyethylene and polystyrene. Can. J. Chem. Eng. 75(6): 1153–1158.

Suvorova, A.I., I.S. Tyukova and E.I. Trufanova. 2000. Biodegradable starch-based polymeric materials. Russ. Chem. Rev. 69(5): 451.

Tan, N.B., S.K. Tai and R.M. Briber. 1996. Morphology control and interfacial reinforcement in reactive polystyrene/amorphous polyamide blends. Polymer 37(16): 3509–3519.

Tang, W., N.S. Murthy, F. Mares, M.E. Mcdonnell and S.A. Curran. 1999. Poly (ethylene terephthalate)–poly (caprolactone) block copolymer. I. Synthesis, reactive extrusion, and fiber morphology. J. Appl. Polym. Sci. 74(7): 1858–1867.

Taniguchi, I., W.A. Kuhlman, A.M. Mayes and L.G. Griffith. 2006. Functional modification of biodegradable polyesters through a chemoselective approach: Application to biomaterial surfaces. Polym. Int. 55(12): 1385–1397.

Temuujin, J., K. Okada, T. Jadambaa, K.J. Mackenzie and J. Amarsanaa. 2002. Effect of grinding on the preparation of porous material from talc by selective leaching. J. Mat. Sci. Let. 21(20): 1607–1609.

Tomasik, P., Y.J. Wang and J.L. Jane. 1995. Facile route to anionic starches. Succinylation, maleination and phthalation of corn starch on extrusion. Starch-Stärke 47(3): 96–99.

Valle, G.D., P. Colonna and J. Tayeb. 1991. Use of a twin-screw extruder as a chemical reactor for starch cationization. Starch-Stärke 43(8): 300–307.

van Duin, M., A.V. Machado and J. Covas. 2001. A look inside the extruder: Evolution of chemistry, morphology and rheology along the extruder axis during reactive processing and blending. *In*: Macromol. Symp., Vol. 170, No. 1, pp. 29–40, Wiley VCH Verlag GmbH.

Van Soest, J.J.G., K. Benes, D. De Wit and J.F.G. Vliegenthart. 1996. The influence of starch molecular mass on the properties of extruded thermoplastic starch. Polymer 37(16): 3543–3552.

Vergnes, B. and F. Berzin. 2006. Modeling of reactive systems in twin-screw extrusion: Challenges and applications. Chimie 9(11): 1409–1418.

von Schenck, H., M. Ryner, A.C. Albertsson and M. Svensson. 2002. Ring-opening polymerization of lactones and lactides with Sn (IV) and Al (III) initiators. Macromolecules 35(5): 1556–1562.

Wang, L., R.L. Shogren and J.L. Willett. 1997. Preparation of starch succinates by reactive extrusion. Starch-Stärke 49(3): 116–120.

Wang, N., J. Yu and X. Ma. 2007. Preparation and characterization of thermoplastic starch/PLA blends by one-step reactive extrusion. Polym. Int. 56(11): 1440–1447.

Wang, N., J. Yu and X. Ma. 2008a. Preparation and characterization of compatible thermoplastic dry starch/poly (lactic acid). Polym. Composite. 29(5): 551–559.

Wang, N., J. Yu, P.R. Chang and X. Ma. 2008b. Influence of formamide and water on the properties of thermoplastic starch/poly (lactic acid) blends. Carbohyd. Polym. 71(1): 109–118.

Wang, S., J. Yu and J. Yu. 2005. Compatible thermoplastic starch/polyethylene blends by one-step reactive extrusion. Polym. Int. 54(2): 279–285.

Wang, S.M., J.M. Bouvier and M. Gelus. 1990. Rheological behaviour of wheat flour dough in twin-screw extrusion cooking. Int. J. Food Sci. Tech. 25(2): 129–139.

Wang, Z., J. Massam and T. Pinnavaia. 2000. Epoxy-clay nanocomposites. pp. 127. *In*: Pinnavaia, T.J. and G. Beall (eds.). Polymer-Clay Nanocomposites. Wiley, Indiana.

Whaling, A., R. Bhardwaj and A.K. Mohanty. 2006. Novel talc-filled biodegradable bacterial polyester composites. Ind. Eng. Chem. Res. 45(22): 7497–7503.

White, J.L., W. Szydlowski, K. Min and M.H. Kim. 1987. Twin screw extruders; development of technology and analysis of flow. Adv. Polym. Tech. 7(3): 295–332.

White, J.L., B.J. Kim, S. Bawiskar and J.M. Keum. 2001. Development of a global computer software for modular self-wiping corotating twin screw extruders. Polym.-Plast. Technol. Eng. 40(4): 385–405.

Willett, J.L. and R.L. Shogren. 2002. Processing and properties of extruded starch/polymer foams. Polymer 43(22): 5935–5947.

Wing, R.E. and J.L. Willett. 1997. Water soluble oxidized starches by peroxide reactive extrusion. Ind. Crop. Prod. 7(1): 45–52.

Wu, Q., H. Sakabe and S. Isobe. 2003. Processing and properties of low cost corn gluten meal/wood fiber composite. Ind. Eng. Chem. Res. 42(26): 6765–6773.

Wurzburg, O.B. 1986. Modified Starches—Properties and Uses. CRC Press Inc., Miami, United States.

Xanthos, M. 1992. Process analysis from reaction fundamentals: Examples of polymerization and controlled degradation in extruders. pp. 33–45. *In*: Xhanthos, M. (ed.). Reactive Extrusion, Hanser Publishers, New York.

Xie, F., L. Yu, H. Liu and L. Chen, 2006. Starch modification using reactive extrusion. Starch-Stärke 58(3–4): 131–139.

Xu, Y., V. Miladinov and M.A. Hanna. 2004. Synthesis and characterization of starch acetates with high substitution 1. Cereal Chem. 81(6): 735–740.

Xu, Y. and M.A. Hanna. 2005. Physical, mechanical, and morphological characteristics of extruded starch acetate foams. J. Polym. Environ. 13(3): 221–230.

Yu, L., K. Dean, Q. Yuan, L. Chen and X. Zhang. 2007. Effect of compatibilizer distribution on the blends of starch/biodegradable polyesters. J. Appl. Polym. Sci. 103(2): 812–818.

Zeng, J.B., L. Jiao, Y.D. Li, M. Srinivasan, T. Li and Y.Z. Wang. 2011. Bio-based blends of starch and poly(butylene succinate) with improved miscibility, mechanical properties, and reduced water absorption. Carbohyd. Polym. 83: 762–768.

Zhu, L., K.A. Narh and K.S. Hyun. 2005. Investigation of mixing mechanisms and energy balance in reactive extrusion using three-dimensional numerical simulation method. Int. J. Heat Mass Transfer 48(16): 3411–3422.

Zou, G.X., P.Q. Jin and L.Z. Xin. 2008. Extruded starch/PVA composites: Water resistance, thermal properties, and morphology. J. Elastom. Plast. 40(4): 303–316.

10

Polyhydroxyalkanoates – A Prospective Food Packaging Material: Overview of the State of the Art, Recent Developments and Potentials

Katrin Jammernegg, Franz Stelzer and Stefan Spirk*

Institute for Chemistry and Technology of Materials, Graz University of Technology, Stremayrgasse 9, 8010 Graz, Austria

Introduction

During the past decades, a tremendous increase of waste derived from packaging caused many ecological problems with far-reaching consequences. Particularly, inappropriate waste management and the fact that most of these materials are made of petrochemical-based, long-lasting polymers contributed significantly to plastics pollution of almost every ecosystem. The role of food packaging should not be underestimated in this context (Avella et al. 2005; Malathi 2014). Plastics demand is continually growing (e.g. about 4-5% each year just in the USA) (Philip et al. 2007) and one of the fastest growing segments is food packaging. To overcome these problems, large efforts in academia and industry have been focused in research on the development of biodegradable polymers. Although there are already a number of convenient materials on the market, the higher production costs compared to fossil-based polymers are the main reason why they are not widely employed so far.

The most common and extensively characterized material for renewable and biodegradable food packaging is probably polylactic acid (PLA). It is already utilized as packaging material for fruits and vegetables, oneway

*Corresponding author: stefan.spirk@tugraz.at

cutlery etc. (Jamshidian et al. 2010). However, much effort was invested on
the improvement of this material by blending with other polymers (Feijoo
et al. 2006) or reinforcing additives (Salmieri et al. 2014; Busolo et al. 2010).
Besides the application of PLA, also other polyesters like polycaprolactone
(PCL) (Joseph et al. 2011; Beltran et al. 2014), polyvinylalcohol (PVA)
(Gaikwad et al. 2015; Cano et al. 2015), polyhydroxyalkanoates (PHA) as
well as their copolymers have been already suggested. Particular attention
is nowadays paid to PHAs since they exhibit quite advantageous properties
for food packaging (Khosravi-Darani 2015; Plackett and Siró 2011). This
review is constructed as follows: after a short overview on PHAs structure
and chemistry we will provide detailed insight into recent developments
and progress in PHAs chemistry with a particular focus on processing
conditions. For other more specified information, particularly on microbial
production of PHB, we refer to other recently published reviews, for example
(Koller 2014).

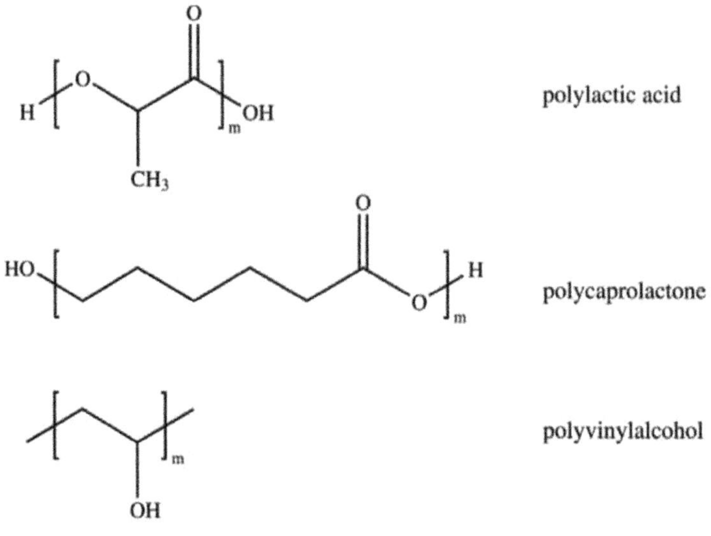

polylactic acid

polycaprolactone

polyvinylalcohol

Fig. 1. Structural formulas of polylactic acid (PLA), polycaprolactone
(PCL) and polyvinylalcohol (PVA).

Polyhydroxyalkanoates

Polyhydroxyalkanoates are a group of renewable and biodegradable
polyesters which are mainly derived from microorganisms. Depending
on the type of microorganism and substrate used, the resulting products
differ in structure, molecular weight and composition (Koller et al. 2013).
PHA's composition was initially revealed by the French scientist Lemoigne
(Lemoigne 1925, 1926) in the middle of the 1920s. However, it was only in the
1960s when scientists realized that this polyester is produced by bacteria as

a kind of intracellular energy and carbon storage polymer when there is an excess of carbon sources present while there are limited nutrients available. (Philip et al. 2007; Halami 2008; Matsumoto et al. 2001).

The general structure of PHA and its most important derivatives are shown in Fig. 2. The side chain residue marked "R" can be, for instance, hydrogen, methyl-, ethyl-, propyl-groups etc. Beside the various side chains, PHAs can be altered through different lengths of the main chain. These two parameters are primarily responsible for the resulting thermal and mechanical properties of the final materials. Related to these two parameters, it became quite common to distinguish between short chain length (scl) and medium chain length (mcl) PHAs. Scl-PHAs are defined to have 3-5 carbon atoms (Rahman et al. 2013) in their main chain, whereas mcl-PHAs feature 6-14 carbon atoms (Rupp et al. 2010; Le Meur et al. 2013). In summary, PHAs can be classified as a linear, isotactic polyester with chiral centres revealing R-configuration (Chodak 2008; Kolahchi and Kontopoulou 2015).

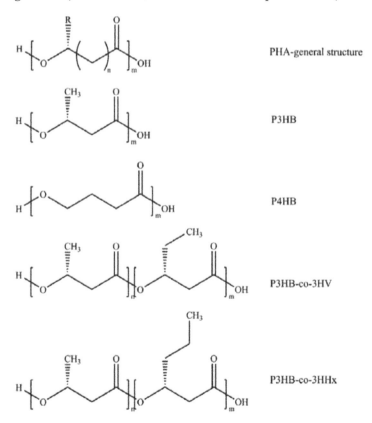

Fig. 2. General structure of polyhydroxyalkanoates (PHA) and its most important representatives poly(3-hydroxybutyrate) (P3HB), poly(4-hydroxybutyrate) (P4HB), poly(3-hydroxybutyrate-co-3-hydroxyvalerate) (P3HB-co-3HV) and poly(3-hydroxybutyrate-co-3-hydroxyhexanoate) (P3HB-co-3HHx).

The most important representative of the PHA family is poly(3-hydroxybutyrate) (P3HB) (Kolahchi and Kontopoulou 2015). It is a semicrystalline polyester with a relatively high melting point in the range 170-180 °C (Arrieta et al. 2014e) compared to other types of PHA. Considering mechanical performance, PHB shows similar values for Young's modulus and tensile strength as commonly used polypropylene (PP) (Raghunatha et al. 2015), but strain at break is significantly lower, in most cases between 3-6% (Plackett and Siró 2011). By means of PHB, the brittle performance typical for scl-PHAs is revealed. An efficient option to overcome stiffness and associated brittleness, a copolymer of PHB, poly(3-hydroxybutyrate-co-3-hydroxyvalerate), PHBV, is readily used. With respect to the amount of 3-hydroxyvalerate incorporated in the copolymer, melting temperature and the degree of crystallinity are decreased whereas ductility is enhanced. A major advantage of PHBV is its larger processing window compared to PHB (Kolahchi and Kontopoulou 2015). Since degradation of PHB is induced at around 180 °C which is barely above its melting point (Malinova and Brozek 2011), processing conditions have to be carefully optimized to avoid thermal decomposition. Until now, PHB and its copolymer PHBV are the only products of PHA's family which are produced at a commercial scale.

Table 1. Comparison of thermal properties of different kinds of PHA and fossil-based polymers (T_m = melting temperature; T_g = glass transition temperature; X = degree of crystallinity).

Polymer*	T_m [°C]	T_g [°C]	X [%]
P(3HB)	175	4	60
P(4HB)	53	-48	~ 34
P(3HB-co-7mol%-3HV)	153	5	51
P(3HB-co-11mol%-3HHx)	112	-8	-
PLLA	175	65	-
PCL	65	-61	67
LDPE	110	-30	50
PP	176	-10	50

*Data derived from (Chodak 2008; Saito and Doi 1994; Gao et al. 2006; Martin and Williams 2003).

Furthermore, poly(4-hydroxybutyrate) (P4HB) (Le Meur et al. 2013, 2014; Kämpf et al. 2014) and poly(3-hydroxybutyrate-co-3-hydroxyhexanoate) (PHB-HHx) (Vandewijngaarden et al. 2016; Díez-Pascual and Díez-Vicente 2016) are also investigated extensively.

A comparison of some crucial parameters for processing of PHAs and synthetic polymers are depicted in Tables 1 and 2 respectively.

Table 2. Comparison of mechanical properties (tensile strength, Young's modulus and strain at break) of P(3HB), P(4HB), P(3HB-3HV), P(3HB-3HHx), PLA, PCL, PP, PE and PET.

Polymer*	Tensile strength [MPa]	Young's modulus [MPa]	Strain at break [%]
P(3HB)	40	1700-3500	3-6
P(4HB)	104	149	1000
P(3HB-3HV)	30-38	700-2900	20
P(3HB-3HHx)	22	309	525
PLA	28-50	1200-2700	7-9
PCL	16	400	120-800
LDPE	10-15	200	300-500
PP	35-40	1700	150
PET	56	2200	70-100

*Data adopted from (Saito and Doi 1994; Plackett and Siró 2011; Lim et al. 2013)

As already mentioned before, PHAs are mainly biotechnologically produced by microorganisms although it is possible to fabricate them via chemical syntheses as well (Chodak 2008). Until now around 150 (Rupp et al. 2010) different types of microorganisms, Gram-positive as well as Gram-negative, are reported to produce PHA. Among others, *Ralstonia eutropha, Alcaligenes latus, Pseudomonas putida* etc. are frequently used (Philip et al. 2007; Khosravi-Darani 2015; Chodak 2008). In general, each of these microorganism strains are able to produce either scl or mcl PHAs, but lately, bacteria able to synthesize both types were reported. By selecting appropriate microorganisms and carbon sources, certain types of PHA (P3HB, P4HB, copolymers etc.) can be specifically manufactured. Under optimal conditions, yields up to 80% PHA of microorganism's dry cell weight can be achieved (Philip et al. 2007; Khosravi-Darani 2015; Plackett and Siró 2011; Chodak 2008).

An overview of microbial PHB synthesis is given in Fig. 3. The reaction involves three main steps. First, the selected carbon source is converted into an acetate which is further attached by coenzyme A (CoA). In a next step, two molecules of acetyl-CoA react via a condensation catalyzed by β-ketothiolase to establish acetoacetyl-CoA (Senior and Dawes 1973). This intermediate product is subsequently reduced to hydroxybutyryl-CoA, catalyzed by a NADPH-dependent acetoacetyl-CoA reductase. Polymerization takes place by reaction of the hydroxybutyryl-CoA monomer units catalyzed by PHA synthase to gain the PHA polyesters (Plackett and Siró 2011; Chodak 2008; Khanna and Srivastava 2005).

As already mentioned, most of the PHAs can be chemically synthesized as well. This is accomplished by a ring opening polymerization of the

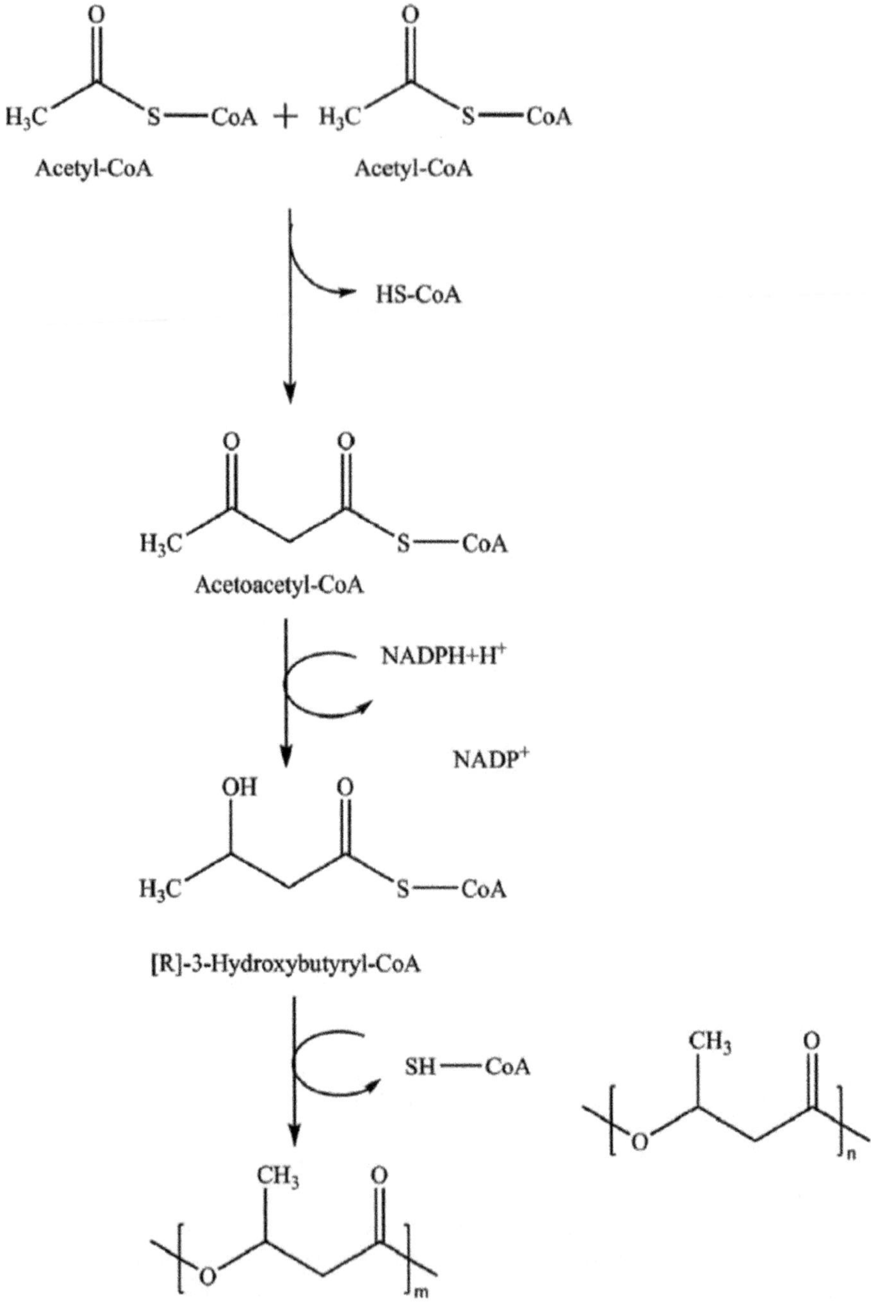

Fig. 3. Simplified overview of PHB's synthesis by microorganisms starting with the condensation of two molecules of acetyl-CoA and three different enzymes involved into the process (Chodak 2008).

respective lactones, for example, butyrolactone and valerolactone to yield PHBV, where aluminium or zinc alkyl compounds are used as catalysts. A very important parameter in this case is stereochemistry since biodegradation is only guaranteed when stereoregularity is accomplished. However, the chemical route is much more cost intensive than the biotechnological one, therefore it is very rarely applied (Chodak 2008; Müller and Seebach 1993).

Probably, the most important parameter of PHA's utilization as food packaging material is its approved biodegradation. Herein, it has the advantage that it is potentially degraded under aerobic as well as under anaerobic conditions (Arrieta et al. 2014d; Abou-Zeid et al. 2001). Furthermore, enzymatic degradation plays a significant role in this process (Renard et al. 2004). In general, the rate of decomposition depends on environmental conditions (compost, marine, sewage, soil, fresh water) (Abou-Zeid et al. 2004) and the polyester's molecular weight, chemical structure, melting temperature and the degree of crystallinity (Renard et al. 2004). Several publications reveal that degradation slows down when the PHAs feature high molecular weight, high crystallinity and high melting temperature.

Degradation in the environment is mainly catalyzed by extracelluar PHA depolymerases which can be produced by microorganisms, fungi and algae (Savenkova et al. 2000). Briefly, these bacteria etc. adsorb onto the polymer surface and secrete related enzymes which promote hydrolysis of PHA. As soon as soluble intermediates are produced, they diffuse through the cell walls where they are further degraded into oligomers and sometimes, depending on the microorganisms involved, into monomer units (Philip et al. 2007). The degradation rate is further strongly influenced by the temperature, humidity, pH and nutrient conditions of the surrounding environment (Abou-Zeid et al. 2001).

The processing of PHAs by extrusion, injection moulding, film blowing or compression moulding is a major challenge compared to other polymers. The operational window for PHB is rather narrow (170-180 °C) making the process rather difficult to control (Chodak 2008; Luef 2015). Thermal degradation is initiated at temperatures above 190 °C resulting in a dramatic decrease of molecular weight and mechanical performance (Arrieta et al. 2014b). This thermal decomposition proceeds via cis-elimination and transesterification reactions (Kim et al. 2006; Kopinke et al. 1996). A simplified scheme of this reaction is shown in Fig. 4. In the case of cis-elimination of PHB, crotonic acid (El-Hadi et al. 2002) and PHB oligomers end-capped with crotonyl moieties are established. A way to avoid or minimize thermal degradation is the usage of copolymers like PHBV or the addition of plasticizers (Fernandes et al. 2004). However, the change of other parameters like mechanical performance, crystallinity or water vapour/oxygen permeability has to be considered and a balance needs to be found which in turn depends on the material requirements. The usage of copolymers and plasticizers usually results in more flexible materials with a lower degree of crystallinity. Especially the water vapour permeability (Follain et al. 2014) is strongly influenced by the

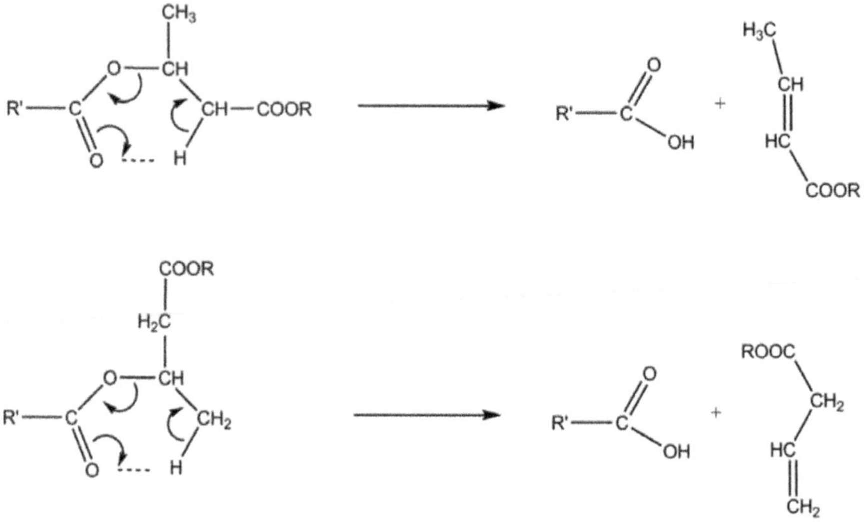

Fig. 4. Simplified reaction of PHA's thermally induced cis-elimination resulting in the corresponding monomer units and crotonic acid (Chodak 2008).

crystallinity of a polymer. The higher the crystalline portion of a material, the greater is the decrease. However, this phenomenon will be discussed in detail later.

Besides food packaging materials, there is a growing interest in using PHAs in the medical and pharmaceutical field too. Therefore, besides biodegradability, PHB also offers biocompatibility and hydrophobicity. Moreover, its degradation product, the monomer 3-hydroxybutyrate, is a harmless compound, produced also by the human organism. Particularly in the last decade, PHB was proposed as excellent material for bone fixation and repair as well as for sutures (Celarek et al. 2012; Chen and Wu 2005). Moreover, several groups are performing research on composite materials with various sorts of apatite (Yang et al. 2014; Sadat-Shojai et al. 2013) to discover bone replacement materials with well-defined degradation and proliferation rates. Therefore, several attempts of scaffold fabrication via 3D-printing have been reported as well (Zhao et al. 2014). In the field of pharmaceutical engineering, the most important application is a matrix material to control drug release (Romero et al. 2015; Michalak et al. 2013; Francis et al. 2011). Other potential applications for instance are disposal bags (Rosa et al. 2004), foils (Musioł et al. 2015), the coating of paper and cardboard (Cyras et al. 2009) and the manufacturing of fibres (Arrieta et al. 2016; Hufenus et al. 2015).

Till today only a few companies all over the world produce PHAs on a large scale. All of these products are biotechnologically fabricated and can be purchased in high purity, but these are still very expensive compared to conventional plastics like polyethylene or polypropylene. The most important suppliers are listed in Table 3. Moreover, products from Metabolix are already approved by the FDA (US Food and Drug Administration) to be used for food packaging purpose (Philip et al. 2007).

Table 3. Overview of worldwide polyhydroxyalkanoate supplying companies.

Company	Product	Country	Type of PHA
Biomer	Biomer®	Germany	PHB
Metabolix	Mirel™	USA	PHB
PHB Industrial Brasil S. A.	Biocycle®	Brazil	PHB, PHBV
Meredian, Inc.	Nodax™	USA	Copolymers
TianAn Biologic Materials Co.	Enmat™	China	PHBV
Bio-on	Minerv-PHA™	Italy	Unknown

PHA as Food Packaging Material

In general, food packaging materials are used to protect food products from external factors such as light, moisture, microorganisms, physical damage etc. and to extend their shelf-life (Rhim et al. 2013). Depending on the type of foodstuff packaged, different criteria have to be fulfilled; for example, high levels of water vapour permeability are desired for the storage of respiring fresh products while low ones are needed for suitable packaging of dry foods like bread or cereals. Therefore the ideal food packaging material should exhibit appropriate permeabilities for oxygen, water vapour, carbon dioxide and odours as well as satisfying mechanical properties, optical appearance, biodegradability and thermal properties. Moreover, antimicrobial activity and renewability are desirable too.

For this reason, several factors influencing the material's performance have to be taken into account when using PHA as packaging material. The most important "tunable" properties are processability, mechanical, thermal and migration properties, biodegradability, morphology, water vapour, carbon dioxide and oxygen permeability, as well as optical appearance. However, three main approaches to obtain materials suitable for food packaging can be identified, namely to use the neat polymer, polymer blends or polymer composites. In the following sections we will discuss these options accordingly.

PHA

Neat PHB and PHBV with various amounts of HV have been investigated extensively. Since there are numerous commercially available PHAs on the market, four different types of these PHAs, a P(3HB-co-8mol%-3HV) called Enmat Y1000P from Tianan Biologic (CN), two types of P(3HB-co-4HB) named Mirel F1006 and Mirel F3002 (both from Telles (USA)) and a P(3HB) called P226 from Biomer (DE), have been characterized by Corre et al. (2012) concerning their suitability as food packaging material. The obtained data were in some cases compared to polypropylene (PP), polyamide 6 (PA), polystyrol (PS) and polylactic acid (PLA). However, thermal analyses showed a rapid crystallization process for PHAs at approximately the same temperatures. Furthermore, the low mechanical performance of PHB was evidenced for all analyzed samples (brittle), which further deteriorates during ageing. However, after some weeks mechanical properties stabilize again. Moreover, the flow behaviour of molten samples was investigated in order to determine the processability with methods like extrusion or injection moulding. It turned out that all samples showed non-Newtonian properties. Nevertheless, the main problem with PHA's processing is its low degradation temperature and hence molecular weight is decreased afterwards. In terms of water vapour and oxygen permeability, the performance of PHA is a bit worse than those of many fossil based polymers (see Table 4), but significantly better than those reported for PLA (Corre et al. 2012).

Table 4. Comparison of values obtained for oxygen and water vapour permeability of commercial available polymers. Data taken from (Corre et al. 2012).

Polymer	Oxygen permeability [$cm^3 \cdot \mu m \cdot m^{-2} \cdot day^{-1}$]	Water vapour permeability [$g \cdot m^{-1} \cdot s^{-1} \cdot Pa^{-1}$]
Enmat Y 1000P (TianAn Biologic)	0.26	$2.7 \cdot 10^{-12}$
P226 (Biomer)	1.93	$1.6 \cdot 10^{-11}$
Mirel F 1006 (Telles)	1.51	$5.6 \cdot 10^{-12}$
Mirel 3002 (Telles)	1.16	$5.2 \cdot 10^{-12}$
PA MXD6 (Mitsubishi Chem.)	0.01	$1.6 \cdot 10^{-12}$
PLA 7001D (Natureworks)	2.95	$1.8 \cdot 10^{-11}$
PP 7712 (Total Chem.)	11.50	$6.2 \cdot 10^{-13}$
PS 1540 (Arkema)	12.85	$7.0 \cdot 10^{-12}$

Moreover, water and gas permeability as well as water sorption behaviour of commercial PHA films have been analyzed by Follain et al. (2014) too. Differences in results regarding the film processing method and the presence of additives were investigated. It turned out that water vapour permeability reduced when the degree of crystallinity increased. Concerning gas permeability, no differences between the samples is evident. Moreover, thermal analyses of the sheets show that the film forming process does not influence the melting behaviour in contrast to water vapour permeability properties, which are lower when fabricated by compression moulding instead of solution casting (Follain et al. 2014). Another study already presented in 1999 by Miguel and Iruin confirms the good barrier

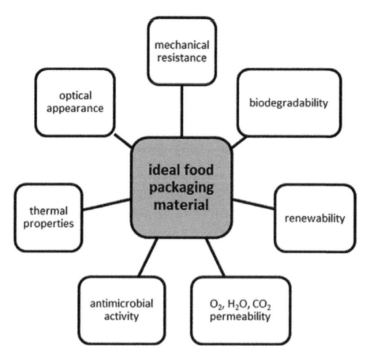

Fig. 5. Outline of desirable properties for ideal food packaging materials.

properties of PHB and PHBV. They compared them to polyvinylchloride (PVC) and polyethylene terephthalate (PET) and obtained similar values for carbon dioxide and water vapour permeation. Moreover, the investigated permeability of organic compounds is reported to be very low (Miguel and Iruin 1999). Migration tests of PHBV sheets were further performed by Chea et al. (2016). For this purpose, PHBV films were exposed to food simulating media including water, acetic acid 3% (w/v), ethanol 20% (v/v) and isooctane. The results indicated that all samples were still stable after storage for 10 days at 40 °C under these conditions. An exception is observed when the samples have been stored in ethanol 95% (v/v) where higher water sorption behaviour and water vapour permeability was observed (Chea et al. 2016).

In order to create more flexible materials for food packaging applications, PHB with 5, 12 and 20% of 3-HV content was synthesized, with different molecular weights, and investigated by Modi et al. (2011). Thermal analyses indicated glass transition temperatures of all films in a range between –10 and 20 °C and the presence of two melting peaks which can be attributed to the formation of crystals during the cooling process. Although the processing window with increasing HV content is getting larger, it is recommended to perform extrusion experiments at temperatures below 160 °C with a low screw speed since molecular weight decreases at elevated temperatures. Furthermore, mechanical analyzes showed high elastic modulus and flexural strength combined with a low tensile strength and elongation at break. Water vapour permeability was found to be quite high, due to increasing hydrophilicity caused by the hydroxyvalerate content (Modi et al. 2011). The suitability of PHB-HHx as possible packaging material was also evaluated. Sheets prepared by compression moulding show oxygen permeability similar to PLA and PET, but 6-25 times lower than polyethylene, polypropylene or polystyrene. Additionally, the influence of temperature and relative humidity (RH) on oxygen barrier properties revealed that increasing temperature and RH leads to higher oxygen permeability. Water vapour permeability values for PHB-HHx were also in the same range as for PLA, EVOH, PET and PA, but a little bit higher than for PP and PE. Regarding carbon dioxide permeability, investigations reveal 2-20 times lower levels of PHB-HHx than for PE, PP or PS but 7 times higher levels than for PET and 650 times higher levels than for PVA (Vandewijngaarden et al. 2014).

The processability of PHB was explored by Bucci et al. (2007). For this purpose, jar-cap-sets of PHB were prepared by an injection moulding process and compared with packaging products made of polypropylene. The results indicated a more difficult manufacturing process for PHB devices than for those made from PP; however, PHB acts as a more efficient light transmission barrier in the range of 250-350 nm and is biodegradable in various environments within 60 days (Bucci et al. 2007).

In another study, Bucci et al. (2005) compared PHB injection moulded devices with those used for PP. The obtained data showed that the deformation value of PHB is about 50% lower than that of PP, but dynamic compression and impact resistance of PP are much better than those of PHB. Differences were also found regarding the temperature: PHB shows a good performance at elevated temperatures, while at low temperatures it performs worse compared to PP. More or less identical results for both polymers were obtained in sensorial experiments with food (Bucci et al. 2005). Moreover, films and cups of commercial or semi-commercial PHB, PLA, wheat starch and corn starch were compared by Petersen et al. (2001). In this study, PHB and PLA were classified as more suitable food packaging materials than wheat starch and corn starch. This claim is confirmed for instance by the O_2 : CO_2 permeability which is in a range between 1:7 and 1:12. Furthermore, mechanical properties of PHB and PLA are comparable to those of PS or PE. Water vapour permeabiltiy of films is 12.6-18.6 $g \cdot m^{-2} \cdot day^{-1}$ and for cups between 2.2-10.5 $g \cdot 700 \ \mu m \cdot m^{-2} \cdot day^{-1}$ (Petersen et al. 2001).

To evaluate the biodegradation, PHBV was exposed in the presence of cellulose and acetate to a noncontrolled inocula consisting of activated sludge from different seasons. Obtained carbon balances were analyzed indicating that a low carbon/inoculum concentration ratio results in high mineralization rates (Vázquez-Rodríguez et al. 2006).

Further, the impact of fungal growth (Bergenholtz and Nielsen 2002) and food quality (Bertelsen et al. 2003) are crucial. Fungal growth on PHB, PLA, wheat, potato and corn starch films was evaluated by two different methods according to ASTM G21-96 and by a new method where 10 μL of a spore suspension are dropped in a spot onto the sample's surface. The obtained data indicated in contrast to starch-based films no significant fungal growth on PHB and PLA test specimen. This phenomenon may be explained by the higher water vapour barriers of PLA and PHB compared to starch (Bergenholtz and Nielsen 2002). For the assessment of the food quality an orange juice simulant and a dressing were stored for 10 weeks at 4 °C under fluorescent light and in the dark in PLA, PHB and HDPE cups. Regarding the orange juice simulant, changes in colour and the loss of ascorbic acid was determined while for the dressing primary and secondary lipid oxidation as well as the reduction of alpha-tocopherols and again the change of the colour was investigated. It turned out that PLA and PHB are as suitable packaging materials as HDPE and alterations in food quality were mainly caused by the exposure to light (Bertelsen et al. 2003).

Polymer Blends

To further improve PHA's properties, several attempts of blending PHA with other polymers were reported. Much effort was invested in the development of PHB/PLA blends (Armentano 2015; Arrieta et al. 2014a, b, c, d, e; Bartczak et al. 2013, Arrieta et al. 2015) and PHBV/PLA blends (Boufarguine et al. 2013; Jost and Kopitzky 2015). Compared to the neat polyester these blends are more flexible and therefore the mechanical performance, especially elongation at break, and the processability are enhanced. Bartczak et al. (2013) investigated the compatibility of PLA/PHB blends with a PHB amount of up to 20 wt%. It turned out that these blends are only partially miscible; phase-separated blends are formed and glass transition temperature of PLA decreases as the amount of PHB increases. Interestingly, PHB induces the formation of submicron domains in the PLA matrix which differ depending on the processing method (compression moulding, foil extrusion) being applied. The authors observed crazes at the surface due to the formation of cavities. Furthermore, mechanical properties changed from brittle to ductile, which can be evidenced by an increase in elongation from 6 ± 2% (neat PLA) to 21 ± 10% for the PLA/PHB blend with a ratio of 80 : 20 wt%. Impact strength improved from 50 kJ·m^{-2} in pure PLA to approximately 120 kJ·m^{-2} in PLA/PHB blends.

To gain a better interaction, ductility and processing performance of PHB/PLA blends a number of plasticizers were tested e.g. acetyl-tri-*n*-butyl citrate (ATBC) (Arrieta et al. 2014a, d, e, 2015, lactic acid oligomer (OLA) (Armentano 2015), polyethylene glycol (PEG) (Arrieta et al. 2014d) and D-limonene (Arrieta et al. 2014c).

The addition of ATBC resulted in less brittle samples. Furthermore degradation tests under composting conditions showed ATBC accelerated decomposition whereas increasing amounts of PHB decreased the degradation rate, because PHB acts as a nucleating agent (compare Fig. 6). However, plasticized and unmodified PLA/PHB samples were completely degraded within one month. As a consequence of decomposition caused by hydrolysis, mechanical properties deteriorated. These experiments were carried out with samples prepared by melt-extrusion followed by compression moulding (Arrieta et al. 2014d) and by an electrospinning technique (Arrieta et al. 2015). In another study, the PLA/PHB/ATBC matrix was modified by the incorporation of catechin (Arrieta et al. 2014a), an antioxidant, to prepare an active food packaging material for fatty food. An enhancement in terms of thermal properties and Young's modulus caused by catechin was reported (Arrieta et al. 2014a). Arrieta et al. (2014d) used ATBC and PEG in one of their studies and compared the influence of the plasticizer on the PLA/PHB blend properties. In general, it turned out that by the incorporation of PHB into a PLA matrix, crystallinity was increased. Furthermore, oxygen permeability improved, while wettability decreased. Concerning plasticizer efficiency, investigations revealed that ATBC plasticized samples showed better results than PEG plasticized ones (Arrieta et al. 2014d).

The usage of lactic acid oligomer as a plasticizer led to a lower glass transition temperature, enhanced water and oxygen barrier properties in combination with improved ductility. A content of 30 wt% OLA in a blend of PLA/PHB (85 : 15 wt%) has been recommended by the authors. Published data showed a good equilibrium between above mentioned parameters. The enhancement of ductility can be revealed by the values obtained for elongation at break which is 100 ± 30% for pure PLA, 140 ± 60% for PLA containing 15 wt% of PHB and 370 ± 20% for PLA+15 wt% PHB plasticized with 30 wt% OLA (Armentano 2015).

D-limonene was added in a concentration of about 15 wt% to a blend consisting of 75 wt% PLA and 25 wt% PHB. Tensile testing measurements show that PHB addition leads to an increase of Young's modulus from 1.15 ± 0.1 GPa to 1.39 ± 0.2 GPa. Furthermore, water vapour and oxygen barrier properties improved whereas latter ones were reported to 44.4 ± 0.9 $cm^3 \cdot mm \cdot m^{-2} \cdot day^{-1}$ for PLA, 11.5 ± 4.7 $cm^3 \cdot mm \cdot m^{-2} \cdot day^{-1}$ for PHB and 53.9 ± 2.3 $cm^3 \cdot mm \cdot m^{-2} \cdot day^{-1}$ for the above described blend film. In general, incorporation of D-limonene increased again elongation at break from 1.5 ± 0.3% (neat PLA) to 8.0 ± 0.2% (PLA / PHB / D-limonene) and accelerated the degradation process under composting conditions.

Similar observations concerning the miscibility between PHB and PLA were found in PHBV/PLA blends. Boufarguine et al. (2013) fabricated blends by incorporating 10 wt% of PHBV by a custom multilayer co-extrusion process into a PLA matrix. Morphological investigations revealed the formation of very thin, long and crystalline lamellas were spread within the PLA matrix. However, gas barrier rates could be successfully decreased by 35-40% compared to neat PLA and elongation at break could be enhanced by 35-52% depending on the number of layers.

Moreover PLA/PHBV blends with a PHBV amount of 20-35 wt% were analyzed by Jost and Kopitzky in 2015. They found that a content of 20-35% of PHBV in a PLA matrix was the optimum regarding miscibility. Higher amounts may lead to co-continuous phase systems which are disadvantageous for these applications. For samples containing 25 wt%

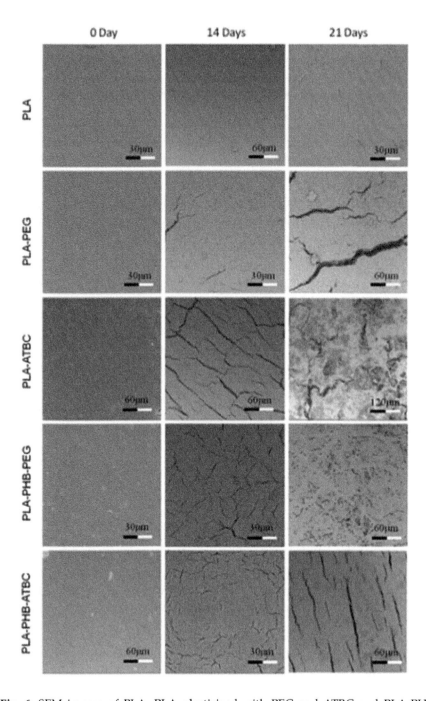

Fig. 6. SEM images of PLA, PLA plasticized with PEG and ATBC and PLA-PHB blends plasticized with PEG and ATBC at zero time, 14 days and 21 days of storage under composting conditions (Arrieta et al. 2014d).

of PHBV, water vapour permeability is decreased by 46% and oxygen permeability by 40% compared to neat PLA.

Blends of PHB-HHx (Lim et al. 2013) with PLA were prepared by direct melt compounding. However, obtained results from DSC measurements revealed the incompatibility of these two polyesters and a decrease in PLA's ability to crystallize at higher PHB-HHx contents. Furthermore, an increase in stiffness compared to neat PLA, determined by DMA analyses was reported. Moreover, the less PHB-HHx was added into the PLA matrix, the less agglomerates were formed and the material's toughness was enhanced which could be seen as a key parameter in food packaging applications (Lim et al. 2013).

Another possible biodegradable blend component of PHB is polycaprolactone (PCL) (Adriana et al. 2006). Compared to PHB, PCL is a rather ductile polyester and has a quite low melting temperature of approximately 56 °C. Investigated PHB/PCL blends were prepared in a 70 : 30 wt% ratio by injection moulding and exhibit improved mechanical performance, indicated by a value of 87 ± 20% for elongation at break. A problem which has to be overcome is again molecular interaction between the two components. A partial miscibility was proven by the shift of glass transition temperature of PHB from around 5 °C for the neat polymer to –10.6 °C for the blend, but nevertheless a second T_g for PCL was observed at –62.9 °C (Adriana et al. 2006).

The preparation of multilayer films of polyvinylalcohol (PVA) as inner layer and PHA as outer layer was realized by Thellen et al. (2013) by means of a co-extrusion process. To improve the compatibility of the two blend components, maleic anhydride combined with a dicumyl peroxide initiator were chosen resulting in a doubling of values for peel strength. Furthermore, oxygen permeability was measured at different relative humidities (RH), more precisely at 0%, 60% and 90% RH giving values of 27 cc·m^{-2}·day^{-1}, 41 cc·m^{-2}·day^{-1} and 52 cc·m^{-2}·day^{-1}. Compared to a pure PVA monolayer which showed data of 60 cc·m^{-2}·day^{-1} at 0% RH and 999 cc·m^{-2}·day^{-1} at 60% RH, oxygen permeation could be significantly enhanced. Nevertheless, biodegradation in marine environment could only be proven for PHA samples but not for PVA ones (Thellen et al. 2013).

A further interesting approach is the blending of natural rubber with PHBV assisted by gamma radiation. By means of gamma radiation, crosslinking and therefore a better adhesion of the two components could be obtained, evidenced by SEM, FT-IR and tensile testing. Gamma doses of 5 kGy, 10 kGy and 15 kGy were applied. However, the radiation dose had an influence on elongation at break, which was decreased when high (15 kGy) gamma doses were used (Maneewong et al. 2013).

Besides already particularized blends consisting of two fully biodegradable and renewable polyesters, several attempts of generating blends made of a decomposable part, namely a PHA, and a "conventional" fossil-based part represented by polymers like polyethylene (PE) or polypropylene (PP) are reported in the literature. In most cases, PHB or PHBV was added as additive in concentration ranges between 10 and 30 wt%. These samples were then analyzed in terms of changes in biodegradation (Masood et al. 2014), recyclability (Norrrahim et al. 2013), gas barrier properties (Norrrahim et al. 2013), mechanical properties (Norrrahim et al. 2013; Ohashi et al. 2009) and miscibility (Pankova et al. 2010).

To determine the impact of PHBV in a PP film, biodegradation was studied in field tests and under controlled laboratory conditions. SEM analyses of treated films revealed agglomerates, holes, pits and grooves due to microbial activity. Furthermore, oxo-biodegradation was confirmed by an increase of the carbonyl peak measured by IR. Also an enhanced thermal stability could be observed after soil burial (Masood et al. 2014).

Moreover, a study regarding the recyclability of PHBV/PE films provides a good insight into the sustainability of such polymer systems. It is suggested to pyrolyze these films; thereby PHBV is converted into its volatile compounds and can easily be separated from PE. However, films with an incorporated PHBV percentage of 20-30 wt% showed improved oxygen permeability as well as similar mechanical properties compared to conventional food packaging materials. Due to these characteristics, such films may be suitable for the packaging of cereals, dried food and bread (Norrrahim et al. 2013). A further approach was reported by Ohashi et al. (2009). They added 0.2-0.4% colloidal silica particles to a PE/PHB blend which resulted in an increase in tensile strength of about 20%, while elongation at break was set between 5-80% (Ohashi et al. 2009). In order to adjust hydrolysis and, therefore, biodegradation of films, PHB was blended with two types of polymers with different polarity, namely LDPE and polyamide (PA). Due to a better resistance against hydrolysis and a lower price compared to pure PHB, PHB/LDPE blends were proposed as a good alternative packaging material, whereas PHB/PA blends are more suitable as a polymer material for drug delivery systems (Pankova et al. 2010).

Polymer Composites

Besides the utilization of pure PHA and PHA blends, composites of PHB and PHBV, respectively, are a third promising option in developing biodegradable food packaging materials. This type of material consists in general of a matrix polymer, here a kind of PHA, and further additives, filler materials etc. These composites are in most cases aimed to improve crucial parameters like water vapour permeability, oxygen barrier properties, mechanical strength but also the introduction of new properties like antibacterial activity is a major aim. Since by the insertion of antimicrobial agents the shelf-life and the quality of the packaged food can be increased (Díez-Pascual et al. 2014a) hopefully less food has to be disposed due to inedibility in the future. One of the main issues is to reduce the production costs compared to neat PHB or PHBV since filler materials for instance are usually very cheap. The cost reduction is necessary to match up with convenient food packaging materials like polyethylene or polypropylene.

Possible filler materials used in composites are, for example, several sorts of fibres. They benefit from their renewability, availability and their low price because they are often obtained as by-products in different types of industries. Water vapour permeability could be highly influenced by the incorporation of wheat straw fibres for example. In general, the usage of smaller particle sizes of those fibres increases the hydrophobic character of the samples and higher amounts of wheat straw fibres can be incorporated into a PHBV matrix. However, the drawback is that higher straw content leads to more brittle samples. Nevertheless, the advantage of increasing straw fibre amount is that water vapour permeability improves from 11 up to 110 $g \cdot m^{-2} \cdot day^{-1}$ which makes these kind of composites a promising candidate for a food packaging material for respiring fresh products (Berthet et al. 2015a). To overcome the brittleness resulting from incorporated wheat straw fibres, ATBC, glycerol triacetate (GTA) and PEG were used as external plasticizers for PHBV. It turned out that an amount of 10 wt% of ATBC or GTA resulted in an increase in elongation at break from 1.8% to 6% when plasticizing pure PHBV. Anyway, when wheat straw fibres were incorporated too, the plasticizing effect was neutralized explainable by the formation of microscopic defects and a poor PHBV/wheat straw fibre adhesion (Martino et al. 2015).

Beside wheat straw fibres, other sorts of lignocellulosic fibres obtained from brewing spent grains and olive mills differing in composition and particle size were also recently investigated (Berthet et al. 2015b). While mechanical properties, fibre-matrix adhesion and crystallinity deteriorate compared to neat PHBV, water vapour permeability of these composites increased by 3.5 fold when wheat straw fibres and decreased by 2.5 fold when fibres from olive mills were used as filler material. The reason for these contrary effects may be the higher amount of lignin in olive mill fibres, which is known to be an effective barrier material (Berthet et al. 2015b). Moreover, Cunha et al. (2015) analyzed the processability of PHBV/beer spent grain fibres by an extrusion process followed by film blowing. It was not possible to process composites containing amounts larger than 10 wt% of beer spent grain fibres. However, investigations of blown films revealed that processing conditions had a big influence on mechanical, barrier and structural characteristics (Cunha et al. 2015). Empty fruit bunch fibres from oil palms were also analyzed in view of potential filler materials for PHBV (Sugama et al. 2011). To gain a better interaction between fibre and polymer, maleic anhydride and benzoyl peroxide were used as compatibilizers. It turned out that the presence of maleic anhydride in PHBV reduced its molecular weight but improved thermal stability. A percentage of 35% empty fruit bunch fibres was found to be the best, because high values for tensile strength could be achieved (Sugama et al. 2011).

An even more promising method in PHA composite preparation suitable for food packaging is the implementation of nanoparticles. Therefore, inorganic substances like zinc oxide (Díez-Pascual and Díez-Vicente 2014 a, b) or silver nanoparticles (Castro-Mayorga et al. 2014, 2015) are often used as well as organic compounds like cellulose nanocrystals (CNC) (Yu et al. 2014a, c; Dhar et al. 2015; Bhardwaj et al. 2014) and carbon nanotubes (Yu et al. 2014b; Sanchez-Garcia et al. 2010).

The incorporation of ZnO nanoparticles into a polymer matrix like PHBV offers several advantages. It seems that ZnO nanoparticles can act as a nucleating agent, which is evidenced by a decreased crystallite size and a raise in crystallinity and crystallization temperature from 77.3 °C for neat PHBV up to 103.4 °C for PHBV containing 8 wt% ZnO. In general, ZnO improved thermal stability, glass transition temperature, E-modulus and tensile strength whereas gas and water barrier properties were reduced. This phenomenon is explained by a strong interfacial adhesion between the matrix and the nanoparticles resulting from hydrogen bonding formation between the components. Incorporation of 4 wt% ZnO was chosen to be the optimum. Moreover, ZnO shows antibacterial activity against *Escherichia coli* and *Staphylococcus aureus* (Díez-Pascual and Díez-Vicente 2014b). The same effects were reported in a second study by Diez-Pascual et al. (2014a) where ZnO nanoparticles were incorporated into a PHB matrix instead of PHBV, by solution casting (Díez-Pascual and Díez-Vicente 2014a).

Castro-Mayorga et al. (2015) prepared nanocomposites consisting of PHB-co-18 mol%-HV *in situ* stabilized silver nanoparticles which was further melt-compounded with a commercially available PHB-co-3 mol%-HV. They reported that the addition of 0.04 wt% of silver nanoparticles dramatically decreased oxygen permeability by 56% compared to pure polyester whereas thermal stability and optical properties remained unchanged. However, the main purpose of incorporating silver nanoparticles into a PHBV matrix was to gain antimicrobial activity which could be successfully proven against *Salmonella enterica* and *Listeria monocytogenes* for a period of seven months (Castro-Mayorga et al. 2015). Moreover, Castro-Mayorga et al. (2014) had already published before a method to avoid well-known agglomeration and instability of silver nanoparticles. Briefly, silver nanoparticles were prepared *in situ* and by physical mixing by a chemical reduction in a suspension of untreated PHBV. Investigations of the obtained nanoparticles revealed a spherical form (7 ± 3 nm) and stability of at least 40 days. The successful prevention of agglomeration can be explained by untreated PHBV being an efficient capping agent. Additionally, silver nanoparticles showed strong antibacterial activity against *Salmonella enterica* at very low concentrations of about 0.1-1 ppm which classifies them as useful additive in food packaging applications (Castro-Mayorga et al. 2014).

Nanocomposite sheets of PHB reinforced with cellulose nanocrystals (CNC) were reported by Dhar et al. (2015). They suggest an ideal CNC amount of 2 wt% when PHB is used as matrix polymer, which was verified by d-spacing from XRD measurements. Migration studies showed that in polar as well as in apolar media the values remained within the predefined limits. It is worth mentioning that after migration experiments, crystallization temperature decreased by about 15 °C and a number of melting peaks appeared. However, there was also a significant influence on oxygen permeability, solubility and diffusivity evidenced by a decrease of about 65%, 57% and 17%, respectively. These changes were attributed to the hydrogen bonding interaction between PHB and CNCs (Dhar et al. 2015). Further, Bhardwaj et al. (2014) summarized among others the role of PHA-CNC nanocomposites in the development of food packaging materials in detail.

Another interesting method is the preparation by solution casting of cellulose nanocrystals/silver (CNC-Ag) nanohybrids embedded in a PHBV matrix described by Yu et al. (2014a). This approach combined antibacterial and reinforcing effects, evidenced for instance in a sample containing 10 wt% CNC-Ag where investigations showed an increase in tensile strength of about 140%, E-modulus was improved by 200% and degradation temperature raised by 24.2 °C compared to pure PHBV. Furthermore, water vapour permeability and water uptake were decreased as well as migration levels in polar and apolar solvents, which may be caused by a higher crystallinity and improved interfacial adhesion. Additionally, antibacterial activity was evidenced against *E. coli* and *S. aureus* (Yu et al. 2014a).

In order to produce transparent food packaging materials, PHBV/CNC-methyl ester films were prepared by solution casting. Because CNC-methyl esters are more hydrophobic than unmodified CNC, a homogenous dispersion throughout the whole sample could be ensured. Crystallization temperature as well as the degree of crystallization could be increased, since CNC acted as nucleating agent. The higher the amount of incorporated CNC-methyl ester, the better mechanical, thermal, barrier and migration properties became. The highest tensile strength and degradation temperature could be reached with an amount of 20 wt% CNCs (Yu et al. 2014c).

Beside cellulose nanocrystals, carbon nanotubes were also embedded into a PHBV matrix for the preparation of nanocomposites suitable for food packaging application. This was for instance demonstrated by Yu et al. (2014b) who fabricated composites of PHBV and PHBV grafted with 1-10 wt% of multi-walled carbon nanotubes by solution casting. The optically transparent and homogenously dispersed samples showed an increase in E-modulus and tensile strength by 88% and 172% when about 7 wt% of PHBV grafted multi-walled carbon nanotubes were incorporated into PHBV matrix. Moreover, degradation temperature raised by 22.3 °C and therefore the melt-processing window got larger. Water vapour permeability and water uptake could be reduced, while overall migration values stayed within the limits for food packaging applications (Yu et al. 2014b).

Another study in this area was performed by Sanchez-Garcia et al. (2010). They prepared composites of PHBV and PCL with carbon nanofibres or carbon nanotubes again by solution casting and compared them with each other. Optical analyses (SEM, TEM and AFM) revealed homogenously dispersed carbon nanofibres and carbon nanotubes within the polymer matrix at low concentrations, but increasing their amounts resulted in the formation of agglomerates. However, thermal analyses revealed that carbon nanofibres as well as carbon nanotubes inhibited crystallization transformation but degradation temperatures were improved. Moreover electrical conductivity was enhanced as well as water uptake, oxygen permeability, Young's modulus and yield stress. The optimum concentration for carbon nanotubes and carbon nanofibres was chosen to be 1 wt% (Sanchez-Garcia et al. 2010).

Another interesting approach is reported by Fabra et al. (2013, 2014a, b, c) in several publications. They extensively investigated composites made of PHA and proteins, mainly zein. Therefore, multilayer systems containing a high barrier interlayer of electrospun zein nanofibres were prepared by compression moulding and solution casting. Thereby, the influence of the preparation method on several properties could be demonstrated, for example, mechanical resistance improved whereas water vapour permeability and transparency decreased more when outer layers were manufactured by compression moulding instead of solution casting. In general, zein nanofibre interlayers enhanced oxygen permeability and water vapour barrier properties as well as limonene depending on the amount of zein added (Fabra et al. 2013). Additionally, further nanostructured high barrier interlayers of whey protein isolate, pullulan, zein and zein blends with whey protein isolate and pullulan were incorporated into PHB-co-3%-HV multilayer systems. Investigations showed that oxygen permeability could be improved by 38-48% and water vapour permeability by 28-35% when fibrillary pullulan and zein fibres were used. Whey protein fibres did not improve these properties because of the formation of bead microstructures. Moreover, mechanical and optical properties changed only slightly, independently of the used interlayer material (Fabra et al. 2014b). In order to determine the performance of self-adhesive, nanostructured high barrier zein interlayers in other multilayer systems, they were assembled in PHB, PHB-co-5%-HV, PLA and PET matrices. It turned out that oxygen and limonene permeability increased independently on the used polymer, but water vapour barrier properties and film flexibility were significantly better in combination with PHA than PLA (Fabra et al. 2014a). Moreover, Fabra et al. (2014c) also investigated the physico-chemical ageing properties of multilayer systems based on PHB or PHB-co-3%-HV with zein nanostructured interlayers and polyethylene glycol as plasticizer (only for PHB-co-3%-HV) over a period of three months. Microbiological tests were therefore performed in an inoculum of *Listeria monocytogenes*. The obtained data showed that the degree of crystallization did not significantly change within three months, except for PHBV-PEG stored at 0% RH. Also, the optical and barrier properties were not affected. Furthermore, tests with *Listeria monocytogenes* showed no bacterial survival, except when stored at 100% RH. Zein interlayer did not influence microbiological activity (Fabra et al. 2014c).

Beside zein, keratin, a fibrous protein, was also added as an active additive into a PHA matrix by melt-mixing. Compatibility of both components could be evidenced by SEM images. Moreover incorporation of 1 wt% of keratin improved material's performance indicated on the one hand by a decrease in water, limonene and oxygen permeability and on the other hand by increased mechanical properties (Pardo-Ibanez et al. 2014). Recently, composites made of PHA and nanokeratin were also reported by Fabra et al. (2016). They focused their investigations on the differences of material properties resulting from the preparation methods used. Therefore, composites were fabricated by direct melt-compounding and by incorporation of nanokeratin into electrospun PHA followed by a melt-compounding step with PHA resulting in improved barrier properties. As a second method, multilayer sheets were prepared by solution casting revealing improved adhesion and barrier properties as well (Fabra et al. 2016).

Further attempts for the preparation of food packaging materials based on PHA were reported by Kovalcik et al. (2015). They fabricated composites of PHBV and Kraft lignin in a concentration range of 0-10 wt% by thermoforming. Investigations revealed that Kraft lignin acts as an active additive, which is highly compatible with PHBV and hence enhances mechanical properties, particularly stiffness. Samples containing 1 wt% of Kraft lignin showed a decrease in oxygen permeability by 77% and in CO_2 permeability by 91% compared to pure PHBV. Moreover, thermo-oxidative stability increased. However, a drawback is that the prepared composites are not transparent (Kovalcik et al. 2015).

Films of PHB and cellulose cardboard were fabricated by solution casting (10 wt% optimum of PHB amount) and compression moulding (15 wt% optimum of PHB amount) by Seoane et al. (2015). It turned out that casted samples showed a decrease in moisture absorption compared to compression moulded ones because hydrophobic PHB solution was able to diffuse into the hydrophilic cellulose cardboard. Therefore, compression moulded samples had better mechanical and permeation performance due to the formation of a continuous layer on the cellulose cardboard (Seoane et al. 2015).

An innovative method was recently reported by Fan et al. (2015) suggesting the fabrication of fibrous membranes by an electrospinning technique of poly[5,5-dimethyl-3-(3'-triethoxysilylpropyl)hydantoin] (PSPH), an *N*-halamine precursor and PHB. While the *N*-halamine precursor had a slight influence on thermal properties, antimicrobial activity could be proven when obtained fibrous membranes were treated with chlorine bleach. A reduction of 92% against *Staphylococcus aureus* and 85% against *Escherichia coli* within 30 min of contact time are indicated (Fan et al. 2015). In order to change the crystalline performance of PHB, it was laminated on glassine paper by hot-pressing. Although distinct bilayers are observed, glassine paper acted as a nucleating agent and therefore a reduction of cracks in the PHB layer was observed. Furthermore, the impact of quenching at crystallization temperature was investigated but the optimum conditions were found by quenching the sample in an ice bath with subsequent crystallization at room temperature. Thereby, a Young's modulus of 2.5 GPa, tensile strength of 80 MPa and water vapour permeability of 20 $g \cdot m^{-2} \cdot day^{-1}$ could be achieved (Safari and van de Ven 2015).

The influence of different kinds of clay on PHAs performance has been the subject of several works in past years. Cabedo et al. (2009), for instance, investigated the impact of montmorillonite and kaolinite on PHBV's properties. Obtained results revealed that the molecular mass of PHBV was already reduced by about 38% after the extrusion process and with added montmorillonite further degraded whereas no further decrease was observed with incorporated kaolinite. Due to the fact that the decrease of PHBV's molecular weight was also observed when the amount of incorporated montmorillonite increased, it is assumed that the vapourized water from the clay's surface is responsible for this phenomenon. Nevertheless, X-ray diffraction measurements showed intercalation of both sorts of clay in the polymer matrix (Cabedo et al. 2009). A further study completed by Gonzalez-Ausejo et al. (2016) deals with the impact of tris(nonylphenyl) phosphite (TNPP), added as a chain extender, on PHBV/clay properties. Two different types of clays, a laminar organomodified montmorillonite (MMT) and a tubular unmodified halloysite (HNT), were compared. Optical analyses of the MMT indicated a partially exfoliated morphology and an appropriate dispersion of the latter one in the PHBV matrix. Furthermore, no changes in crystallinity could be observed, but decomposition temperature raised when TNPP was added. Also mechanical performance improved due to the presence of clay and chain extension caused by TNPP (Gonzalez-Ausejo et al. 2016).

Conclusion and Outlook

In summary, it can be concluded that PHAs bear a high potential for a variety of applications, especially in the field of food packaging. They provide good performance in terms of oxygen and water vapour permeability in combination with satisfying mechanical indicators. Despite many efforts in the past, the processing of PHB remains a major unresolved task due to its thermal lability (depolymerization) as well as its still rather high price. In this context, PHBV is a more promising material for food packaging than PHB, since the processing window in terms of temperature is wider. The hydrophilic character of PHBV is larger compared to PHB which results in higher water vapour transmission rates. Since the transmission rate is related to the amount of the HV in the main chain, this allows for a control of the vapour permeability in a certain range. This could be particularly interesting if the packaged goods need to be equilibrated with the environmental humidity. If the materials properties of neat PHB are aimed at, this can be accomplished by the usage of plasticizers. However, the major prerequisite is to ensure that the whole material is still biodegradable. Another promising field to improve the cost performance of PHB derived materials is to incorporate fibrous materials. However, in most cases, the properties of PHB cannot be retained.

It is still unclear which approach will lead to wider applications of PHB in the future. However, it is clear that both major problems, the price for the final material and the processing to achieve the desired materials properties must be tackled. If these problems can be circumvented in the future, PHAs are certainly serious candidates to replace synthetic packaging materials in the sector of food packaging.

References

Abou-Zeid, D.-M., R.-J. Müller and W.-D. Deckwer. 2001. Degradation of natural and synthetic polyesters under anaerobic conditions. J. Biotechnol. 86: 113–126.

Abou-Zeid, D.M., R.-J. Müller and W.D. Deckwer. 2004. Biodegradation of aliphatic homopolyesters and aliphatic-aromatic copolyesters by anaerobic microorganisms. Biomacromolecules 5: 1687–1697.

Adriana, M., T. Duarte, R. Goulart, E. Sant and A. Martins. 2006. Thermal and mechanical behavior of injection molded poly (3-hydroxybutyrate)/poly (ε-caprolactone) blends 2. Material and Methods 9: 25–27.

Armentano, I. 2015. Processing and characterization of plasticized PLA/PHB blends for biodegradable multiphase systems. Express Polym. Lett. 9: 583–596.

Arrieta, M.P., M. del M. Castro-López, E. Rayón, L.F. Barral-Losada, J.M. López-Vilariño, J. López, et al. 2014a. Plasticized poly(lactic acid)–poly(hydroxybutyrate) (PLA–PHB) blends incorporated with catechin intended for active food-packaging applications. J. Agric. Food Chem. 62: 10170–10180.

Arrieta, M.P., E. Fortunati, F. Dominici, E. Rayón, J. López and J.M. Kenny. 2014b. Multifunctional PLA–PHB/cellulose nanocrystal films: Processing, structural and thermal properties. Carbohydr. Polym. 107: 16–24.

Arrieta, M.P., J. López, A. Hernández and E. Rayón. 2014c. Ternary PLA–PHB–Limonene blends intended for biodegradable food packaging applications. Eur. Polym. J. 50: 255–270.

Arrieta, M.P., J. López, E. Rayón and A. Jiménez. 2014d. Disintegrability under composting conditions of plasticized PLA-PHB blends. Polym. Degrad. Stab. 108: 307–318.

Arrieta, M.P., M.D. Samper, J. Lopez and A. Jimenez. 2014e. Combined effect of poly(hydroxybutyrate) and plasticizers on polylactic acid properties for film intended for food packaging. J. Polym. Environ. 22: 460–470.

Arrieta, M.P., J. López, D. López, J.M. Kenny and L. Peponi. 2015. Development of flexible materials based on plasticized electrospun PLA–PHB blends: Structural, thermal, mechanical and disintegration properties. Eur. Polym. J. 73: 433–446.

Arrieta, M.P., J. López, D. López, J.M. Kenny and Peponi, L. 2016. Biodegradable electrospun bionanocomposite fibers based on plasticized PLA–PHB blends reinforced with cellulose nanocrystals. Ind. Crops Prod. (in press).

Avella, M., J.J. De Vlieger, M.E. Errico, S. Fischer, P. Vacca and M.G. Volpe. 2005. Biodegradable starch/clay nanocomposite films for food packaging applications. Food Chem. 93: 467–474.

Bartczak, Z., A. Galeski, M. Kowalczuk, M. Sobota and R. Malinowski. 2013. Tough blends of poly(lactide) and amorphous poly([R,S]-3-hydroxy butyrate) – morphology and properties. Eur. Polym. J. 49: 3630–3641.

Beltran, A., A.J.M. Valente, A. Jimenez and M.C. Garrigos. 2014. Characterization of poly(ε-caprolactone)-based nanocomposites containing hydroxytyrosol for active food packaging. J. Agric. Food Chem. 62: 2244–2252.

Bergenholtz, K. P. and P.V. Nielsen. 2002. New improved method for evaluation of growth by food related fungi on biologically derived materials. J. Food Sci. 67: 2745–2749.

Bertelsen, G., V.K. Haugaard and B. Danielsen. 2003. Impact of polylactate and poly (hydroxybutyrate) on food quality. Eur. Food Res. Technol. 216: 233–240.

Berthet, M.-A., H. Angellier-Coussy, V. Chea, V. Guillard, E. Gastaldi and N. Gontard. 2015a. Sustainable food packaging: Valorising wheat straw fibres for tuning PHBV-based composites properties. Compos. Part A Appl. Sci. Manuf. 72: 139–147.

Berthet, M.-A., H. Angellier-Coussy, D. Machado, L. Hilliou, A. Staebler, A. Vicente, et al. 2015b. Exploring the potentialities of using lignocellulosic fibres derived from three food by-products as constituents of biocomposites for food packaging. Ind. Crops Prod. 69: 110–122.

Bhardwaj, U., P. Dhar, A. Kumar and V. Katiyar. 2014. Polyhydroxyalkanoates (PHA)-cellulose based nanobiocomposites for food packaging applications. ACS Symp. Ser. 1162: 275–314.

Boufarguine, M., A. Guinault, G. Miquelard-Garnier and C. Sollogoub. 2013. PLA/PHBV films with improved mechanical and gas barrier properties. Macromol. Mater. Eng. 298: 1065–1073.

Bucci, D.Z., L.B.B. Tavares and I. Sell. 2005. PHB packaging for the storage of food products. Polym. Test. 24: 564–571.

Bucci, D.Z., L.B.B. Tavares and I. Sell. 2007. Biodegradation and physical evaluation of PHB packaging. Polym. Test 26: 908–915.

Busolo, M.A., P. Fernandez, M.J. Ocio and J.M. Lagaron. 2010. Novel silver-based nanoclay as an antimicrobial in polylactic acid food packaging coatings. Food Addit. Contam. Part A. 27: 1617–1626.

Cabedo, L., D. Plackett, E. Gimenez and J.M. Lagaron. 2009. Studying the degradation of polyhydroxybutyrate-co-valerate during processing with clay-based nanofillers. J. Appl. Polym. Sci. 112: 3669–3676.

Cano, A.I., M. Chafer, A. Chiralt and C. Gonzalez-Martinez. 2015. Physical and microstructural properties of biodegradable films based on pea starch and PVA. J. Food Eng. 167: 59–64.

Castro-Mayorga, J.L., A. Martínez-Abad, M.J. Fabra, C. Olivera, M. Reis and J.M. Lagarón. 2014. Stabilization of antimicrobial silver nanoparticles by a polyhydroxyalkanoate obtained from mixed bacterial culture. Int. J. Biol. Macromol. 71: 103–110.

Castro-Mayorga, J.L., M.J. Fabra and J.M. Lagaron. 2015. Stabilized nanosilver based antimicrobial poly(3-hydroxybutyrate-co-3-hydroxyvalerate) nanocomposites of interest in active food packaging. Innov. Food Sci. Emerg. Technol. (in press).

Celarek, A., T. Kraus, E.K. Tschegg, S.F. Fischerauer, S. Stanzl-Tschegg, P.J. Uggowitzer, et al. 2012. PHB, crystalline and amorphous magnesium alloys: Promising candidates for bioresorbable osteosynthesis implants? Mater. Sci. Eng. C. 32: 1503–1510.

Chea, V., H. Angellier-Coussy, S. Peyron, D. Kemmer and N. Gontard. 2016. Poly(3-hydroxybutyrate-co-3-hydroxyvalerate) films for food packaging: Physical-chemical and structural stability under food contact conditions. J. Appl. Polym. Sci. 133: 41850.

Chen, G.Q. and Q. Wu. 2005. The application of polyhydroxyalkanoates as tissue engineering materials. Biomaterials 26: 6565–6578.

Chodak, I. 2008. Polyhydroxyalkanoates: Origin, properties and applications. pp. 451–477. *In*: Belgacem, M.N. and A. Gandini (eds.). Monomers, Polymers and Composites from Renewable Resources. Elsevier, Oxford, UK.

Corre, Y.M., S. Bruzaud, J.L. Audic and Y. Grohens. 2012. Morphology and functional properties of commercial polyhydroxyalkanoates: A comprehensive and comparative study. Polym. Test. 31: 226–235.

Cunha, M., M.-A. Berthet, R. Pereira, J.A. Covas, A.A. Vicente and L. Hilliou. 2015. Development of polyhydroxyalkanoate/beer spent grain fibers composites for film blowing applications. Polym. Compos. 36: 1859–1865.

Cyras, V.P., C.M. Soledad and V. Analía. 2009. Biocomposites based on renewable resource: Acetylated and non acetylated cellulose cardboard coated with polyhydroxybutyrate. Polymer (Guildf). 50: 6274–6280.

Dhar, P., U. Bhardwaj, A. Kumar and V. Katiyar. 2015. Poly (3-hydroxybutyrate)/cellulose nanocrystal films for food packaging applications: Barrier and migration studies. Polym. Eng. Sci. 55: 2388–2395.

Díez-Pascual, A.M. and A.L. Díez-Vicente. 2014a. ZnO-reinforced poly(3-hydroxybutyrate-co-3-hydroxyvalerate) bionanocomposites with antimicrobial function for food packaging. ACS Appl. Mater. Interfaces 6: 9822–9834.

Díez-Pascual, A.M. and A.L. Díez-Vicente. 2014b. Poly(3-hydroxybutyrate)/ZnO bionanocomposites with improved mechanical, barrier and antibacterial properties. Int. J. Mol. Sci. 15: 10950–10973.

Díez-Pascual, A.M. and A.L. Díez-Vicente. 2016. Electrospun fibers of chitosan-grafted polycaprolactone/poly(3-hydroxybutyrate-co-3-hydroxyhexanoate) blends. J. Mater. Chem. B. 4: 600–612.

El-Hadi, A., R. Schnabel, E. Straube, G. Müller and M. Riemschneider. 2002. Effect of melt processing on crystallization behavior and rheology of poly (3-hydroxybutyrate) (PHB) and its blends. Macromol. Mater. Eng. 287: 363–372.

Fabra, M.J., A. Lopez-Rubio and J.M. Lagaron. 2013. High barrier polyhydroxyalcanoate food packaging film by means of nanostructured electrospun interlayers of zein. Food Hydrocoll. 32: 106–114.

Fabra, M. J., A. Lopez-Rubio and J.M. Lagaron. 2014a. Nanostructured interlayers of zein to improve the barrier properties of high barrier polyhydroxyalkanoates and other polyesters. J. Food Eng. 127: 1–9.

Fabra, M.J., A. López-Rubio and J.M. Lagaron. 2014b. On the use of different hydrocolloids as electrospun adhesive interlayers to enhance the barrier properties of polyhydroxyalkanoates of interest in fully renewable food packaging concepts. Food Hydrocoll. 39: 77–84.

Fabra, M.J., G. Sánchez, A. López-Rubio and J.M. Lagaron. 2014c. Microbiological and ageing performance of polyhydroxyalkanoate-based multilayer structures of interest in food packaging. LWT-Food Sci. Technol. 59: 760–767.

Fabra, M.J., P. Pardo, M. Martinez-Sanz, A. Lopez-Rubio and J.M. Lagaron. 2016. Combining polyhydroxyalkanoates with nanokeratin to develop novel biopackaging structures. J. Appl. Polym. Sci. 133: 42695.

Fan, X., Q. Jiang, Z. Sun, G. Li, X. Ren, J. Liang, et al. 2015. Preparation and characterization of electrospun antimicrobial fibrous membranes based on polyhydroxybutyrate (PHB). Fibers Polym. 16: 1751–1758.

Feijoo, L., P. Villanueva and E. Gime. 2006. Optimization of biodegradable nanocomposites based on aPLA/PCL blends for food packaging applications. Macromol. Symp. : 191–197.

Fernandes, E.G., M. Pietrini and E. Chiellini. 2004. Thermo-mechanical and morphological characterization of plasticized poly[(R)-3-hydroxybutyric acid]. Macromol. Symp. 218: 157–164.

Follain, N., C. Chappey, E. Dargent, F. Chivrac, R. Crétois and S. Marais. 2014. Structure and barrier properties of biodegradable polyhydroxyalkanoate films. J. Phys. Chem. C. 118: 6165–6177.

Francis, L., D. Meng, J. Knowles, T. Keshavarz, A.R. Boccaccini and I. Roy. 2011. Controlled delivery of gentamicin using poly(3-hydroxybutyrate) microspheres. Int. J. Mol. Sci. 12: 4294–4314.

Gaikwad, K.K., J.Y. Lee and Y.S. Lee. 2015. Development of polyvinyl alcohol and apple pomace bio-composite film with antioxidant properties for active food packaging application. J. Food Sci. Technol. (in press).

Gao, Y., L. Kong, L. Zhang, Y. Gong, G. Chen, N. Zhao, et al. 2006. Improvement of mechanical properties of poly(dl-lactide) films by blending of poly(3-hydroxybutyrate-co-3-hydroxyhexanoate). Eur. Polym. J. 42: 764–775.

Gonzalez-Ausejo, J., E. Sanchez-Safont, J. Gamez-Perez and L. Cabedo. 2016. On the use of tris(nonylphenyl) phosphite as a chain extender in melt-blended poly(hydroxybutyrate-co-hydroxyvalerate)/clay nanocomposites: Morphology, thermal stability, and mechanical properties. J. Appl. Polym. Sci. 133: 42390.

Halami, P.M. 2008. Production of polyhydroxyalkanoate from starch by the native isolate Bacillus cereus CFR06. World J. Microbiol. Biotechnol. 24: 805–812.

Hufenus, R., F.A. Reifler, M.P. Fernández-Ronco and M. Heuberger. 2015. Molecular orientation in melt-spun poly(3-hydroxybutyrate) fibers: Effect of additives, drawing and stress-annealing. Eur. Polym. J. 71: 12–26.

Jamshidian, M., E.A. Tehrany, M. Imran, M. Jacquot and S. Desobry. 2010. Poly-lactic acid: Production, applications, nanocomposites, and release studies. Compr. Rev. Food Sci. Food Saf. 9: 552–571.

Joseph, C.S., K.V.H. Prashanth, N.K. Rastogi, A.R. Indiramma, S.Y. Reddy and K.S.M.S. Raghavarao. 2011. Optimum blend of chitosan and poly-(ε-caprolactone) for fabrication of films for food packaging applications. Food Bioprocess Technol. 4: 1179–1185.

Jost, V. and R. Kopitzky. 2015. Blending of polyhydroxybutyrate-co-valerate with polylactic acid for packaging applications – Reflections on miscibility and effects on the mechanical and barrier properties. Chem. Biochem. Eng. Q. 29: 221–246.

Kämpf, M.M., L. Thöny-Meyer and Q. Ren. 2014. Biosynthesis of poly(4-hydroxybutyrate) in recombinant Escherichia coli grown on glycerol is stimulated by propionic acid. Int. J. Biol. Macromol. 71: 8–13.

Khanna, S. and A.K. Srivastava. 2005. Recent advances in microbial polyhydroxyalkanoates. Process Biochem. 40: 607–619.

Khosravi-Darani, K. 2015. Application of poly(hydroxyalkanoate) in food packaging: Improvements by nanotechnology. Chem. Biochem. Eng. Q. 29: 275–285.

Kim, K.J., Y. Doi and H. Abe. 2006. Effects of residual metal compounds and chain-end structure on thermal degradation of poly(3-hydroxybutyric acid). Polym. Degrad. Stab. 91: 769–777.

Kolahchi, A.R. and M. Kontopoulou. 2015. Chain extended poly(3-hydroxybutyrate) with improved rheological properties and thermal stability, through reactive modification in the melt state. Polym. Degrad. Stab. 121: 222–229.

Koller, M., A. Salerno, A. Muhr, A. Reiterer and G. Braunegg. 2013. Polyhydroxyalkanoates: Biodegradable polymers and plastics from renewable resources. Mater. Technol. 47: 5–12.

Koller, M. 2014. Poly (hydroxyalkanoates) for food packaging: Application and attempts towards implementation. Appl. Food Biotechnol. 1: 1.

Kopinke, F.-D., M. Remmler and K. Mackenzie. 1996. Thermal decomposition of biodegradable polyesters—I: Poly(β-hydroxybutyric acid). Polym. Degrad. Stab. 52: 25–38.

Kovalcik, A., M. Machovsky, Z. Kozakova and M. Koller. 2015. Designing packaging materials with viscoelastic and gas barrier properties by optimized processing of poly(3-hydroxybutyrate-co-3-hydroxyvalerate) with lignin. React. Funct. Polym. 94: 25–34.

Le Meur, S., M. Zinn, T. Egli, L. Thöny-Meyer and Q. Ren. 2013. Poly(4-hydroxybutyrate) (P4HB) production in recombinant Escherichia coli: P4HB synthesis is uncoupled with cell growth. Microb. Cell Fact. 12: 123.

Le Meur, S., M. Zinn, T. Egli, L. Thöny-Meyer and Q. Ren. 2014. Improved productivity of poly (4-hydroxybutyrate) (P4HB) in recombinant Escherichia coli using glycerol as the growth substrate with fed-batch culture. Microb. Cell Fact. 13: 131.

Lemoigne, M. 1925. Études sur l'autolyse microbienne acidification par formation d'acide. Ann. Inst. Pasteur Paris. 39: 144.

Lemoigne, M. 1926. Produit de déshydratation et de polymérisation de l'acide β-oxybutyrique. Bull. Soc. Chim. Biol. 8: 770–782.

Lim, J.S., K. Park, G.S. Chung and J.H. Kim 2013. Effect of composition ratio on the thermal and physical properties of semicrystalline PLA/PHB-HHx composites. Mater. Sci. Eng. C. Mater. Biol. Appl. 33: 2131–2137.

Luef, K.P. 2015. Poly(hydroxy alkanoate)s in medical applications. Chem. Biochem. Eng. Q. 29: 287–297.

Malathi, A.N. 2014. Recent trends of biodegradable polymer: Biodegradable films for food packaging and application of nanotechnology in biodegradable food packaging. Curr. Trends Technol. Sci. 3: 73–79.

Malinova, L. and J. Brozek. 2011. Mixtures poly((R)-3-hydroxybutyrate) and poly(l-lactic acid) subjected to DSC. J. Therm. Anal. Calorim. 103: 653–660.

Maneewong, C., J. Sunthornvarabhas and K. Sriroth. 2013. Evaluation of gamma radiation on NR/PHBV blends. Appl. Mech. Mater. 300-301: 1325–1329.

Martin, D.P. and S.F. Williams. 2003. Medical applications of poly-4-hydroxybutyrate: A strong flexible absorbable biomaterial. Biochem. Eng. J. 16: 97–105.

Martino, L., M.-A. Berthet, H. Angellier-Coussy and N. Gontard. 2015. Understanding external plasticization of melt extruded PHBV–wheat straw fibers biodegradable composites for food packaging. J. Appl. Polym. Sci. 132: 41611/1–41611/11.

Masood, F., T. Yasin and A. Hameed. 2014. Comparative oxo-biodegradation study of poly-3-hydroxybutyrate-co-3-hydroxyvalerate/polypropylene blend in controlled environments. Int. Biodeterior. Biodegradation 87: 1–8.

Matsumoto, K., S. Nakae, K. Taguchi and H. Matsusaki. 2001. Biosynthesis of from sugars by recombinant Ralstonia eutropha harboring the phaC1 Ps and the phaG Ps genes of Pseudomonas sp . 61-3. Biomacromolecules 2: 934–939.

Michalak, M., A.A. Marek, J. Zawadiak, M. Kawalec and P. Kurcok. 2013. Synthesis of PHB-based carrier for drug delivery systems with pH-controlled release. Eur. Polym. J. 49: 4149–4156.

Miguel, O. and J.J. Iruin 1999. Evaluation of the transport properties of poly(3-hydroxybutyrate) and its 3-hydroxyvalerate copolymers for packaging applications. Macromol. Symp. 144: 427–438.

Modi, S., K. Koelling and Y. Vodovotz. 2011. Assessment of PHB with varying hydroxyvalerate content for potential packaging applications. Eur. Polym. J. 47: 179–186.

Müller, H.M. and D. Seebach. 1993. Poly(hydroxyfettsäureester), eine fünfte Klasse von physiologisch bedeutsamen organischen Biopolymeren? Angew. Chemie. 105: 483–509.

Musioł, M., W. Sikorska, G. Adamus, H. Janeczek, M. Kowalczuk and J. Rydz. 2016. (Bio)degradable polymers as a potential material for food packaging: Studies on the (bio)degradation process of PLA/(R,S)-PHB rigid foils under industrial composting conditions. Eur. Food Res. Technol. 242: 815–823.

Norrrahim, M.N.F., H. Ariffin, M.A. Hassan, N.A. Ibrahim and H. Nishida. 2013. Performance evaluation and chemical recyclability of a polyethylene/poly(3-hydroxybutyrate-co-3-hydroxyvalerate) blend for sustainable packaging. RSC Adv. 3: 24378.

Ohashi, E., W.S. Drumond, N.P. Zane, P.W. de Faria Barros, M.G. Lachtermacher, H. Wiebeck, et al. 2009. Biodegradable poly(3-hydroxybutyrate) nanocomposite. Macromol. Symp. 279: 138–144.

Pankova, Y.N., A.N. Shchegolikhin, A.L. Iordanskii, A.L. Zhulkina, A.A. Ol'khov and G.E. Zaikov. 2010. The characterization of novel biodegradable blends based on polyhydroxybutyrate: The role of water transport. J. Mol. Liq. 156: 65–69.

Pardo-Ibanez, P., A. Lopez-Rubio, M. Martinez-Sanz, L. Cabedo and J.M. Lagaron. 2014. Keratin-polyhydroxyalkanoate melt-compounded composites with improved barrier properties of interest in food packaging applications. J. Appl. Polym. Sci. 131: 39947/1–39947/10.

Petersen, K., P.V. Nielsen and M.B. Olsen. 2001. Physical and mechanical properties of biobased materials – Starch, polylactate and polyhydroxybutyrate. Starch/Staerke. 53: 356–361.

Philip, S., T. Keshavarz and I. Roy. 2007. Polyhydroxyalkanoates: Biodegradable polymers with a range of applications. J. Chem. Technol. Biotechnol. 82: 233–247.

Plackett, D. and I. Siró. 2011. Polyhydroxyalkanoates (PHAs) for food packaging. pp. 498–526. *In*: Lagarón, J.M. (ed.). Multifunctional and Nanoreinforced Polymers for Food Packaging. Woodhead Publishing Ltd., Cambridge, UK.

Raghunatha, K., H. Sato and I. Takahashi. 2015. Intermolecular hydrogen bondings in the poly (3-hydroxybutyrate) and chitin blends: Their effects on the crystallization behavior and crystal structure of poly (3-hydroxybutyrate). Polymer (Guildf). 75: 141–150.

Rahman, A., E. Linton, A.D. Hatch, R.C. Sims and C.D. Miller. 2013. Secretion of polyhydroxybutyrate in Escherichia coli using a synthetic biological engineering approach. J. Biol. Eng. 7: 24.

Renard, E., M. Walls, P. Guerin and V. Langlois. 2004. Hydrolytic degradation of blends of polyhydroxyalkanoates and functionalized polyhydroxyalkanoates. Polym. Degrad. Stab. 85: 779–787.

Rhim, J.-W., H.-M. Park and C.-S. Ha 2013. Bio-nanocomposites for food packaging applications. Prog. Polym. Sci. 38: 1629–1652.

Romero, A.I., J.M. Bermudez, M. Villegas, M.F.D. Ashur, M.L. Parentis and E.E. Gonzo. 2015. Modeling of progesterone release from poly (3-Hydroxybutyrate) (PHB) membranes. AAPS Pharm. Sci. Tech. (in press).

Rosa, D., N. Lotto, D. Lopes and C.G. Guedes. 2004. The use of roughness for evaluating the biodegradation of poly-β-(hydroxybutyrate) and poly-β-(hydroxybutyrate-co-β-valerate). Polym. Test. 23: 3–8.

Rupp, B., C. Ebner, E. Rossegger, C. Slugovc, F., Stelzer and F. Wiesbrock. 2010. UV-induced crosslinking of the biopolyester poly(3-hydroxybutyrate)-co-(3-hydroxyvalerate). Green Chem. 12: 1796–1802.

Sadat-Shojai, M., M.T. Khorasani, A. Jamshidi and S. Irani. 2013. Nano-hydroxyapatite reinforced polyhydroxybutyrate composites: A comprehensive study on the structural and in vitro biological properties. Mater. Sci. Eng. C. 33: 2776–2787.

Safari, S. and T.G.M. van de Ven. 2015. Effect of crystallization conditions on the physical properties of a two-layer glassine paper/polyhydroxybutyrate structure. J. Mater. Sci. 50: 3686–3696.

Saito, Y. and Y. Doi. 1994. Microbial synthesis and properties of poly(3-hydroxybutyrate-co-4-hydroxybutyrate) in Comamonas acidovorans. Int. J. Biol. Macromol. 16: 99–104.

Salmieri, S., F. Islam, R.A. Khan, F.M. Hossain, H.M.M. Ibrahim, C. Miao, et al. 2014. Antimicrobial nanocomposite films made of poly(lactic acid)-cellulose nanocrystals (PLA-CNC) in food applications: Part A: Effect of nisin release on the inactivation of Listeria monocytogenes in ham. Cellulose 21: 4271–4285.

Sanchez-Garcia, M.D., J.M. Lagaron and S.V. Hoa. 2010. Effect of addition of carbon nanofibers and carbon nanotubes on properties of thermoplastic biopolymers. Compos. Sci. Technol. 70: 1095–1105.

Savenkova, L., Z. Gercberga, V. Nikolaeva, A. Dzene, I. Bibers and M. Kalnin. 2000. Mechanical properties and biodegradation characteristics of PHB-based films. Process Biochem. 35: 573–579.

Senior, P.J. and E.A. Dawes. 1973. The regulation of poly-beta-hydroxybutyrate metabolism in Azotobacter beijerinckii. Biochem. J. 134: 225–238.

Seoane, I.T., L.B. Manfredi and V.P. Cyras. 2015. Properties and processing relationship of polyhydroxybutyrate and cellulose biocomposites. Procedia Mater. Sci. 8: 807–813.

Sugama, Y., A.A. Abdullah, C. Stephen, S. Mohd, M. Nasir and M. Ibrahim. 2011. Bioresource technology biosynthesis of poly(3-hydroxybutyrate-co-3-hydroxyvalerate) and characterisation of its blend with oil palm empty fruit bunch fibers. Bioresour. Technol. 102: 3626–3628.

Thellen, C., S. Cheney and J.A. Ratto. 2013. Melt processing and characterization of polyvinyl alcohol and polyhydroxyalkanoate multilayer films. J. Appl. Polym. Sci. 127: 2314–2324.

Vandewijngaarden, J., M. Murariu, P. Dubois, R. Carleer, J. Yperman, P. Adriaensens, et al. 2014. Gas permeability properties of poly(3-hydroxybutyrate-co-3-hydroxyhexanoate). J. Polym. Environ. 22: 501–507.

Vandewijngaarden, J., R. Wauters, M. Murariu, P. Dubois, R. Carleer, J. Yperman, et al. 2016. Poly(3-hydroxybutyrate-co-3-hydroxyhexanoate)/organomodified montmorillonite nanocomposites for potential food packaging applications. J. Polym. Environ. (in press).

Vázquez-Rodríguez, G.A., A. Calmon, F. Silvestre, G. Goma and J.-L. Rols. 2006. Effect of the inoculation level in aerobic biodegradability tests of polymeric materials. Int. Biodeterior. Biodegradation 58: 44–47.

Yang, S., J. Wang, L. Tang, H. Ao, H. Tan, T. Tang, et al. 2014. Mesoporous bioactive glass doped-poly(3-hydroxybutyrate-co-3-hydroxyhexanoate) composite scaffolds with 3-dimensionally hierarchical pore networks for bone regeneration. Colloids Surf. B. Biointerfaces 116: 72–80.

Yu, H., B. Sun, D. Zhang, G. Chen, X. Yang and J. Yao. 2014a. Reinforcement of biodegradable poly(3-hydroxybutyrate-co-3-hydroxyvalerate) with cellulose nanocrystal/silver nanohybrids as bifunctional nanofillers. J. Mater. Chem. B. 2: 8479–8489.

Yu, H., Z.Y. Qin, B. Sun, X.G. Yang and J.M. Yao. 2014b. Reinforcement of transparent poly(3-hydroxybutyrate-co-3-hydroxyvalerate) by incorporation of functionalized carbon nanotubes as a novel bionanocomposite for food packaging. Compos. Sci. Technol. 94: 96–104.

Yu, H., C. Yan and J. Yao. 2014c. Fully biodegradable food packaging materials based on functionalized cellulose nanocrystals/poly(3-hydroxybutyrate-co-3-hydroxyvalerate) nanocomposites. RSC Adv. 4: 59792–59802.

Zhao, S., M. Zhu, J. Zhang, Y. Zhang, Z. Liu, Y. Zhu, et al. 2014. Three dimensionally printed mesoporous bioactive glass and poly(3-hydroxybutyrate-co-3-hydroxyhexanoate) composite scaffolds for bone regeneration. J. Mater. Chem. B. 2: 6106.

Active Biopackaging Based on Proteins

Mariana Pereda[1], María R. Ansorena[2] and Norma E. Marcovich[1]*

[1] Institute of Material Science and Technology (INTEMA),
National Research Council, Mar del Plata, Argentina
[2] Chemical Engineering Department, Food Engineering Group, Engineering Faculty,
National University of Mar del Plata, Mar del Plata, Argentina

Introduction

The globalization of food markets and increasing concerns on food stability and control lead researchers and industrial companies to develop active and intelligent packaging materials. Food packaging technology is continuously evolving in response to growing challenges from a modern society. Major current and future challenges to fast-moving consumer goods packaging include legislation, global markets, longer shelf life, convenience, safer and healthier food, environmental concerns, authenticity and food waste (Kerry 2014). Every year a growing amount of edible food is lost along the entire food supply chain. Annual food waste generation estimated in Europe is around 89 million tonnes varying considerably between individual countries and the various sectors (EU No 528/2012). Spoilage of raw meat along the food supply chain (production, retailers, and consumers) represents a loss which could be as high as 40% (Sperber 2010). Packaging optimization strategies such as varying pack sizes to help consumers to buy the right amount, and designing packaging to maintain food quality and increase its shelf life have been proposed to reduce food waste (EU No 528/2012). In addition, the growing consumer demand for minimally processed, more natural, fresh and convenient food products as well as continuous changes at industrial, retail and distribution levels associated with globalization pose major challenges to food safety and quality (Realini and Marcos 2014).

---
*Corresponding author: marcovic@fi.mdp.edu.ar

Active food packaging is currently one of the most studied areas, stressing the development of new techniques capable of improving conservation and food quality in terms of their interaction with the packaging (González and Alvarez Igarzabal 2015). The dynamic packaging technologies have been modernized to provide more reliable quality and food safety, and also to minimize package-related environmental contamination and disposal problems (Ozdemir and Floros 2004). According to Regulation (CE) N450/2009 of the European Union Commission on active and intelligent materials, an active food packaging comprises the group of materials intended to prolong shelf life or maintain or improve the condition, security and quality of the packaged food (Vermeiren et al. 1999). Thus, an active packaging can be defined as a type of material that changes its packaging conditions to extend shelf life, interacting directly with the food, enhancing security and maintaining quality. The expansion of active packaging has led to advances in many applications incorporating properties such as antioxidant and antimicrobial activities, controlled respiration rates, and reducing water vapour permeability. Other active packaging technologies include carbon dioxide absorber/emitters, odour absorbers, ethylene absorber and aroma emitters (Restuccia et al. 2010). Among these techniques, oxygen scavengers, moisture absorbers and antimicrobial packaging constitute more than 80% of the market (Robinson and Morrison 2010). In particular, the antimicrobial packaging is one of the most innovative and promising active packaging type developed over the last decade, which includes systems capable of inhibiting microorganism action and avoiding loss of food quality (Seydim and Sarikus 2006; Pérez et al. 2011; González and Alvarez Igarzabal 2015). Additionally, antioxidant active packaging seeks to prevent or slow down the oxidation of certain food components, like lipids and proteins, which lead to the deterioration of physical characteristics (such as flavour and colour) of food products. This active material approach requires the intentional incorporation of antioxidants within the packaging materials and their further migration to those foods (Robertson 2012).

Polymers, and in particular biomass-derived polymers, are the preferred materials for active packaging because of their intrinsic properties (Han and Gennadios 2005; Colak et al. 2015) which includes good film forming properties (Van de Velde and Kiekens 2002; Arancibia et al. 2013) and the possibility of combining several polymers through blending or multilayer assembly in order to tailor the application, constituting an ideal carrier for active agents, with the advantage of being adaptable in terms of controlled release. The drawbacks that initially characterized these biopolymers in terms of poor barrier properties and high instability have, in turn, resulted in novel applications, making highly permeable and water- plasticizable biopolymers an ideal partner for active packaging where the package is no longer a passive barrier, but actively contributes to the preservation of food by controlled release of the substances (Martínez-Abad et al. 2014). Active films based on biopolymers provide not only the opportunity to extend shelf-life of foods by preventing contamination or growth of unwanted microorganisms, but also offer potential as a way to overcome the limitations of synthetic plastics, since they are biodegradable and thus, could limit the accumulation of residues (Bucci et al. 2005), reducing therefore the environmental impact of conventional packaging materials (Van de Velde and Kiekens 2002).

Films, casings and coatings made from polysaccharides and proteins are effective barriers against oxygen (O_2), carbon dioxide (CO_2) and low polarity compounds. In comparison with polysaccharide films, protein films show considerably lower O_2 and CO_2 permeabilities and CO_2/O_2 permeability ratio (Bourlieu et al. 2009; Kaewprachu and Rawdkuen 2014; Taylor et al. 2014). Proteins are transparent, odourless and tasteless and also have a good aroma (Van de Velde and Kiekens 2002), barrier properties under dry conditions, excellent film-forming properties and high nutritional value. An additional advantage is that proteins can be processed by diverse methods such as dissolution-solvent evaporation or thermo-mechanical processing to produce films with suitable mechanical properties (Ciannamea et al. 2014). The unique structure of proteins (based on 20 different monomers) provides a wider range of functional properties, especially a high intermolecular binding potential since their secondary, tertiary and quaternary structures result in various interactions and bondings, differing in position, type and energy (González et al. 2011; Silva et al. 2014). Moreover, literature have shown that protein coatings or films require relatively lower amounts of added antimicrobial agents to reach the desired effect as compared to the synthetic polymer or to other biopolymer films (Cha et al. 2003; Zhao et al. 2013).

Active Additives

Antimicrobials

In order to control undesirable microorganisms in food surfaces, natural or synthetic antimicrobial agents can be incorporated into polymer coatings (Pérez et al. 2011). Several compounds have been proposed as antimicrobial agents in food packaging including organic acids, enzymes, fungicides and natural compounds such as spices and essential oils (Tharanathan 2003; Seydim and Sarikus 2006; Pérez et al. 2011). However, additives in the polymer matrix such as fillers, antifogging and antistatic agents, lubricants, stabilizers and plasticizers can affect the activity of antimicrobial agents. These additives may change the conformation of the polymer and alter the diffusion of the antimicrobial, or they may interact directly with it. Further considerations in antimicrobial packaging choice are the effect of film thickness on the activity and the concentration of antimicrobials in polymeric films. The effect of the antimicrobials on the mechanical, barrier and optical properties of a polymeric matrix must also be considered.

Potassium sorbate (PS), the potassium salt of sorbic acid, has a long history and is generally recognized as a safe food preservative, being widely used to inhibit or retard the growing of a number of recognized food pathogens (Pérez et al. 2011). PS is used in dairy, meat and bakery products as a yeast, mould and bacteria inhibitor (Kristo et al. 2008). Moreover, its effectiveness in preventing microbial growth and release properties from several solvent-cast films have been reported (Shen et al. 2010; Sayanjali et al. 2011). The use of PS is effective up to pH 6.5 but its effectiveness increases as the pH decreases; so changes in the pH of the films and the foods should be monitored because this compound, being a weak acid, is most effective in the undissociated form due to its increased ability to penetrate the cytoplasmic membrane of bacteria (Pérez et al. 2011).

Essential oils (EOs) extracted from plants are rich sources of biologically active compounds such as terpenoids and phenolic, which exhibit a wide range of biological effects including antioxidant and antimicrobial properties (Ansorena et al. 2016). Spices are rich in phenolic compounds, such as flavonoids and phenolic acids (Dadalioglu and Evrendilek 2004; Seydim and Sarikus 2006). Generally, the EOs possessing the strongest antibacterial properties against foodborne pathogens contain high concentrations of phenolic compounds such as carvacrol, eugenol (2-methoxy-4-(2-propenyl) phenol), citral and thymol (Burt 2004; Seydim and Sarikus 2006). According to FDA, thymol and carvacrol are considered as food additives; thus, they have lower potential than eugenol and citral which are in GRAS status (Suppakul et al. 2003). The mode of action of these phenolic compounds is generally considered to be the disturbance of the cytoplasmic membrane, disrupting the proton motive force, electron flow, active transport and/or coagulation of cell contents (Burt 2004). Some spice essential oils incorporated into packaging materials can control microbial contamination in beef muscle by reducing the growth of *Escherichia coli* O157:H7 and *Pseudomonas* spp. (Oussallah et al. 2004; Seydim and Sarikus 2006). Essential oil fractions of oregano and pimento are efficient against various foodborne bacteria such as *Salmonella* and *E. coli* O157:H7 (Burt and Reinders 2003; Seydim and Sarikus 2006). Spice extracts from oregano, clove, sage, rosemary, garlic, thyme and pimento are also reported to possess strong antioxidant and antimicrobial properties (Seydim and Sarikus 2006). Carvacrol, a major constituent of oregano essential oil, was found to be highly active against strains of *Staphylococcus aureus* and *Escherichia coli* (Benarfa et al. 2007), including the pathogen *E. coli* O157:H7 (Lim et al. 2010). The antioxidant activity and antimicrobial properties of thyme essential oil have been demonstrated and attributed to its major components, thymol and carvacrol (Ansorena et al. 2016). Citronella essential oil possesses insect repellent, herbicidal and antimicrobial properties and can, therefore, be used to replace chemical fungicides (Lim et al. 2010). Mosquito repellent and weed control formulations in organic agriculture look promising (Arancibia et al. 2013). The principal components of the essential oil extracted from the flowers of

Matricaria recutita (also known as Matricaria chamomilla) are the terpenoids α- bisabolol and its oxides and azulenes, including chamazulene, which is responsible for the strong blue color of the extract. The components of chamomile, believed to have antimicrobial properties, include α- bisabolol, luteolin, quercetin and apigenin. Chamomile oil, at a concentration of 25 mg/mL, demonstrates antibacterial activity against gram-positive bacteria such as *Bacillus subtilis, Staphylococcus aureus, Streptococcus mutans* and *Streptococcus salivarius*, as well as some fungicidal activity against *Candida albicans*. In addition, chamomile extracts block aggregation of *Helicobacter pylori* and various strains of *Escherichia coli*.

Essential oils can be trapped and released slowly from a film matrix to make "active packaging" (Lim et al. 2010). However, EOs have a distinctive odour and taste that is incompatible with most food systems. Thus, it was suggested that the application of essential oils in active packaging should be combined with use of suitable flavour compounds (Gutierrez et al. 2009). Gutierrez et al. (2009) found that acceptable organoleptic profiles could be obtained by combining carvacrol and thymol with vanilla aroma and strawberry aroma with thymol.

Copaiba oil is extracted directly from the trunk of its tree and its correct denomination is oleoresin. In Brazilian popular traditional medicine, copaiba oil has been used as a healing agent, anti-inflammatory and antibiotic for a long time (Veiga et al. 2001; Santos et al. 2008), being swallowed or applied directly on the skin surface. Copaiba oil is approved by the Food and Drug Administration (FDA) since 1972, being suitable for food contact (Veiga and Pinto 2002). Copaiba oil was incorporated for the first time by Lipparelli Morelli et al. (2015) on paper sheets and plastic films, and it was found to be efficient against gram-positive bacteria *B. subtilis* when the copaiba oil content was about 20 wt%.

Citric acid (CA) is a natural organic acid found in citrus foods. It is used as a food additive in various food production processes for its antibacterial and acidulant effects. It reinforces the antioxidant and antidarkening actions of other substances and improves flavour and aroma (Park et al. 2005; Azevedo et al. 2015).

Hen egg white lysozyme (E1105) is a widely used enzyme authorized for food preservation in European Union under 2008/1333/EC Regulation on food additives (Colak et al. 2015). It has also been accepted as an antimicrobial agent in casings and in ready-to-eat products (FDA 2000). Lysozyme exhibits a strong antimicrobial activity against most Gram-positive and some Gram negative bacteria. It damages peptidoglycans in bacterial cell wall by catalyzing hydrolysis of β-1–3 glycosidic linkages between Nacetylmuramic acid and N-acetylglucosamine. In particular, lysozyme is effective on a very critical pathogen, *L. Monocytogenes* (Min et al. 2005; Duan et al. 2007; Ünalan et al. 2011) which is a food pathogen for dairy products due to its incidence in raw milk and its capacity to grow at refrigeration temperatures (Kozak et al. 1996). Thus, its current antimicrobial film applications are concentrated mainly on fighting against listeriosis.

Nisin, a well-known bacteriocin obtained from lactic acid bacteria, is a cationic peptide. It shows antimicrobial activity by interacting with the anionic phospholipids at the bacterial surfaces, forming pores and dissipating proton motive forces at the bacterial membrane (Sudağidan and Yemenicioğlu 2012). Similar to lysozyme, nisin cannot overcome the protective LPS of Gram-negative bacteria, and shows antimicrobial activity mainly on Gram-positive bacteria. However, the advantage of using nisin in active packaging comes from its potency against *S. aureus*, which is quite resistant to the action of lysozyme.

ε-polylysine is formed by 25-35 l-lysine residues, and it is produced commercially from aerobic fermentation by *Streptomyces albulus*, which is a non-pathogenic microorganism (Hiraki et al. 2003; FDA 2004; Geornaras and Sofos 2005). The major advantage of using ε-polylysine as an antimicrobial agent is that it is effective on major Gram-positive and Gram-negative food pathogenic bacteria such as *L. monocytogenes, E. coli* O157:H7, and *Salmonella typhimurium* (Geornaras and Sofos 2005; Geornaras et al. 2007). The antibacterial action of ε-polylysine is attributed to its polycationic and surface active nature that enables its interaction with bacterial membranes (Ho et al. 2000). Moreover, a recent study made by Liu et al. (2015) suggested that ε-polylysine is also capable of interacting with bacterial DNA when it penetrates into cells and also showed that the combination of ε-polylysine with nisin creates a synergetic antimicrobial activity against *E. coli* and *S. aureus*.

TiO_2 photocatalyst forms have been developed for antimicrobial purposes since they are non-toxic and exhibit a notable disinfecting activity against a broad spectrum of microorganisms, due to the high redox potential of the surface species formed by photo excitation (under UV illumination) affording a non-selective oxidative attack on microorganisms (Gumy et al. 2006). The FDA approved TiO_2 as a "food contact substance" as well, meaning that it is safe to incorporate it into food packaging (Martirosyan and Schneider 2014).

Nano-sized particles can be incorporated into biopolymers to improve the mechanical and barrier properties of biopolymer-based films (Shankar et al. 2014a, b, 2015a, b) and to provide extra functional properties, such as antioxidant and antimicrobial functions. Metallic nanoparticles such as gold, silver, copper, zinc, etc. have become important for food packaging applications. Silver nanoparticles (AgNPs) have attracted enormous attention for their use in food packing due to an outstanding and broad spectrum of antimicrobial effects against foodborne pathogens. Copper-based nanoparticles have advantages over silver nanoparticles because of the low cost of the source materials, insignificant sensitivity to human tissues and high sensitivity to microorganisms (Hostynek and Maibach 2004). Copper ions and copper complexes have been used as effective materials for sterilizing liquids, and as antibacterials, antifungals and antivirals since ancient times (Borkow and Gabbay 2009). However, till today, no report of copper nanoparticles with proteins films is available. This area of research is yet to be exploited for use in the food packaging area. Zinc nanoparticles (ZnO NP) have gained an excellent ability for interfacial interactions and particle dispersion in a protein matrix, owing to their high surface energy and large surface area to volume ratio (Rouhi et al. 2013). ZnO is listed as a "generally recognized as safe" (GRAS) material by the Food and Drug Administration. ZnO is reported to have strong antimicrobial activity against foodborne pathogens and spoilage bacteria (Espitia et al. 2013).

Antioxidants

Owing to their strong antioxidant effects, antioxidants are beneficial compounds as they reduce the risk of diseases associated with oxidative stress and also protect the human body from the harmful effects of free radicals and reactive oxygen species (Lu and Foo 2000; Sanchez-Moreno et al. 2000; Malik et al. 2011). Antioxidants also slow down the progress of many chronic diseases and lipid eroxidation (Lai et al. 2001; Gülçin et al. 2004). Polyphenols have been widely used as food additives to protect food nutrients against oxidative degradation (Gülçin et al. 2005). The molecular weight and structure of phenolic compounds can show great variation, and they may contain different numbers of hydroxyl groups, which are capable of forming H-bonds with peptide carbonyl groups of proteins (Damodaran 1996).

Tannic acid (TA), which possesses antioxidant activity, is one example of a polyphenol (Andrade et al. 2005; Gülçin et al. 2010) abundantly present in several beverages, including red wine, beer, coffee, black and green tea, and many foods such as grapes, pear, banana, sorghum, black-eyed peas, lentils and chocolates (Helal et al. 2012). Grape seed procyanidins and green tea polyphenols have been used recently as active substances in edible films (Wu et al. 2013; Li et al. 2014; Nie et al. 2015).

Flavonoids are polyphenols that are present in vegetables and fruits. Catechins (CAT) compose an interesting class of flavonoids ubiquitously found in fruits such as plum and apple, as well as in green and black tea, chocolate, and red wine (Yilmaz 2006). The major catechins include catechin, epicatechin, epicatechin gallate and epigallocatechin gallate (Yilmaz 2006). The catechins are the main active phenolic compounds in green tea extracts and up to 50% of the catechins in the green tea is formed by epigallocatechin gallate that might show antimicrobial effect on both Gram-positive (*Staphylococcus* spp.) and Gram-negative (*Salmonella* spp.) bacteria (Perumalla and Hettiarachchy 2011). CAT are recognized as potent anticancer, anti-allergy and antioxidant agents (Kondo et al. 2000; Ku et al. 2008a, b) and are regarded as the most powerful antioxidants among plant phenols.

Natural tannins are comprised of a wide range of oligomeric and polymeric phenols, which are categorized as GRAS food additives commonly used to protect food nutrients against oxidative degradation

(Wang et al. 2016). They are usually classified in hydrolysable tannins and condensed tannins. Hydrolysable tannins are composed of one molecule of sugar, generally glucose, joined to phenolic acid, while condensed tannins (i.e. proanthocyanidins) are polymers and oligomers comprising flavan-3-ol units.

Chestnut bur/skin extracts contain hydrolysable gallotannins (gallic acid esters of glucose) (Fernández-Agulló et al. 2014), ellagic acid and vescalagin/castalagin (Maria do Carmo et al. 2010) and have high antioxidant activity values, particularly for lipid peroxidation inhibition in the thiobarbituric acid-reactive substances (Barreira et al. 2008).

Phlorotannins are phenolic compounds present in brown seaweeds (e.g. *Ascophyllum nodosum*) which have reported biological activities including antioxidant, angiotensin-I-converting enzyme (ACE-I) inhibition, and bactericidal and anticancer (Li et al. 2011).

Chlorophyllins are semi-synthetic porphyrins derived from chlorophyll with antioxidative, anticarcinogenic and antimutagenic properties that are used as food colorants, dietary supplements, cosmetics, an internal deodorant and an accelerant in wound healing (Ferruzzi et al. 2002; Kapiotis et al. 2005).

Active Packaging

Several studies have shown that protein films are good vehicles for antimicrobial substances, for example, essential oils (Zinoviadou et al. 2009; Shakeri et al. 2011; Fernández-Pan et al. 2012; Matan 2012), organic acids salts (Kristo et al. 2008), glucose oxidase (Murillo-Martínez et al. 2013), nisin (Kristo et al. 2008; Cao-Hoang et al. 2010b; Murillo-Martínez et al. 2013), thymol (Li et al. 2012) and viable cells of *Lactobacillus sakei* (Beristain-Bauza et al. 2016). Therefore, numerous proteins such as corn zein, wheat gluten, soy, peanut, cottonseed, sunflower, rice bran, serum albumin, egg white, collagen, gelatin, myofibrils, casein and whey proteins have been studied as potential film-forming agents (Kumar and Gupta 2012). Also, these protein films were blended with other proteins (Bai et al. 2013), polysaccharides (Li et al. 2013) and lipids (Guerrero et al. 2011) to form composite films. In addition, a new class of materials represented by bio-nanocomposites with enhanced barrier, mechanical and thermal properties has also been considered as a promising option for active packaging materials.

The main characteristic of selected proteins and a summary of the last findings regarding their corresponding active packaging are reviewed in the following sections.

Films Based on Milk Proteins

Milk proteins are common sources to formulate edible films and coatings (Beristain-Bauza et al. 2016) owing to their excellent nutritional value and their numerous functional properties, such as solubility in water and ability to act as emulsifiers (Aliheidari et al. 2013). The typical composition of milk proteins is 80% caseins and 20% whey proteins (Beristain-Bauza et al. 2016).

Caseins are a blend of four fractions of proteins named αs1-, αs2-, β- and κ-caseins which represents approximately 37, 10, 35 and 12%, respectively (Beristain-Bauza et al. 2016). Sodium caseinate (SC) is the water-soluble form of casein, obtained by acid precipitation of the casein (Aliheidari et al. 2013) and it is one of the most studied proteins in edible films due to its excellent film-forming properties (Fabra et al. 2009; Pereda et al. 2010; Pereda et al. 2011; Jimenez et al. 2012). In addition, due to the high number of polar groups in their structure, caseinates also show good adhesion to different substrates making them excellent barrier to non-polar substances, such as oxygen, carbon dioxide and aromas (Arrieta et al. 2013).

Numerous antimicrobial agents including essential oils (Atares et al. 2010), organic acids (Arrieta et al. 2013; Arrieta et al. 2014), bacteriocins (Cao-Hoang et al. 2010a) and lysozyme (Mendes de Souza et al. 2010) have been incorporated into SC films to obtain biodegradable edible packaging films with enhanced protective functions. However, almost all these antimicrobial SC films were prepared by a solvent-casting method, while most of the commercial plastic films are prepared by extrusion-blowing. Interestingly, Belyamani et al. (2014) recently reported for the first time the possibility to obtain SC-based edible thin films (~30 μm) by blown-film extrusion. More recently, Colak et al. (2015) obtained active films by blown-film-extrusion of pellets prepared by twin-screw extrusion of sodium caseinate, lysozyme and glycerol. The challenge was to define conditions limiting inactivation of lysozyme by temperature and shear stress prevailing during thermo-mechanical treatments used to prepare pellets and subsequently films. A twin screw extrusion temperature of 65 °C and a glycerol content of 25 or 20% allowed getting films with up to 26.4% of the initial activity of lysozyme. Moreover, it was checked that they effectively inhibited the growth of *M. luteus* even below lysozyme minimal inhibitory concentration. The enzymatic activity of films was preserved for at least five weeks even when stored at room temperature. The authors also verified that lysozyme neither impaired films biodegradability, nor affected their mechanical properties.

Aliheidari et al. (2013) compared the antimicrobial effectiveness, as well as the water vapour permeability and thermal and mechanical properties, of SC-incorporated films with two lipids – stearic acid (SA) and oleic acid (OA) – and Matricaria recutita essential oil (MEO), which was chosen as antimicrobial agent in the formulation for its positive characteristics. Incorporation of OA/SA and MEO improved the barrier properties of casein-based films; WVP was reduced the most by the incorporation of MEO. Films containing MEO exhibited a high inhibitory effect on *L. monocytogenes*, *S. aureus* and *E. coli*. The films developed in this study can have applications in packaging in a wide range of food products, particularly those that are highly oxidative and microbial-sensitive.

Helal et al. (2012) chose caseins and caseinates to protect polyphenols against rapid oxidation. These authors investigated the effect of protecting TA, as an example of phenols, and CAT, as an example of flavonoids, against oxidation during storage at different relative humidities by entrapping these molecules in macromolecular films of different casein (C)/SC ratios. The SC films containing phenolic compounds had surface Radical-scavenging activity (RSA) adaptable to the needs of food products that are more susceptible to alteration of its properties, including oxidation, during storage at high relative humidity. Indeed, the surface RSA increased with storage time due to plasticizing and perhaps alteration of the SC and C network. Antioxidants can then migrate easily to the surface and any oxidant molecule can better migrate from food to the film/coating, which limits food oxidation. The structural changes of the C/polyphenols complexes have nevertheless to be investigated to better explain the early changes in antioxidant capacity and the excellent stabilization of the complexes for up to 90 days storage.

Arrieta et al. (2014) evaluated the functional properties of high transparent antimicrobial bio-films based on plasticized sodium and calcium caseinates (CC) with carvacrol and their suitability for food active packaging applications, particularly for the meat industry. The incorporation of carvacrol into plasticized caseinates was successfully performed to obtain transparent active films by solvent casting. Caseinate–carvacrol interactions suggested that the resulting materials could have more hydrophobic character than the materials without the active agent. The antimicrobial activity of SC and CC films containing carvacrol was clearly demonstrated against two indicator bacteria (Gram negative and Gram positive). Barrier properties to oxygen were excellent and diffusion of dyes through films was dependent on the caseinate (SC or CC), since CC resulted in less permeable composites due to the ability to promote cross-linking.

Kadam et al. (2015) investigated the effect of the incorporation of *Ascophyllum nodosum* extract at different ratios on the total phenolic content, antioxidant activity, appearance, structural and other physico-chemical properties of SC films. Seaweed extracts, being rich in phycocolloids, phenolics and other antioxidative compounds, are a very good source for fortification of biopolymer-based films. SC films were casted with incorporation of *A. nodosum* extract at 25 and 50% (w/w of base material) with glycerol added as a plasticizer. The total phenolic content and antioxidant activity of the films increased at higher levels of seaweed extract incorporation.

Whey proteins are soluble proteins present in milk serum after caseinate coagulation during cheese processing. Utilization of cheese whey, which is produced in large quantities as a by-product in the cheese making could effectively alleviate the whey disposal problem by the conversion of whey into value-added products, such as edible films and coatings (Javanmard 2009; Pérez et al. 2011). Whey proteins contain five main proteins: α-lactalbumin, β-lactoglobulin, bovine serum albumin (BSA), immunoglobulin and proteose peptones. The β-lactoglobulin monomer comprises approximately 57% of the protein in whey (Dybing and Smith 1991). The industrial processes used for whey protein recovery are ultrafiltration, reverse osmosis, gel filtration, electrodialysis, ion-exchange chromatography and diafiltration (Letendre et al. 2002). These processes are generally used to produce whey protein concentrate (WPC, 25 to 80% protein) or whey protein isolate (WPI, 90% protein). Formation of intact and insoluble whey protein films could be realized by heat denaturation of the proteins. Heating modifies the three-dimensional structure of the protein, exposing internal SH and hydrophobic groups (Shimada and Cheftel 1998), which promote intermolecular S-S bonds and hydrophobic interactions upon drying (McHugh and Krochta 1994).

Pérez et al. (2011) incorporated PS into WPC plasticized (WPC/Glycerol 3:1) films to determine the inhibitory effects of these films against eight non-O157 STEC strains isolated from ready-to-eat food samples. The active films were prepared by casting of film forming solutions adjusted to pH 5.2 or 6.0, to study also the effect of the pH on the antimicrobial performance. In protein stabilized solutions, the net charge is highly dependent on pH values. When pH is close to the isoelectric point (pI) of the protein (pI ≈ 5 for whey protein), its net charge approaches to zero, electrostatic repulsions become weak and attractive interactions become important. They found that increasing the concentration of PS into the film at both pH levels increased the antimicrobial activity (Table 1). Control films without antimicrobials were non-inhibitory. Acidic films (pH 5.2) containing PS were generally more inhibitory since the diffusion of PS from the films is affected by the WPC background in a way that at a pH value near the isoelectric point of the proteins it will be less retained due to a decreased capability to generate dipolar interactions with the amide groups present in the matrix. On the other hand, films obtained at pH 5.2 were more opaque than those prepared at pH 6 due to partial protein aggregation, which means that its acceptability to the consumers would be affected.

Table 1. Antimicrobial activity of WPC edible films containing PS against non-O157 STEC strains and ATTCC 43895 strain. From Pérez et al. (2011).

| Strain number | Diameter of inhibition zone (mm) | | | | | |
| | WPC films (pH 5.2) | | | WPC films (pH 6.0) | | |
	PS 0.5%	PS 1.0%	PS 1.5%	PS 0.5%	PS 1.0%	PS 1.5%
ARG 4827	4.3 ± 0.6	8.3 ± 1.5	11.3 ± 0.6			
ARG 2379	4.7 ± 1.2	8.3 ± 1.5	13.0 ± 1.0	-	7.3 ± 1.5	9.0 ± 1.0
ARG 5266	4.0 ± 1.0	8.7 ± 0.6	10.7 ± 0.6	3.7 ± 0.6	6.3 ± 0.6	6.3 ± 0.6
ARG 4824	3.3 ± 1.0	8.7 ± 0.6	12.7 ± 1.2	-	5.7 ± 1.2	7.0 ± 1.0
ARG 4627	3.0 ± 1.0	8.3 ± 1.2	9.0 ± 1.0	-	2.0 ± 1.0	7.0 ± 1.0
ARG 20	4.3 ± 1.5	8.7 ± 1.2	10.3 ± 0.6	3.0 ± 1.0	3.7 ± 0.6	7.7 ± 0.6
ARG 4823	2.7 ± 1.2	5.3 ± 1.2	9.0 ± 1.0	-	6.0 ± 1.0	5.0 ± 1.0
ARG 5468	3.0 ± 1.0	6.7 ± 1.2	10.3 ± 0.6	3.3 ± 1.5	6.0 ± 0.0	7.00 ± 0.0

Seydim and Sarikus (2006) evaluated the antimicrobial performance of whey protein isolate (WPI) films prepared by casting method, containing 1.0–4.0% (wt/vol) ratios of oregano, rosemary and garlic essential oils by testing them against *Escherichia coli* O157:H7 (ATCC 35218), *Staphylococcus aureus* (ATCC 43300), *Salmonella enteritidis* (ATCC 13076), *Listeria monocytogenes* (NCTC 2167) and *Lactobacillus plantarum* (DSM 20174). The films containing oregano essential oil was the most effective against these bacteria at 2% level ($p < 0.05$). The use of rosemary essential oil incorporated into WPI films did not exhibit any antimicrobial activity whereas inhibitory effect of WPI film containing garlic essential oil was observed only at 3% and 4% level ($p < 0.05$). The different inhibitory effects of essential oils were attributed to the differences in the biological properties of the main compounds in the essential oils.

Vonasek et al. (2014) developed bacteriophage-based antimicrobial edible coatings using a model T4 bacteriophage in edible WPI films. Confocal imaging measurements show that fluorescently labeled phages were homogenously distributed within the WPI film matrix. The ability of WPI films to stabilize phages at ambient and refrigerated conditions without significant loss in phage infectivity over a period of one month was demonstrated. Additionally, the WPI films were able to release significant concentration of phages in an aqueous environment and leaf surface within 3 h of incubation. Antimicrobial activity measurements demonstrate that the phage encapsulating WPI film could effectively inhibit the microbial growth: results showed an approximately 5 log difference in microbial levels between the control and the treatment samples. This study demonstrates integration

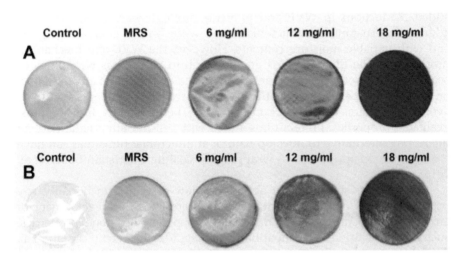

Fig. 1. Protein films of whey protein isolate (A) and calcium caseinate (B) added with 0 (control), MRS (de Man, Rogosa and Sharpe broth), 6, 12, or 18 mg/ml of cell-free supernatant of *Lactobacillus rhamnosus* NRRL B-442. From Beristain-Bauza et al. (2016).

of phages with edible packaging materials to develop novel active packaging materials for biocontrol applications.

According to Barba et al. (2015), control release of eugenol and carvacrol is difficult to obtain even when WPI is used as a carrier and thus they decided to use active compounds cyclodextrin (CD) inclusion complexes to encourage molecular interaction intervening in their release. In their work the release of eugenol and carvacrol in a WPI matrix with and without the use of β-CD was compared. Specifically, release for β-CD:carvacrol inclusion complexes allowed the extension of the carvacrol time delivery in the food simulant (ethanol:water, 1:1, v/v) tried, which was explained by the higher retention that carvacrol has in β-CD due to its preference for the apolar compounds. This fact shows significance for delivery of antimicrobial compounds that must release gradually in time.

Films Based on Soy Protein

Beristain-Bauza et al. (2016) evaluated the antimicrobial and physical properties of whey protein isolate or calcium caseinate films added with cell-free supernatant of *Lactobacillus rhamnosus* NRRL B-442. Cell-free supernatant was obtained by centrifugation and filtration from cultures of *L. rhamnosus* in MRS broth, freeze-dried, and rehydrated to add 6, 12 or 18 mg/ml to protein film solution. Films' inhibition zones were determined by agar disk diffusion assay against tested microorganisms. Noticeable antimicrobial activity (about 3 mm) was observed against *E. coli, L. monocytogenes, S. aureus* or *S. Typhimurium* when 18 mg/ml of cell-free supernatant were added. Gradual and significant ($p < 0.05$) increments were obtained in thickness, colour and solubility when different increasing concentrations of supernatant were added. Reductions in WVP and puncture strength were observed when adding supernatant. Antimicrobial films were brown coloured (Fig. 1) and with variable moisture contents. However, the WVP and mechanical properties of the films resulted good enough to be used as wrappings or coatings. On the other hand, heat treatment had an important adverse effect on antimicrobial activity of the films, thus severe heat treatments should be avoided. WPI films were significantly ($p < 0.05$) better than CC films for most evaluated properties. Protein films added with cell-free supernatant are an interesting alternative to develop natural antimicrobial films that can have a potential food application as wrappings or coatings (Beristain-Bauza et al. 2016).

Owing to its sustainability, abundance, low cost and functionality, soy protein (SP) has attracted great research interest for the development of environment friendly protein materials with potentially good properties, such as regeneration, biocompatibility, biodegradability, etc. (Garrido et al. 2013; Echeverría et al. 2014; Félix et al. 2014; Silva et al. 2014; Gupta and Nayak 2015). Soy protein is extracted from soybean seeds that are used to obtain soy oil and can be used for food packaging or edible films since it meets food grade standards (Garrido et al. 2013). SPs are composed of a mixture of albumins and globulins, 90% of which are storage proteins with globular structure, consisting mainly in 7S (β-conglycinin) and 11S (glycinin) globulins (Ciannamea et al. 2014; Hsieh et al. 2014; Nishinari et al. 2014; Yuan et al. 2014). Globulins are protein fractions in which the subunits are associated via hydrophobic, hydrogen bonding and disulphide bonds. Soy proteins consist of 20 different amino acids, including lysine, leucine, phenylalanine, tyrosine, aspartic, glutamic acid, etc. (Félix et al. 2014). Soy protein is commercially available in three different forms from soybean processing plants, namely, soy flour (SF, 56% protein, ~34% carbohydrates), soy protein concentrate (SPC, 65-72% protein, >18% carbohydrates) and soy protein isolate (SPI, ≥ 90% protein) (Saenghirunwattana et al. 2014). SF is made by grinding soybeans into a fine powder; SPC is prepared by eluting soluble components from defatted soy flour while SPI is a highly refined or purified form of soy protein made from defatted soy flour which has had most of the non-protein components, fats and carbohydrates removed (Kalman 2014). In a recent study by Vasconcellos et al. (2014), antimicrobial and antioxidant properties of β-conglyicinin and glycinin from SPI were analyzed and it was found that glycinin peptides have better activity against almost all microbial strain; however β-conglycinin and glycinin peptides have equivalent microbial activity against *Escherichia coli*. Assays showed that antioxidant activity was dependent on the concentration of glycinin and β-conglycinin peptides. Moreover, SPI has shown advantages over other sources due to its exceptional film-forming properties and low cost (for its extensive production in several countries and for being isolated from oil industry waste). Chemical treatments can be used to modify properties to produce films with good physicochemical characteristics and capability to act as carriers for active compounds. Arancibia et al. (2014) developed active biodegradable bilayer films based on soy protein isolate, lignin and formaldehyde. These films showed high water resistance and malleability, making them suitable for use in extreme environmental conditions. The presence of lignin in the formulation provided greater protection against light, especially UV. With the addition of 3% w/w citronella essential oil, the films showed good antifungal activity against *Fusarium oxysporum* pathogen microorganism in bananas. When bananas were covered by the films during storage, there was a noticeable reduction in total aerobic mesophiles and moulds and yeasts. As the films aged (1, 3, 6 months), in both controlled and ambient conditions, the main active compounds in the citronella essential oil, citronellal and geraniol, were progressively released from the film matrix, citronellal more than geraniol. Over 30 days of soil degradation, the films lost around 30% of weight irrespective of the conditioning time.

González and Alvarez Igarzabal (2013) developed active biodegradable bilayer films from SPI and polylactic acid (PLA) without adding compatibilizing agents, adhesives or without chemically modifying film surfaces. It was found that the addition of PLA markedly improved fundamental physical and mechanical properties of the films. The films were evaluated as active packaging by incorporation of an antifungal and an antibacterial agent (natamycin and thymol, respectively) to the SPI layer, showing a marked growth inhibition of mould, yeast and two strains of bacteria by *in vitro* microbiological assays. Figure 2 shows a slice of soft cheese partially coated with the natamycin-containing bilayer film (SPI-PLA 60/40); an initial growth of mould on the free surface (not in contact with the coating) was observed after six days of storage at room temperature. By contrast, no presence of mould was found in the coated surface due to the inhibitory action of the active agent. These results showed that these materials may be highly suitable as a biodegradable material for active food coatings.

The physical and antioxidant properties of SPI films were investigated incorporated with chestnut bur extract (CBE) at different levels, 20, 50, 80 and 100 g/kg (based on SPI content) (Wang et al. 2016). Increased protection against UV light at wave-lengths <400 nm and oxygen barriers were observed in the SPI films containing CBE, which would help to reduce food deterioration. On the other hand, CBE contributed to limiting the light transmission of films at visible ranges. The antioxidant properties of SPI films were improved with the addition of CBE as determined by 2, 2-diphenylpicrylhydrazyl radical scavenging activities. Temperature also affected the release of CBE from SPI films: the DPPH scavenging activity of the films at 25°C and 35°C was significantly ($p < 0.05$) higher than that at 15°C. The tensile strength of SPI/CBE film reached the maximum value of 2.1 ± 0.1 MPa when 80 g/kg CBE was added, which was associated with cross-linking interactions via hydrogen bonding between the CBE and SPI, as well as the distributions of the secondary protein structure in the film network. The cross-section of the SPI/CBE films viewed using scanning electron microscopy became more compact as the CBE level increased. These results suggest that SPI/CBE film is an ideal choice for food packaging and preservation.

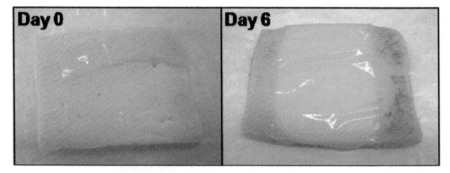

Fig. 2. Visual analysis of the behaviour of a slice of soft cheese partially coated with metamycin-containing SPI-PLA 60/40 film before and after six days of storage. From González and Alvarez Igarzabal (2013).

González and Alvarez Igarzabal (2015) obtained SPI films reinforced with starch nanocrystals (SNC) by a simple casting method. The films were transparent and homogeneous and became more resistant and less elongated as the SNC amount increased. In addition, β-cyclodextrins (β-CD)-containing SPI-SNC films were prepared by adding β-CD (14, 71 and 143% of β-CD in mass with respect to SPI) in powder to SPI-SNC dispersions (containing 5 or 20% of SNC). Assays performed demonstrated that β-CD-containing SPI-SNC films were able to sequester cholesterol when brought into contact with cholesterol-rich food such as milk (Table 2). This effect was more marked as the amount of β-CD into the films increased, and it was explained by the fact that β-CD has cavities available to form inclusion complexes with cholesterol. Furthermore, variation in the amount of the SNC added into the films (5 or 20%) did not significantly affect the cholesterol retention. These methodologies allowed yielding active biodegradable films with optimized physical and mechanical properties.

Table 2. Amount of cholesterol and its decrease in percentage (%) for milk samples after contact with the active films. From González and Alvarez Igarzabal (2015).

Films	Milk (mL)	Cholesterol (ppm)	Cholesterol decrease (%)
-	50	4.7 ± 0.1^A	-
SPI-SNC 5%	50	4.7 ± 0.2^A	-0.50
SPI-SNC 20%	50	4.6 ± 0.2^A	0.15
SPI-SNC 5%-βCD 14%	50	4.6 ± 0.4^A	1.31
SPI-SNC 20%-βCD 14%	50	4.6 ± 0.1^A	0.77
SPI-SNC 5%-βCD 71%	50	3.7 ± 0.2^B	20.10
SPI-SNC 20%-βCD 71%	50	3.5 ± 0.3^B	25.79
SPI-SNC 5%-βCD 143%	50	2.8 ± 0.1^C	40.52
SPI-SNC 20%-βCD 143%	50	2.6 ± 0.0^D	43.30

Any two means in the same column followed by the same letter are not significantly ($p \geq 0.05$) different according to the Turkey test.

Water uptake by protein films might be undesirable for certain applications. However, it could be exploited as an advantageous property aiding in the release of active principles for "in package" treatments of foods. In this sense, Ortiz et al. (2013) developed and tested soy protein biodegradable films releasing the inhibitor of ethylene action 1-methylcyclopropene (1-MCP). Soy protein pads were prepared by casting from formulations of different glycerol concentrations (20, 40 or 60% on protein basis) and pHs (2.0, 7.0 or 10.0). For this, protein films obtained with the different formulations and conditioned for two days at 58% RH were cut into squares. One milligram of Smart-Fresh® (AgroFresh, Springhouse, PA), containing 0.14% 1-MCP in a cyclodextrin matrix was weighed over each film and subsequently covered with a second protein film layer. The film surfaces were then thermo-sealed to generate 8 cm^2 protein film pads containing the 1-MCP (Fig. 3A). For each formulation, six trays with a releasing pad and covered with PVC were prepared (Fig. 3B). Based on the better mechanical properties, lower initial water content and efficacy to delay ripening, soy protein films with 20% glycerol at pH 7.0 were selected as the most suitable for 1-MCP release. These pads delayed tomato softening and pectin solubilization and reduced decay (Fig. 4) and lycopene accumulation, without causing negative quality changes in sugar content, acidity or antioxidants and thus, could be useful for postharvest "in package" treatments.

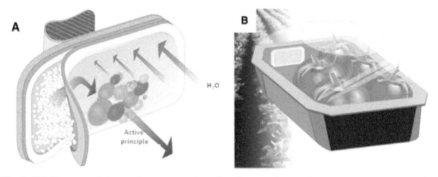

Fig 3. (A) Design of the soybean protein-releasing pad containing an active principle retained in a cyclodextrin matrix. (B) Intended use of the soybean protein-releasing pads containing a cyclodextrin matrix for "in package" treatment of tomato fruit. From Ortiz et al. (2013).

Control   1 - MCP

Fig. 4. Appearance of control or "in package" 1-MCP treated fruits after eight days of storage at 10 °C. From Ortiz et al. (2013).

Films Based on Gelatin

Gelatin is a water soluble protein derived from the partial hydrolysis of the fibrous insoluble protein, collagen. Collagen is the major protein constituent in the bones, cartilages and skins of animals and fish. Therefore, the properties of the gelatin are greatly influenced by the source, age of the animal, type of collagen and processing method (Gómez-Guillén et al. 2011). For the production of gelatin, various mammalian sources are used, such as pig skin (46%), bovine hide (29.4%) and pork and cattle bones (23.1%). Gelatin is classified into two types, depending on the processing methods: type A and type B gelatins (Arvanitoyannis 2002). Type A gelatin is obtained by acid treatment of collagen and it has an isoelectric point of 7-9, whereas type B gelatin is derived from the alkali treatment of collagen with an isoelectric point of 4-5 (Gómez-Guillén et al. 2011). Gelatin is composed of 50.5% carbon, 25.2% oxygen, 17% nitrogen and 6.8% hydrogen (Smith 1921), and it is a mixture of α-chains (one polymer/single chain), β-chains (two α-chains covalently cross-linked), and γ-chains (three α-chains covalently cross-linked) (Papon et al. 2007). Extreme pH and high temperature denature the gelatin and change its properties by disrupting its three dimensional structures and forming a random coil nature. Therefore, the processing conditions of gelatin need to be carefully controlled to get a high gelling strength. Gelatin has been widely used industrially, not only for food packaging materials, but also in various applications such as emulsifiers, cosmetics, photography, hydrogels and colloid stabilizers (Arora and Padua 2010; Shankar et al. 2010). In the food industry, gelatin is used in food packaging and food coating materials, and also as gelling, stabilizer, texturizing, and emulsifier for bakery, beverages, confectionery and dairy products (Djagny et al. 2001). Gelatin is a promising protein biopolymer, owing to its remarkable film forming property, biodegradability, abundance and cost effectiveness (Farris et al. 2011).

Ahmad et al. (2012) incorporated lemongrass essential oil to increase the shelf-life of sea bass slices stored at 4 °C. They found that the incorporation of the lemongrass in a gelatin film resulted in a delay in microbial spoilage of sea bass slices. Lemongrass oil enhanced the antimicrobial and antioxidant properties of the films, which resulted in the extended shelf-life of the products stored at a refrigerated temperature. Kavoosi et al. (2014b) prepared 10% (w/v) gelatin film incorporated with Zataria multiflora essential oil (ZMO). They found that gelatin/ZMO exhibited profound antioxidant and antimicrobial activity against both gram-positive and gram-negative bacteria. Wu et al. (2014) prepared silver carp skin gelatin-chitosan composite films blended with different concentrations of oregano essential oil (OEO). The incorporation of OEO into gelatin-chitosan film distinctively improved the antimicrobial activity oil against *E. coli, S. aureus, B. subtilis, B. enteritidis* and *Shigabacillus*. Further, Wu et al. (2014) prepared gelatin films incorporated with cinnamon essential oil (CEO) nanoliposomes, synthesized by thin film ultrasonic dispersion method. They observed that the gelatin film incorporated with CEO solution exhibited slightly higher antimicrobial activity compared with the gelatin film with CEO nanoliposomes.

Hong et al. (2009) incorporated grapefruit seed and green tea extracts in Gelidium corneum/gelatin blend films and monitored the quality of pork loin packed with the film during storage. The initial population of *E. coli* O157:H7 and *L. monocytogenes* was 5.16 and 4.80 log CFU/g, respectively, on inoculated pork loins. The populations of the pathogenic bacteria in the pork loins packed with both active blend films decreased significantly during storage, compared with the control.

Kadam et al. (2015) also incorporated *Ascophyllum nodosum* extract at different ratios (25 and 50% w/w of base material) in glycerol plasticized gelatin films obtained by casting and found that the total phenolic content and antioxidant activity and the hydrophilicity and glass transition temperature of the films increased at higher levels of the active compound.

The introduction of nanoparticles into gelatin matrices greatly enhances the overall properties, such as the biocompatibility, physical and mechanical properties, and biodegradability of gelatin-nanocomposite films (Duncan 2011). Kanmani and Rhim (2014) synthesized gelatin-based nanocomposite films reinforced with AgNPs by a solution casting method. They tested the antimicrobial activity of gelatin/AgNPs nanocomposite films against the foodborne pathogens *E. coli* O157:H7 ATCC 43895, *L. monocytogenes* ATCC 15313, *Salmonella typhimurium* ATCC 14028, *S. aureus* ATCC 29213 and *Bacillus cereus* ATCC 21366 by agar diffusion and colony count methods. It was found that the antimicrobial activity of AgNPs was in a dose-dependent manner. Additionally, gram-negative bacteria were more susceptible to AgNPs than gram-positive bacteria. The authors proposed that the higher susceptibility of gram-negative bacteria might be due to a thin peptidoglycan layer present in the cell wall of gram-negative bacteria. Gram-positive bacteria, composed of a 20-80 nm thick peptidoglycan layer, consists of linear polysaccharide chains cross-linked with short peptides. These form a complex structure that resists the AgNPs, and prevents the AGNPs from penetrating inside the cell. However, gram-negative bacteria possess a negatively charged outer membrane and a 7-8 nm thin peptidoglycan layer, which facilitates the penetration of AgNPs inside the cell (Priyadarshini et al. 2013). Further, Kanmani and Rhim (2014) prepared an antimicrobial nanocomposite of gelatin, AgNPs and organoclay by a solvent casting method. They measured the antimicrobial activity of the gelatin/AgNPs/clay nanocomposite films against *E. coli* and *L. monocytogenes* by agar diffusion and colony count methods. The gelatin film that included AgNPs showed high antimicrobial activity against both gram-positive (*L. monocytogenes*) and gram-negative (*E. coli*) bacteria, whereas the gelatin film with organoclay showed antimicrobial activity only against gram-positive bacteria. The combined addition of nanoclay and AgNPs into the gelatin film exhibited more pronounced antimicrobial activity against *L. monocytogenes*.

Nafchi et al. (2014) prepared a gelatin/ZnO-nanorod (nr) nanocomposite film by incorporating various concentrations of ZnO-nr into bovine gelatin matrices. They determined the antimicrobial activity of gelatin/ZnO-nr film against *S. aureus* by measuring the inhibition zone on the solid media and found that the inhibition zones of the nano-incorporated films increased significantly with an increase in ZnO-nr content. Shankar et al. (2015b) prepared gelatin-based ZnO nanocomposite films with various shapes and sizes of ZnO nanoparticles (NPs) synthesized from two different zinc salts (zinc nitrate and zinc acetate) with or without a capping agent (carboxymethyl cellulose). They examined the antimicrobial activity of gelatin/ZnO nanocomposite films against *L. monocytogenes* and *E. coli* using colony count methods and found that the log CFU/mL value of *L. monocytogenes* decreased from 6.5 to 2 in 6 h when incubated with gelatin/ZnO nanocomposite films, while in the control film (neat gelatin film), it increased from 6.5 to 8.8 in 6 h. However, against *E. coli*, gelatin/ZnO nanocomposite films were less active and the log CFU values decreased from 7.2 to 5.2 in 12 h. In contrast, in the control film, the values increased from 7.2 to 9 in 12 h. Further, Shankar et al. (2014b) tested the effects of ZnO NPs concentration on the properties of gelatin/ZnO NPs composite films. They found that the antimicrobial activity of gelatin/ZnO NPs against *L. monocytogenes* and *E. coli* was dependent on the ZnO NPs concentration. At low concentrations of less than 0.5 wt% of ZnO NPs, the films showed only a bacteriostatic effect, but they exhibited a distinctive bactericidal effect above this concentration. Arfat et al. (2014) prepared composite films based on a fish skin gelatin (FSG) and fish protein isolates (FPI) blend, incorporated with 50% and 100% (w/w, protein) basil leaf essential oil (BEO) in the absence and presence of 3% (w/w, protein) ZnO NPs. They found that the FPI/FSG blend films incorporated with 100% BEO in combination with ZnO NP exhibited strong antibacterial activity against foodborne pathogenic and spoilage bacteria. All the mentioned films could be used as active food packaging materials to ensure food safety and extend the shelf-life of packaged foods. Some proposed mechanisms of action of ZnO NPs against bacteria are reported, but the clear mechanism for the antimicrobial activity of ZnO NPs has not been established yet. The ZnO NPs may release Zn^{2+} ions, which could penetrate through the cell wall of bacteria and react to the cytoplasmic content and kill bacteria. ZnO NPs have been known to generate a strong oxidizing agent, hydrogen peroxide (H_2O_2), which damages the cell membrane of bacteria (Tayel et al. 2011). Zhang et al. (2010) suggested that ZnO NPs are able to slow down bacterial growth due to the disorganization of the bacterial membrane, which causes increased membrane permeability and leads to the accumulation of nanoparticles in the bacterial membrane and cytoplasmic regions of the cells. It has been believed that both production of reactive oxygen species and accumulation or deposition of ZnO NPs within the cytoplasm or on the surface of bacteria lead to either inhibition or killing of bacterial cells. However, further studies are required to clarify the exact mechanism of antibacterial action of ZnO NPs.

Kavoosi et al. (2014a) prepared 10 wt% gelatin film reinforced with different percentages of multi-walled carbon nanotubes (MWCN, 0.5, 1, 1.5

and 2% w/w gelatin) and found that all gelatin/MWCNT films exhibited significant antibacterial activities against gram-positive and gram-negative bacteria. MWCNTs are relatively flexible, interact with cell membranes and penetrate various microorganisms and consequently cause cell death (Meredith et al. 2013).

López-Carballo et al. (2008) prepared gelatin film that was reinforced with photosensitizer, sodium magnesium chlorophyllin (E-140) and sodium copper chlorophyllin (E-141). They tested the antimicrobial activity of chlorophyllin-gelatin films against *S. aureus* and *L. monocytogenes* at different settings of light intensities and time periods. The cell viability of both microorganisms decreased as the light intensity and the exposure time increased. The chlorophyllin E-140 exerted higher antimicrobial activity than chlorophyllin E-141. The difference in antimicrobial activity might be due to the greater ability of generating singlet oxygen by E-140 than E-141; the exposure time plays a key role in the photodynamic effect. Ma et al. (2013) prepared a gelatin film using genipin as a cross-linker and studied the lysozyme release kinetics from the film. The release of lysozyme was dose-dependent and decreased as the genipin concentration increased, leading to slow release of lysozyme and extending the antimicrobial effects during food storage. Moreover, the release was retarded at pH 7.0, while it was rapidly released from the films at pH 3.8.

Films Based on Corn Zein Protein

The main advantage of working with zein comes from its compatibility with different natural antimicrobial compounds that are mostly classified as bio-preservatives. This is because of the hydrophobicity of zein, which aids in the formation of an inert film matrix that does not interact with the incorporated bio-preservatives, which are mostly hydrophilic. This helps maintaining the solubility and activity of the incorporated bio-preservatives in the film. It also enables the use of zein as a reservoir for different bio-preservatives to achieve critical microbial inhibitory concentrations at the food surface easily.

Lysozyme is the antimicrobial candidate with the highest potential for use in active zein packaging. The extreme stability of this GRAS status agent, both in ethanolic zein film making solutions and in dried pre-cast zein films kept under refrigeration, has been reported by Mecitoglu et al. (2006). These authors reported that the ethanol that is used in the preparation of zein films causes the activation of the enzyme and increases its activity up to threefold. Moreover, recent findings of Ünalan et al. (2013), who employed lysozyme-containing zein films and zein-based composite films, showed a great potential of such films to suppress the growth of *L. monocytogenes* in fresh cheeses. Zein films have a very fast release profile for most bio-preservatives, including lysozyme and phenolic compounds (Mecitoglu et al. 2006; Ünalan et al. 2013; Arcan and Yemenicioğlu 2014). Zein films have a porous structure and it still has some hydrophilic and amphiphilic constituents that cause its limited swelling. This is a very important problem when zein films incorporate highly hydrophilic lysozyme, which rapidly diffuses from films that show swelling. The incorporation of waxes, such as beeswax, carnauba wax or candelilla wax into zein film serves to increase film hydrophobicity and to reduce film swelling in aqueous media. At the same time, it will increase film tortuosity and could help reducing antimicrobial's diffusion coefficients (Ünalan et al. 2013; Arcan and Yemenicioğlu 2013). Ünalan et al. (2013) reported that the use of zein-wax composite films is highly effective in sustaining lysozyme release rates. However, these authors also noted that the beneficial effects of composites on sustained lysozyme release become sound when films are plasticized using catechin. The catechin is not only an effective plasticizer for the zein, but it also improves sustained release properties of zein films by reducing their pore size (Ünalan et al. 2013; Arcan and Yemenicioğlu 2013). In contrast, the use of low molecular weight phenolic acids like gallic, ferulic and p-hydroxybenzoic acids in plasticization of zein films increases the pore size or number of pores of films and impairs their sustained antimicrobial release properties (Alkan et al. 2011; Arcan and Yemenicioğlu 2013). The type of wax used in composite materials also affects the sustained release properties of films. Waxes that have low melting points (MP) are more easily mixed with zein and more homogenously distributed within the zein film matrix. Thus, a wax with a very low MP impairs the composite structure by reducing the amount of tiny wax particles and aggregates within the films. This causes a reduction in film tortuosity that has a major impact on the sustained release of antimicrobials. The effect of the MP of waxes on the sustained release profiles of zein-wax composites was clearly observed by Arcan and Yemenicioğlu (2011). These authors reported that the use of beeswax (MP: 62-66°C) in zein composite films, instead of candelilla wax (MP: 68.5-72.5°C), causes a 1.7-fold increase in the release rate of lysozyme. However, both composite films still showed 1.8-2.5-fold lower lysozyme release rates than zein control films (Arcan and Yemenicioğlu 2011). Ünalan et al. (2013) tested the antimicrobial and antioxidant potential of zein and zein-wax composite films, which have different release profiles, for lysozyme and mixtures of lysozyme, catechin and gallic acid. The films were tested on cold-stored fresh Kashar cheese that was inoculated with *L. monocytogenes*. The authors reported that all lysozyme-containing films prevented the increase of *L. monocytogenes* counts in Kashar cheese for eight weeks at 4 °C. However, only the zein-wax composite films, with sustained lysozyme release rates, caused a significant reduction in the initial microbial load of the inoculated cheese samples. Another alternative to control the release profiles of antimicrobials from edible zein films consists in adding different concentrations of suitable fatty acids into zein film forming solutions. Arcan and Yemenicioğlu (2011) reported that the hydrophobicity and morphology of zein films can successfully be modified by blending oleic, linoleic or lauric acids at 10% (w/w) in the presence of lecithin as emulsifier. These authors determined that the zein blend films show 2-8.5 and 1.6-2.9-fold lower initial release rates for the model active compounds, lysozyme and catechin, than the zein control films, respectively. Moreover, they reported that the increase of fatty acid chain length reduces the release rates of active compounds considerably. The controlled release properties of zein-fatty acid blend films are attributed mainly to the microspheres formed within their film matrix and the encapsulation of active compounds (Wang et al. 2008; Arcan and Yemenicioğlu 2013, 2014).

GRAS antimicrobial peptides like nisin and polylysin have also been successfully tested in zein film systems (Ünalan et al. 2011). Edible zein coatings incorporating nisin (54.4 AU/cm^2) or nisin/ethylene diamine tetra acetic acid (568 μg/cm^2) controlled the growth of microbial load (less than 1-log increase) in fish balls better as compared to the control, which showed a 3-log increase during a 15-day refrigeration storage period (Ünalan et al. 2011). Janes et al. (2002) found that a zein coating with nisin and calcium propionate was able to prevent growth (at 4°C) of *L. monocytogenes* inoculated on cooked chicken breast meat with an inoculum level of 3 log/g; however, the combination was not able to prevent growth at an inoculum level of 3 or 6 log/g if the incubation temperature was 8 °C. Lungu and Johnson (2005) demonstrated that using nisin alone or in combination with zein, ethanol-glycerol or propylene glycol showed potential as a barrier against the growth of *L. monocytogenes* on frankfurters at 4 °C. They found that initial counts decreased for all the treatments containing nisin and no cells were detected for the low inoculum test (4 log) by day 21. For a high inoculum level (6 log), bacterial counts in the treatment by nisin alone were lower than in the control by 6.1 logs over 28 days. In a recent study, Sudağidan and Yemenicioğlu (2012) showed that the presence of 25 μg/ml nisin is sufficient to inactivate all of the 25 *S. aureus* strains isolated from raw milk and cheese samples. The active packaging conducted by zein films that incorporate nisin could be employed as an important part of a hurdle to reduce risks associated not only with cheese obtained from heated milk, but also from traditional cheeses which are still produced locally from unheated milk.

Moreover, to overcome bacterial resistance problems, nisin can be combined with lysozyme. It has been reported that this combination exhibits synergy against Gram-positive bacteria, including pathogenic ones like *S. aureus* (Chung and Hancock 2000; Sobrino-López and Martin-Belloso 2008). Gill and Holley (2000) also employed this strategy in bologna sausages and reduced the growth of *L. monocytogenes* during a two-week period. The combined application of nisin and lysozyme also gave promising results against *L. monocytogenes* in ready-to-eat seafood products as minced tuna and salmon roe (Takahashi et al. 2012) and in ready-to-eat turkey bologna (Mangalassary et al. 2008). Both nisin and lysozyme are highly compatible with the zein film system; thus, combined incorporation of nisin and lysozyme into zein films might be an alternative option to improve effectiveness of antimicrobial zein packaging.

Ünalan et al. (2011) attributed the good antimicrobial potential of ε-polylysine in the zein film system to its hydrophobic nature, and the limited charged groups of zein that prevent the complexation and immobilization of ε-polylysine within the film matrix by charge-charge interactions.

Zein also provides an excellent opportunity to use pure phenolic compounds, phenolic extracts and phenolic-rich essential oils in active packaging, because its films are prepared in ethanol, an effective solvent for most phenolic compounds. The incorporation of phenolic compounds such as catechin, gallic acid, p-hydroxy benzoic acid and ferulic acid at 3 mg/cm^2 eliminated the brittleness problem of zein films and increased their flexibility considerably. Films containing phenolic compounds demonstrated antioxidant activity, and films containing gallic acid showed *in vitro* antimicrobial activity against *L. monocytogenes* and *Campylobacter jejuni* (Arcan and Yemenicioğlu 2011). Thus, the incorporation of phenolic compounds offers a new avenue for the use of zein films as flexible bioactive packaging materials. However, Arcan and Yemenicioglu (2011) also reported that the phenolic compounds, particularly hydroxyl cinamic acids like gallic acid and ferulic acid, hydroxybenzoic acids like p-hydroxy benzoic acid and flavonoids like (+) catechin, act as natural plasticizers for the zein film matrix, since their hydroxyl groups form hydrogen bonds with the biopolymer to increase the free volume of the film matrix (Sothornvit and Krochta 2005). The hydrophilic hydroxyl groups of phenolic compounds also decrease the hydrophobic interaction among zein molecules, which contributes to their increased mobility and flexibility (Alkan et al. 2011; Arcan and Yemenicioğlu 2011). On the other hand, the effects of phenolic compounds on structural properties of zein films show great variation. For example, catechin reduces the zein film porosity, while gallic acid reduces pore size, but increases the number of pores, and p-hydroxy benzoic acid and ferulic acid mainly increase pore size (Arcan and Yemenicioğlu 2011). These changes in film morphology gain a particular importance when the controlled release of the active substance is a critical factor to obtain a benefit from active packaging.

Regarding essential oils, thymol has been extensively tested in zein film system and its antimicrobial effectiveness on important pathogenic bacteria, yeast and mould and release kinetics from these films have been studied in detail (Del Nobile et al. 2008; Gutierrez et al. 2009; Park et al. 2012). Park et al. (2012) employed eugenol, thymol and carvacrol incorporated in zein films for lamination of low-density polyethylene films intended for antioxidant packaging. Khalil and Deraz (2015) employed eugenol to develop antimicrobial zein films and to improve their mechanical properties.

Films Based on Wheat Gluten Protein (WG)

Among the proteins, wheat gluten (high protein content, >75 wt%) is one of the most frequently studied proteins in edible films due to its interesting viscoelastic properties, good filming properties, ability to cross-link upon heating, low water solubility, low cost and availability as a by-product from the wheat starch industry (Cho et al. 2010; Jansens et al. 2013; Zubeldía et al. 2015). Moreover, wheat gluten films can be obtained by thermoplastic processing, which consists of mixing proteins and plasticizer by a combination of heat and shear followed by an additional stage involving further thermo-mechanical treatments (e.g. compression moulding) (Pommet et al. 2005; Sun et al. 2008), which, from economical and environmental viewpoints, is the most viable way to produce rigid gluten-based materials since it is fast and requires no solvent (Gaellstedt et al. 2004; Jansens et al. 2011). However, only a few studies have been reported on the development of antimicrobial and antioxidant biodegradable films using thermoplastic processes (Dawson et al. 2002), probably because of the inactivation of the incorporated active agents due to the high temperature and pressure associated with these processing methods (Del Nobile et al. 2009).

Türe et al. (2012) reported a successful system for wheat gluten containing potassium sorbate by compression moulding without any loss of antimicrobial properties. It was the first study on thermoplastically produced antimicrobial wheat gluten materials. The antimicrobial efficiency was tested against *Aspergillus niger* and *Fusarium incarnatum* by the agar diffusion assay. The results indicated that films containing more than 10 wt.% PS showed antimicrobial activity against *A. niger* while films containing 2.5 wt.% or more of PS showed antimicrobial activity against *F. incarnatum*. Moreover, most of the PS was released when the film was exposed to the agar solution, which is considered as an interesting feature for edible active packaging. What is more, without seeding of spores, the films resisted microbial growth for at least one week when left in the agar solution. X-ray diffraction and field emission scanning electron microscopy revealed that the PS crystals were dissolved in the wheat gluten material and thus, they acted as plasticizer in the wheat gluten film.

Ansorena et al. (2016) prepared active films based on glycerol-plasticized wheat gluten protein containing different thyme oil (TO) concentrations (0-15 wt.%) by a thermoplastic process involving relatively high temperature and pressure. The addition of the oil leads to heterogeneous films containing hydrophobic discontinuities that reduce the mechanical performance in terms of strength and modulus, but enhance their flexibility. Samples containing lower amounts of TO presented larger WVP and equilibrium moisture content values than control sample, but they showed the opposite behaviour at higher concentrations. Antimicrobial *in vitro* results indicated that at least 10 wt.% TO was necessary to obtain a significant and clear inhibition zone in the case of Gram-positive bacteria and only films containing 15 wt.% TO showed significant antimicrobial activity against both types of bacteria and native microflora of two selected vegetables. Most of the thymol and carvacrol added remained in the formulations after processing, which resulted in a significant antioxidant activity, as indicated by the high percentage of inhibition obtained using DPPH. El-Wakil et al. (2015) developed bio-nanocomposite films consisting of WG, glycerol, cellulose nanocrystals (CNC) and TiO_2 nanoparticles by casting/evaporation. Optimal CNC (7.5%) and TiO_2 nanoparticles (0.6%) contents were established for improving the functional properties of WG-based materials on the basis of tensile strength and water resistance of the bio-nanocomposites. Moreover, coated kraftpaper with WG/CNC 7.5%/0.6% TiO_2 exhibited excellent antimicrobial activities that is, 100, 100 and 98.5% against *S. cervisiae*, *E. coli* and *S. aureus*, respectively, for three layers coated paper after 2 h of exposure to UVA light illumination.

Films Based on Fish Protein

In the seafood processing industry, a substantial amount of by-products are generated, that can be used to recover proteins to prepare films and restructured seafood products. Protein films have been successfully prepared using fish proteins, including myofibrillar and sarcoplasmic proteins (Benjakul et al. 2008). Myofibrillar protein, the main protein component in fish muscle, has excellent film-forming capacities under acidic and alkaline conditions (Shiku et al. 2004; Chinabhark et al. 2007). Fish protein isolate (FPI) prepared by prior washing followed by alkaline solubilisation has been shown as a promising starting material with lower haem protein and lipid contents, leading to the improved mechanical and physical properties of film with negligible discolouration (Tongnuanchan et al. 2011).

Hake proteins recovered from by-products of seafood processing industries were used by Teixeira et al. (2014) for the preparation of biodegradable films. The incorporation of essential oils (garlic, clove and origanum) in these films reduced film thickness as well as the solubility in water, affected film mechanical properties (breaking force and elongation) and improved antioxidant activity. Clove films showed the lowest water vapour permeability and the highest antibacterial activity (against *S. putrefaciens*); garlic films were the most yellowish and had the highest antioxidant activity while origanum films were rather similar to control films.

Sliver carp (Hypophthalmichthys molitrix Val.) is one of the main freshwater fish species in China, still underutilized due to the muddy flavour and tiny intramuscular bones. Therefore, using sliver carp myofibrillar protein to make edible films is an attractive option. In this sense, Nie et al. (2015) prepared myofibrillar protein-based films incorporated with grape seed procyanidins (GSPC) and green tea polyphenol (GTP). Incorporation of GSPC and GTP markedly decreased water solubility, water vapour permeability and elongation at break, whereas increased tensile strength as well as moisture content. Loss in protein solubility and decrease in protein pattern suggested the formation of covalent-crosslinking between phenol and protein molecules.

Other Novel Active Systems

Fabra et al. (2016) used an electro-hydrodynamic processing as a novel route to develop active packaging systems for food packaging applications. Specifically, alpha-tocopherol, as a model antioxidant, was encapsulated in three different hydrocolloid matrices (WPI, SPI and zein) which were directly applied as coatings onto a thermoplastic wheat gluten film. The antioxidant activity of the active compound was preserved during the encapsulation process (up to 95%, Table 3) and, interestingly, the active coatings improved the barrier properties of the bilayer systems. Furthermore, it was shown that the hydrocolloid matrices were able to protect the alpha-tocopherol from degradation during a typical sterilization process (more than 70-85% depending on the type of coating) and, thus, this type of coatings could be used to increase the shelf life of the alpha-tocopherol when incorporated within packaging systems. The release in aqueous media slightly changes depending on the encapsulation matrix used, which could be explained by the different chemical and morphological characteristics of the matrices used. Therefore, this proof-of-concept study demonstrates that hybrid structures of hydrocolloids containing active compounds can play a major role in the design of novel multifunctional materials for food packaging. For instance, one of the potential applications of these types of materials could be a food contact layer of a multilayer film which could act as a package for liquid products (juice, milk, etc). In this case, the most appropriate terminology will be bioactive since the alpha-tocopherol (a hydrophobic vitamin) will be released into the liquid improving health benefits while avoiding lipid oxidation.

Wang et al. (2015) prepared ternary blends agar/alginate/collagen (A/A/C) hydrogel films with silver nanoparticles (AgNPs) and grapefruit seed extract (GSE). The A/A/C film was highly transparent, and both AgNPs and GSE incorporated ternary blend composite films exhibited high UV screening function with improved mechanical and barrier properties. Moreover, the A/A/C blend films formed efficient hydrogel film with the water holding capacity of 23.6 times of their weight. Both A/A/CAgNPs and A/A/CGSE composite films exhibited strong antimicrobial activity against both Gram-positive (*Listeria monocytogenes*) and Gram-negative (*E. coli*) food-borne pathogenic bacteria (Fig. 5). The test results of fresh potatoes packaging revealed that all the A/A/C ternary blend films prevented forming of condensed water on the packaged film surface and both A/A/CAgNPs and and A/A/CGSE composite films prevented greening of potatoes during storage. The results indicate that the ternary blend hydrogel films incorporated with AgNPs or GSE can be used not only as antifogging packaging films for highly respiring fresh agriculture produce, but also as an active food packaging system by means of their strong antimicrobial activity.

Table 3. Antioxidant activity, according to the ABTS methodology, of electrospun non-sterilized coatings with and without the antioxidant, expressed as % of inhibition of alpha-tocopherol·mg antioxidant⁻¹ and the corresponding μM Trolox mg antioxidant⁻¹. Antioxidant activity of bilayer films prepared without alpha-tocopherol was also carried out for comparative purposes, and expressed as % inhibition of alpha-tocopherol·mg antioxidant⁻¹. The encapsulation efficiency (EE, %) estimated from the antioxidant activity studies of non-sterilized samples (μM Trolox mg antioxidant⁻¹) when they were compared to the neat alpha-tocopherol, is also included in this table. From Fabra et al. (2016).

Sample	μM Trolox mg coating⁻¹		μM Trolox mg antioxidant⁻¹		% inhibition mg antioxidant⁻¹		EE
	Non-sterilized samples	Sterilized samples	Non-sterilized samples	Sterilized samples	Non-sterilized samples	Sterilized samples	
α-tocopherol			62.53 (1.5)[a1]	21.49 (2.5)[a2]	3.65 (0.15)[a1]	1.48 (0.40)[a2]	
WPI	0.20 (0.03)[a]						
WPI-T	18.67 (1.35)[b1]	11.29 (1.02)[a2]	59.75 (2.0)[b1]	40.11 (2.8)[b2]	3.29 (0.16)[b1]	2.25 (0.18)[a2]	95%
SPI-GG	0.23 (0.02)[a]						
SPI-GG-T	20.16 (1.05)[b]	14.01 (1.22)[ab2]	60.49 (1.2)[ab1]	42.02 (2.2)[b2]	3.32 (0.12)[b1]	2.35 (0.15)[a2]	96%
Zein	0.23 (0.04)[a]						
Zein-T	20.39 (1.55)[b]	17.45 (2.05)[b2]	63.61 (1.8)[a1]	54.46 (1.9)[a1]	3.53 (0.15)[ab1]	3.08 (0.21)[b2]	100%

Different letters (a-c) within the same column indicate significant differences among antioxidant activity ($p < 0.05$) due to encapsulation matrices.
Different numbers (1-2) within the same file indicate significant differences among antioxidant activity ($p < 0.05$) due to sterilization process.

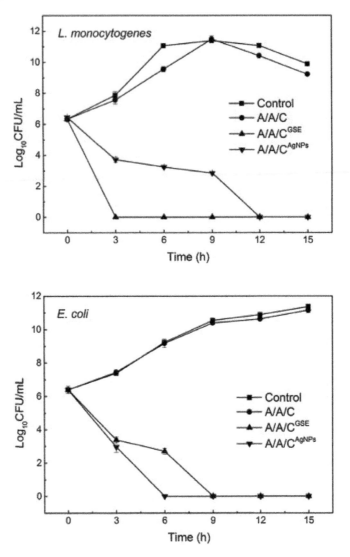

Fig. 5. Antimicrobial activity of agar/alginate/collagen ternary blend
films against foodborne pathogenic bacteria, *L. monocytogenes*
and *E. coli*. From Wang et al. (2015).

Coatings are mostly continuous layers formed on the base packaging
materials. Various novel properties of nanocoating materials, such as optical,
mechanical, chemical, electronic, magnetic and thermal properties, are used
industrially. A variety of methods, such as physical vapour deposition,
chemical vapour deposition, electronic precipitation/electronic coating,
sol-gel process, electrodeposition, rotating coating, spray coating and self-
assembling methods, have been used for the production of nano-thin films
or nanocoatings (Aliofkhazraei 2011). The antimicrobial activity of silver

nanoparticles, ZnO nanoparticles and TiO_2 nanoparticles could be exploited to develop noncytotoxic coatings using gelatin as a matrix. Such biocompatible antimicrobial polymeric films containing antimicrobial activity have high potential for application in antimicrobial active food packaging systems. Self-cleaning smart nanocoatings that destroy bacteria, isolate pathogens or fluoresce under certain conditions are under development (Carneiro et al. 2011).

Final Remarks

The previous sections demonstrated that there is a vast variety of active compounds with the potential to improve functional properties of protein-based films, in particular antimicrobial and antioxidant ones. In addition and due to their intrinsic properties, protein films are suitable alternatives to develop natural active films that can have potential food applications as wrappings, coatings or other novel preservation packaging systems. All the active systems discussed in this chapter could have applications in packaging in a wide range of food products, particularly those that are highly oxidative and microbial-sensitive, although further investigations are still needed to test their effectiveness on selected food systems. Moreover, there are still some problems to overcome, for example, many active agents are thermally sensitive and, thus, their incorporation during typical processing methods used for polymeric materials as casting or melt-blending results in the evaporation of most of the agent during film formation. Therefore, encapsulation techniques ranging from spray drying or coarcervation, to the electro-hydrodynamic process (also called electrospinning) that are also simple and straightforward methods of generating submicron encapsulation structures for a variety of bioactive molecules could be further tested to produce more efficient active protein based packaging systems.

Additionally, the controlled release of the active substance is a critical factor to obtain a benefit from the active packaging. The antimicrobial packaging mainly targets the food surface on which microbiological changes occur most intensively. However, a sufficient antimicrobial effect could not be achieved unless the release rate of antimicrobial compounds from the packaging materials to the food surface is adjusted considering several complex factors. Some of these complex factors include the physical and chemical properties of the food, the growth kinetics of target pathogenic or spoilage microorganisms and the expected shelf life of the food. This is a multifaceted problem that was not completely solved yet and thus still deserves too much research.

Acknowledgements

The authors thank the National Research Council of Republic Argentina (CONICET), the Science and Technology National Promotion Agency

(ANPCyT) and the National University of Mar del Plata (UNMdP) for the financial support.

References

Ahmad, M., S. Benjakul, P. Sumpavapol and N.P. Nirmal. 2012. Quality changes of sea bass slices wrapped with gelatin film incorporated with lemongrass essential oil. Int. J. Food Microbiol. 155: 171–178.

Aliheidari, N., M. Fazaeli, R. Ahmadi, M. Ghasemlou and Z. Emam-Djomeh. 2013. Comparative evaluation on fatty acid and Matricaria recutita essential oil incorporated into casein-based film. Int. J. Biol Macromol. 56: 69–75.

Aliofkhazraei, M. 2011. Synthesis, processing and application of nanostructured coatings. pp. 1–28. *In*: Nanocoatings: Size Effects in Nanostructured Films. Springer-Verlag. Berlin, Germany.

Alkan, D., L.Y. Aydemir, I. Arcan, H. Yavuzdurmaz, H.I. Atabay, C. Ceylan, et al. 2011. Development of flexible antimicrobial packaging materials against Campylobacter jejuni by incorporation of gallic acid into zein-based films. J. Agric. Food Chem. 59: 11003–11010.

Andrade, R.G., L.T. Dalvi, J.M.C. Silva, G.K.B. Lopes, A. Alonso and M. Hermes Lima. 2005. The antioxidant effect of tannic acid on the in vitro coppermediated formation of free radicals. Arch. Biochem. Biophys. 437: 1–9.

Ansorena, M.R., F. Zubeldía and N.E. Marcovich. 2016. Active wheat films obtained by thermoplastic processing. Food Sci. Technol. 69: 47–54.

Arancibia, M., A. Rabossi, P.A. Bochicchio, S. Moreno, M.E. López-Caballero, M. del C. Gómez-Guillén, et al. 2013. Biodegradable films containing clove or citronella essential oils against the Mediterranean fruit fly Ceratitis capitate (Diptera: Tephritidae). J. Agr. Food Technol. 3: 1–7.

Arancibia, M.Y., M.E. López-Caballero, M.C. Gómez-Guillén and P. Montero. 2014. Release of volatile compounds and biodegradability of active soy protein lignin blend films with added citronella essential oil. Food Control 44: 7–15.

Arcan, I. and A. Yemenicioğlu. 2011. Incorporating phenolic compounds opens a new perspective to use zein films as flexible bioactive packaging materials. Food Res. Int. 44: 550–556.

Arcan, I. and A. Yemenicioğlu. 2013. Development of flexible zein-wax composite and zein-fatty acid blend films for controlled release of lysozyme. Food Res. Int. 51: 208–216.

Arcan, I. and A. Yemenicioğlu. 2014. Controlled release properties of zein-fatty acid blend films for multiple bioactive compounds. J. Agric. Food Chem. 62: 8238–8246.

Arfat, Y.A., S. Benjakul, T. Prodpran, P. Sumpavapol and P. Songtipya. 2014. Properties and antimicrobial activity of fish protein isolate/fish skin gelatin film containing basil leaf essential oil and zinc oxide nanoparticles. Food Hydrocolloid. 41: 265–273.

Arora, A. and G.W. Padua. 2010. Review: Nanocomposites in food packaging. J. Food Sci. 75: R43–R49.

Arrieta, M.P., M.A. Peltzer, M. del C. Garrigó and A. Jiménez. 2013. Structure and mechanical properties of sodium and calcium caseinate edible active films with carvacrol. J. Food Eng. 114: 486–494.

Arrieta, M.P., M.A. Peltzer, J. López, M. del C. Garrigós, A.J.M. Valente and A. Jiménez. 2014. Functional properties of sodium and calcium caseinate antimicrobial active films containing carvacrol. J. Food Eng. 12: 94–101.

Arvanitoyannis, I.S. 2002. Formation and properties of collagen and gelatin films and coatings. pp. 275–304. *In*: Gennadios, A. (ed.). Protein-Based Films and Coatings. CRC Press LLC, New York.

Atares, L., J. Bonilla and A. Chiralt. 2010. Characterization of sodium caseinate-based edible films incorporated with cinnamon or ginger essential oils. J. Food Eng. 100: 678–687.

Azevedo, V.M., M.V. Dias, S.V. Borges, A.L.R. Costa, E.K. Silva, É.A.A. Medeiros, et al. 2015. Development of whey protein isolate bio-nanocomposites: Effect of montmorillonite and citric acid on structural, thermal, morphological and mechanical properties. Food Hydrocolloid. 48: 179–188.

Bai, H., J. Xu, P. Liao and X. Liu. 2013. Mechanical and water barrier properties of soy protein isolate film incorporated with gelatin. J. Plast. Film Sheet. 29: 174–188.

Barba, C., A. Eguinoa and J.I. Maté. 2015. Preparation and characterization of β-cyclodextrin inclusion complexes as a tool of a controlled antimicrobial release in whey protein edible films. LWT-Food Sci.Technol. 64: 1362–1369.

Barreira, J., I. Ferreira, M. Oliveira and J. Pereira. 2008. Antioxidant activities of the extracts from chestnut flower, leaf, skins and fruit. Food Chemistry 107(3): 1106–1113.

Belyamani, I., F. Prochazka and G. Assezat. 2014. Production and characterization of sodium caseinate edible films made by blown-film extrusion. J. Food Eng. 121: 39–47.

Benarfa, A., L. Presiozi-Belloy, P. Chalier and N. Gontard. 2007. Antimicrobial paper based on soy protein isolate or modified starch coating including carvacrol and cinnamaldehyde. J. Agr. Food Chem. 55: 2155–2162.

Benjakul, S., A. Artharn and T. Prodpran. 2008. Properties of protein-based film from round scad (Decapterus maruadsi) muscle as influenced by fish quality. LWT-Food Sci. Technol. 41: 753–763.

Beristain-Bauza, S.C., E. Mani-López, E. Palou and A. López-Malo. 2016. Antimicrobial activity and physical properties of protein films added with cell-free supernatant of Lactobacillus rhamnosus. Food Control. 62: 44–51.

Borkow, G. and J. Gabbay. 2009. Copper, an ancient remedy returning to fight microbial, fungal and viral infections. Curr. Chem. Biol. 3: 272–278.

Bourlieu, C., V. Guillard, B. Valles-Pamies, S. Guilbert and N. Gontard. 2009. Edible moisture barriers: How to assess of their potential and limits in food products shelf-life extension? Crit. Rev. Food Sci. Nutr. 49: 474–499.

Bucci, D.Z., L.B.B. Tavares and I. Sell. 2005. PHB packaging for the storage of food products. Polym. Test. 5: 564–571.

Burt, S.A. and R.D. Reinders. 2003. Antibacterial activity of selected plant essential oils against Escherichia coli O157:H7. Lett. Appl. Microbiol. 36: 162–167.

Burt, S.A. 2004. Essential oils: Their antibacterial properties and potential applications in foods: A review. Int. J. Food Microbiol. 94: 223–253.

Cao-Hoang, L., A. Chaine, L. Grégoire and Y. Waché. 2010a. Potential of nisin-incorporated sodium caseinate films to control Listeria in artificially contaminated cheese. Food Microbiol. 27: 940–944.

Cao-Hoang, L., L. Grégoire, A. Chaine and Y. Waché. 2010b. Importance and efficiency of in-depth antimicrobial activity for the control of Listeria development with nisin-incorporated sodium caseinate films. Food Control 21: 1227–1233.

Carneiro, J.O., V. Texeira, P. Carvalho and S. Azevedo. 2011. Self-cleaning smart nanocoatings. pp. 397–413. *In*: Makhlouf, A.S.H. and I. Tiginyanu (eds.). Nanocoatings and Ultra-Thin Films. Woodhead Publishing Ltd., Cambridge, UK.

Cha, D.S., K. Cooksey, M.S. Chinnan and H.J. Park. 2003. Release of nisin from various heat-pressed and cast films. Lebensm. Wiss. Technol. 36: 209–213.

Chinabhark, K., S. Benjakul and T. Prodpran. 2007. Effect of pH on the properties of protein-based film from bigeye snapper (Priacanthus tayenus) surimi. Bioresource Technol. 98: 221–225.

Cho, S.W., M. Gällstedt and M.S. Hedenqvist. 2010. Properties of wheat gluten/poly (lactic acid) laminates. J. Agric. Food Chem. 58: 7344–7350.

Chung, W. and R.E.W. Hancock. 2000. Action of lysozyme and nisin mixtures against lactic acid bacteria. Int. J. Food Microbiol. 60: 25–32.

Ciannamea, E.M., P.M. Stefani and R.A. Ruseckaite. 2014. Physical and mechanical properties of compression molded and solution casting soybean protein concentrate based films. Food Hydrocolloid. 38: 193–204.

Colak, B.Y., P. Peynichou, S. Galland, N. Oulahal, G. Assezat, F. Prochazka, et al. 2015. Active biodegradable sodium caseinate films manufactured by blown-film extrusion: Effect of thermo-mechanical processing parameters and formulation on lysozyme stability. Ind. Crop. Prod. 72: 142–151.

Dadalioglu, I. and G. Evrendilek. 2004. Chemical compositions and antibacterial effects of essential oils of Turkish oregano (Origanum minutiflorum), bay laurel (Laurus nobilis), Spanish lavender (Lavandula stoechas L.), and fennel (Foeniculum vulgare) on common foodborne pathogens. J. Agr. Food Chem. 52: 8255–8260.

Damodaran, S. 1996. Amino acids, peptides and proteins. pp. 217–330. *In*: Fennema, O. (ed.). Food Chemistry. CRC Press, New York, USA.

Dawson, P.L., G.D. Carl, J.C. Acton and I.Y. Han. 2002. Effect of lauric acid and nisin-impregnated soy-based films on the growth of Listeria monocytogenes on turkey bologna. Poultry Sci. 81: 721–726.

Del Nobile, M.A., A. Conte, A.L. Incoronato and O. Panza. 2008. Antimicrobial efficacy and release kinetics of thymol from zein films. J. Food Eng. 89: 57–63.

Del Nobile, M.A., A. Conte, G.G. Buonocore, A.L. Incoronato, A. Massaro and O. Panza. 2009. Active packaging by extrusion processing of recyclable and biodegradable polymers. J. Food Eng. 93: 1–6.

Djagny, K.B., Z. Wang and S. Xu. 2001. Gelatin: A valuable protein for food and pharmaceutical industries: Review. Crit. Rev. Food Sci. Nutr. 41: 481–492.

Duan, J., S.L. Park, M.A. Daeschel and Y. Zhao. 2007. Antimicrobial chitosan-lysozyme (CL) films and coatings for enhancing microbial safety of Mozzarella cheese. J. Food Sci. 72: 355–362.

Duncan, T.V. 2011. Applications of nanotechnology in food packaging and food safety: Barrier materials, antimicrobials and sensors J. Colloid Interface Sci. 363: 1–24.

Dybing, S.T. and D.E. Smith. 1991. Relation of chemistry and processing procedures to whey protein functionality: A review. Cult. Dairy Prod. J. 26: 4–12.

Echeverría, I., P. Eisenberg and A.N. Mauri. 2014. Nanocomposites films based on soy proteins and montmorillonite processed by casting. J. Membrane Sci. 449: 15–26.

El-Wakil, N.A., E.A. Hassan, R.E. Abou-Zeid and A. Dufresne. 2015. Development of wheat gluten/nanocellulose/titanium dioxide nanocomposites for active food packaging. Carbohydr. Polym. 124: 337–346.

Espitia, P.J.P., N.F.F. Soares, R.F. Teófilo, J.S.R. Coimbra, D.M. Vitor, R.A. Batista, et al. 2013. Physical-mechanical and antimicrobial properties of nanocomposite films with pediocin and ZnO nanoparticles. Carbohydr. Polym. 94: 199–208.

European Commission. Regulation (EU) No 528/2012 of the European Parliament and of the Council of 22 May 2012 concerning the making available on the market and use of biocidal products. Official Journal of the European Union, L 167, 1–123. 2012.

Fabra, M.J., P. Talens and A. Chiralt. 2009. Microstructure and optical properties of sodium caseinate films containing oleic acid–beeswax mixtures. Food Hydrocolloid. 23: 676–683.

Fabra, M.J., A. López-Rubio and J.M. Lagaron. 2016. Use of the electrohydrodynamic process to develop active/bioactive bilayer films for food packaging applications. Food Hydrocolloid. 55: 11–18.

Farris, S., K.M. Schaich, L.S. Liu, P.H. Cooke, L. Piergiovanni and K.L. Yam. 2011. Gelatin-pectin composite films from polyion-complex hydrogels. Food Hydrocolloid. 25: 61–70.

FDA. 2004. Office of Food Additive Safety, Agency letter, GRASS Notice No GRN 000135. http://www.cfsan.fda.gov/~rdb/Opa-g135.html.

Félix, M., J.E. Martín-Alfonso, A. Romero and A. Guerrero. 2014. Development of albumen/soy biobased plastic materials processed by injection molding. J. Food Eng. 125: 7–16.

Fernández-Agulló, A., M. Sonia Freire, G. Antorrena, J.A. Pereira and J. González-Álvarez. 2014. Effect of the extraction technique and operational conditions on the recovery of bioactive compounds from chestnut (Castanea sativa) bur and shell. Separ. Sci. Technol. 49(2): 267–277.

Fernández-Pan, I., M. Royo and J.I. Maté. 2012. Antimicrobial activity of whey protein isolate edible films with essential oils against food spoilers and foodborne pathogens. J. Food Sci. 77(7): M383–M390.

Ferruzzi, M.G., V. Bohm, P.D. Courtney and S.J. Schwartz. 2002. Antioxidant and antimutagenic activity of dietary chlorophyll derivatives determined by radical scavenging and bacterial reverse mutagenesis assays. J. Food Sci. 67: 2589–2595.

Gaellstedt, M., A. Mattozzi, E. Johansson and M.S. Hedenqvist. 2004. Transport and tensile properties of compression-molded wheat gluten films. Biomacromolecules 5: 2020–2028.

Garrido, T., A. Etxabide, M. Peñalba, K. de la Caba and P. Guerrero. 2013. Preparation and characterization of soy protein thin films: Processing properties correlation. Mater. Lett. 105: 110–112.

Geornaras, I. and J.N. Sofos. 2005. Activity of ε-polylysine against Escherichia coli O157:H7, Salmonella Typhimurium, Listeria monocytogenes. J. Food Sci. 70: 404–408.

Geornaras, I., Y. Yoon, K.E. Belk, G.C. Smith and J.N. Sofos. 2007. Antimicrobial activity of ε-polylysine against Escherichia coli O157:H7, Salmonella Typhimurium, Listeria monocytogenes in various food extracts. J. Food Sci. 72: 330–334.

Gill, A.O. and R.A. Holley. 2000. Inhibition of bacterial growth on ham and bologna by lysozyme, nisin and EDTA. Food Res. Int. 33: 83–90.

Gómez-Guillén, M.C., B. Giménez, M.E. López-Caballero and M.P. Montero. 2011. Functional and bioactive properties of collagen and gelatin from alternative sources: A review. Food Hydrocolloid. 25: 1813–1827.

González, A., M.C. Strumia and C.I. Alvarez Igarzabal. 2011. Cross-linked soy protein as material for biodegradable films: Synthesis, characterization and biodegradation. J. Food Eng. 106(4): 331–338.

González, A. and C.I. Alvarez Igarzabal. 2013. Soy protein – Poly (lactic acid) bilayer films as biodegradable material for active food packaging. Food Hydrocolloid. 33: 289–296.

González, A. and C.I. Alvarez Igarzabal. 2015. Nanocrystal-reinforced soy protein films and their application as active packaging. Food Hydrocolloid. 43: 777–784.

Guerrero, P., Z.A. Nur Hanani, J.P. Kerry and K. de la Caba. 2011. Characterization of soy protein-based films prepared with acids and oils by compression. J. Food Eng. 107: 41–49.

Gülçin, I., V. Mshvildadze, A. Gepdiremen and R. Elias. 2004. Antioxidant activity of saponins isolated from ivy: a-hederin, hederasaponin-C, hederacolchiside-E and hederacolchiside. F. Planta Medica 70: 561–563.

Gülçin, I., D. Berashvilli and A. Gepdiremen. 2005. Antiradical and antioxidant activity of total anthocyanins from Perilla pankinrensis decne. J. Ethanopharamcol. 101: 287–293.

Gülçin, I., Z. Huyut, M. Elmastas and H.Y. Aboul-Enein. 2010. Radical scavenging and antioxidant activity of tannic acid. Arab. J. Chem. 3: 43–53.

Gumy, D., C. Morias, P. Bowen, C. Pulgarin, S. Giraldo, R. Hajidu, et al. 2006. Catalytic activity of commercial of TiO2 powders for the abatement of the bacteria (E. coli) under solar simulated light: Influence of the isoelectric point. Appl. Catal. B-Environ. 63: 76–84.

Gupta, P. and K.K. Nayak. 2015. Characteristics of protein-based biopolymer and its application. Polym. Eng. Sci. 55: 485–498.

Gutierrez, L., A. Escudero, R. Batlle and C. Nerin. 2009. Effect of mixed antimicrobial agents and flavours in active packaging films. J. Agric. Food Chem. 57: 8564–8571.

Han, J.H. and A. Gennadios. 2005. Edible films and coatings: A review. pp. 239–259. *In*: Han, J.H. (ed.). Innovations in Food Packaging. Elsevier Academic Press, Amsterdam, The Netherlands.

Helal, A., D. Tagliazucchi, A. Contea and S. Desobry. 2012. Antioxidant properties of polyphenols incorporated in casein/sodium caseinate films. Int. Dairy J. 25: 10–15.

Hiraki, J., T. Ichikawa, S. Ninomiya, H. Seki, K. Uohama, H. Seki, et al. 2003. Use of ADME studies to confirm the safety of ε-polylysine as a preservative in food. Regul. Toxicol. Pharmacol. 37: 328–340.

Ho, Y.T., S. Ishizaki and M. Tanaka. 2000. Improving emulsifying activity of ε-polylysine by conjugation with dextran through the Maillard reaction. Food Chem. 68: 449–455.

Hong, Y., G. Lim and K.B. Song. 2009. Physical properties of Gelidium corneum/gelatin blend films containing grapefruit seed extract or green tea extract and its application in the packaging of pork loins. J. Food Sci. 74: C6–C10.

Hostynek, J.J. and H.I. Maibach. 2004. Copper hypersensitive: Dermatologic aspects. Dermatol. Ther. 17: 328–333.

Hsieh, J.F., C.J. Yu, J.Y. Chang, S.T Chen and H.Y. Tsai. 2014. Microbial transglutaminase-induced polymerization of β-conglycinin and glycinin in soymilk: A proteomics approach. Food Hydrocolloid. 35: 678–685.

Janes, M.E., S. Kooshesh and M.G. Johnson. 2002. Control of Listeria monocytogenes on the surface of refrigerated, ready-to-eat chicken coated with edible zein film coatings containing nisin and/or calcium propionate. J. Food Sci. 67: 2754–2757.

Jansens, K.J.A., B. Lagrain, I. Rombouts, M. Smet and J.A. Delcour. 2011. Effect of temperature, time and wheat gluten moisture content on gluten network formation during thermomolding. J. Cereal Sci. 54: 434–441.

Jansens, K.J.A., N. Vo Hong, L. Telen, K. Brijs, A.W. Van Vuure, K. Van Acker, et al. 2013. Effect of molding conditions and moisture content on the mechanical properties of compression molded glassy, wheat gluten bioplastics. Ind. Crop. Prod. 44: 480–487.

Javanmard, M. 2009. Biodegradable whey protein edible films as new biomaterials for food and drug packaging. Iranian J. Pharm. Sci. 5: 129–134.

Jimenez, A., M.J. Fabra, P. Talens and A. Chiralt. 2012. Effect of sodium caseinate on properties and ageing behavior of corn starch based films. Food Hydrocolloid. 29: 265–271.

Kadam, S.U., S.K. Pankaj, B.K. Tiwari, P.J. Cullen and C.P. O'Donnell. 2015. Development of biopolymer-based gelatin and casein films incorporating brown seaweed Ascophyllum nodosum extract. Food Packaging and Shelf Life. 6: 68–74.

Kaewprachu, P. and S. Rawdkuen. 2014. Mechanical and physico-chemical properties of biodegradable protein-based films: A comparative study. Food Appl. Biosc. J. 2(1): 14–29.

Kalman, D. 2014. Amino acid composition of an organic brown rice protein concentrate and isolate compared to soy and whey concentrates and isolates. Foods 3: 394–402.

Kanmani, P. and J.W. Rhim. 2014. Physicochemical properties of gelatin/silver nanoparticle antimicrobial composite films. Food Chem. 148: 162–169.

Kapiotis, S., M. Hermann, M. Exner, H. Laggner and B.M.K. Gmeiner. 2005. Copper- and magnesium protoporphyrin complexes inhibit oxidative modification of LDL induced by hemin, transition metal ions and tyrosyl radicals. Free Radic. Res. 39: 1193–1202.

Kavoosi, G., S.M.M. Dadfar, S.M.A. Dadfar, F. Ahmadi and M. Niakosari. 2014a. Investigation of gelatin/multi-walled carbon nanotube nanocomposite films as packaging materials. Food Sci. Nutr. 2: 65–73.

Kavoosi, G., A. Rahmatollahi, S.M.M. Dadfar and A.M. Purfard. 2014b. Effects of essential oil on the water binding capacity, physicomechanical properties, antioxidant and antibacterial activity of gelatin films. LWT-Food Sci. Technol. 57: 556–561.

Kerry, J.P. 2014. New packaging technologies, materials and formats for fast-moving consumer products. pp. 549–584. In: Han, J.H. (ed.). Innovations in Food Packaging (2nd ed.). Academic Press, San Diego, USA.

Khalil, A.A. and S.F. Deraz. 2015. Enhancement of mechanical properties, microstructure, and antimicrobial activities of zein films cross-linked using succinic anhydride, eugenol, and citric acid. Prep. Biochem. Biotechnol. 45: 551–567.

Kondo, K., M. Kurihara, K. Fukuhara, T. Tanaka, T. Suzuki, N. Miyata, et al. 2000. Conversion of procyanidin B-type (catechin dimer) to A-type: Evidence for abstraction of C-2 hydrogen in catechin during radical oxidation. Tetrahedron Lett. 41: 485–488.

Kozak, J., T. Balmer, R. Byrne and K. Fisher. 1996. Prevalence of Listeria monocytogenes in foods: Incidence in dairy products. Food Control 7: 215–221.

Kristo, F., K.P. Koutsoumanis and C.G. Biliaderis. 2008. Thermal, mechanical and water vapor barrier properties of sodium caseinate films containing antimicrobials and their inhibitory action on Listeria Monocytogenes. Food Hydrocolloid. 22: 373–386.

Ku, K., Y. Hong and K.B. Song. 2008a. Mechanical properties of a Gelidium corneum edible film containing catechin and its application in sausages. J. Food Sci. 73: 217–221.

Ku, K., Y. Hong and K.B. Song. 2008b. Preparation of a silk fibroin film containing catechin and its applications. Food Sci. Biotechnol. 17: 1203–1206.

Kumar, S. and S.K. Gupta. 2012. Applications of biodegradable pharmaceutical packaging materials: A review. Middle East J. Sci. Res. 12: 699–706.

Lai, L.S., S.T. Chou and W.W. Chao. 2001. Studies on the antioxidative activities of Hsian-tsao (Mesona procumbens Hemsl) leaf gum. J. Agr. Food Chem. 49: 963–968.

Letendre, M., G. D'Aprano, M. Lacroix, S. Salmieri and D. St.-Gelais. 2002. Physicochemical properties and bacterial resistance of biodegradable milk protein films containing agar and pectin. J. Agric. Food Chem. 50: 6017–6022.

Li, J.H., J. Miao, J.L. Wu, S.F. Chen and Q.Q. Zhang. 2014. Preparation and characterization of active gelatin-based films incorporated with natural antioxidants. Food Hydrocolloid. 37(6): 166–173.

Li, K.K., S.W. Yin, X.Q. Yang, C.H. Tang and Z.H. Wei. 2012. Fabrication and characterization of novel antimicrobial films derived from thymolloaded zein–sodium caseinate (SC) nanoparticles. J. Agr. Food Chem. 60: 11592–11600.

Li, S., Y. Wei, Y. Fang, W. Zhang and B. Zhang. 2013. DSC study on the thermal properties of soybean protein isolates/corn starch mixture. J. Therm. Anal. Calorim. 115: 1633–1638.

Li, Y.X., I. Wijesekara, Y. Li and S.K. Kim. 2011. Phlorotannins as bioactive agents from brown algae. Process Biochem. 46(12): 2219–2224.

Lim, G.O., Y.H. Hong and K.B. Song. 2010. Application of gelidium corneum edible films containing carvacrol for ham packages. J. Food Sci.75(1): C90–C93.

Lipparelli Morelli, C., M. Mahrous, M. Naceur Belgacem, M.C. Branciforti, R.E. Suman Bretas and J. Bras. 2015. Natural copaiba oil as antibacterial agent for bio-based active packaging. Ind. Crop Prod. 70: 134–141.

Liu, H., H. Pei, Z. Han, G. Feng and D. Li. 2015. The antimicrobial effects and synergistic antibacterial mechanism of the combination of ϵ-polylysine and nisin against Bacillus subtilis. Food Control 47: 444–450.

López-Carballo, G., P. Hernández-Muñoz, R. Gavara and M.J. Ocio. 2008. Photoactivated chlorophyllin-based gelatin films and coatings to prevent microbial contamination of food products. Int. J. Food Microbiol. 126: 65–70.

Lu, Y. and L.Y. Foo. 2000. Antioxidant and radical scavenging activities of polyphenols from apple pomace. Food Chem. 68: 81–85.

Lungu, B. and M.G. Jhonson. 2005. Potasium sorbate does not increase control of Listeria Monocytogenes when added to zein coating with nicin on the surface of full fat turkey frankfurter pieces in a model system at 4°C. J. Food Sci. 70: M95–M99.

Ma, W., C.H. Tang, S.W. Yin, X.Q. Yang and J.R. Qi. 2013. Genipin-crosslinked gelatin films as controlled releasing carriers of lysozyme. Food Res. Int. 51: 321–324.

Malik, A., A. Kushnoor, V. Saini, S. Singhal, S. Kumar and Y. Chand Yadav. 2011. In vitro antioxidant properties of Scopoletin. J. Chem. Pharm. Res. 3: 659–665.

Mangalassary, S., I. Han, J. Rieck, J. Acton and P. Dawson. 2008. Effect of combined nisin and/or lysozyme with in-package pasteurization for control of Listeria monocytogenes in ready-to-eat turkey bologna during refrigerated storage. Food Microbiol. 25: 866–870.

Maria do Carmo, B., R.N. Bennett, S. Quideau, R. Jacquet, E.A. Rosa and J.V. Ferreira-Cardoso. 2010. Evaluating the potential of chestnut (Castanea sativa Mill.) fruit pericarp and integument as a source of tocopherols, pigments and polyphenols. Ind. Crop. Prod. 31(2): 301–311.

Martínez-Abad, A., M.J. Ocio and J.M. Lagaron. 2014. Morphology, physical properties, silver release, and antimicrobial capacity of ionic silver-loaded poly(L-lactide) films of interest in food-coating applications. J. Appl. Polym. Sci. 131: 41001.

Martirosyan, A. and Y.J. Schneider. 2014. Engineered nanomaterials in food: Implications for food safety and consumer health. Int. J. Environ. Heal. R. 11: 5720–5750.

Mastromatteo, M., G. Barbuzzi, A. Conte and M.A. Del Nobile. 2009. Controlled release of thymol from zein based film. Innovative Food Sci. Emerg. Technol. 10: 222–227.

Matan, N. 2012. Antimicrobial activity of edible film incorporated with essential oils to preserve dried fish (Decapterus maruadsi). Int. Food Res. J. 19(4): 733–1738.

McHugh, T.H. and J.M. Krochta. 1994. Water vapor permeability properties of edible whey protein-lipid emulsion films. J. Am. Oil Chem. Soc. 71: 307–312.

Mecitoglu, Ç., A. Yemenicioğlu, A. Arslanoglu, Z.S. Elmacı, F. Korel and A.E. Çetin. 2006. Incorporation of partially purified hen egg white lysozyme into zein films for antimicrobial food packaging. Food Res. Int. 39: 12–21.

Mendes de Souza, P., A. Fernández, G. López-Carballo, R. Gavara and P. Hernández-Muñoz. 2010. Modified sodium caseinate films as releasing carriers of lysozyme. Food Hydrocolloid. 24: 300–306.

Meredith, J.R., C. Jin, R.J. Narayan and R. Aggarwal. 2013. Biomedical applications of carbon-nanotube composites. Front. Biosci. 5: 610–621.

Min, S., L.J. Harris, J.H. Han and J.M. Krochta. 2005. Listeria monocytogenes inhibition by whey protein films and coatings incorporating lysozyme. J. Food Prot. 68: 2317–2325.

Murillo-Martínez, M.M., S.R. Tello-Solís, M.A. García-Sánchez and E. Ponce-Alquicira. 2013. Antimicrobial activity and hydrophobicity of edible whey protein isolate films formulated with nisin and/or glucose oxidase. J. Food Sci. 78(4): M560–M566.

Nafchi, A.M., M. Moradpour, M. Saeidi and A.K. Alias. 2014. Effects of nanorod-rich ZnO on rheological, sorption isotherm, and physicochemical properties of bovine gelatin films. LWT-Food Sci. Technol. 58: 142–149.

Nie, X., Y. Gong, N. Wang and X. Meng. 2015. Preparation and characterization of edible myofibrillar protein-based film incorporated with grape seed procyanidins and green tea polyphenol. LWT-Food Sci. Technol. 64: 1042–1046.

Nishinari, K., Y. Fang, S. Guo and G.O. Phillips. 2014. Soy proteins: A review on composition, aggregation and emulsification. Food Hydrocolloid. 39: 301–318.

Ortiz, C.M., A.N. Mauri and A.R. Vicente. 2013. Use of soy protein based 1-methylcyclopropene-releasing pads to extend the shelf life of tomato (Solanum lycopersicum L.) fruit. Innov. Food Sci. Emerg. 20: 281–287.

Oussallah, M., S. Caillet, S. Salmieri, L. Saucier and M. Lacroix. 2004. Antimicrobial and antioxidant effects of milk protein based film containing essential oils for the preservation of whole beef muscle J. Agr. Food Chem. 52: 5598–5605.

Ozdemir, M. and J.D. Floros. 2003. Film composition effects on diffusion of potassium sorbate through whey protein films. J. Food Sci. 68: 511–516.

Ozdemir, M. and J.D. Floros. 2004. Active food packaging technology. Crit. Rev. Food Sci. Nutr. 44: 185–193.

Papon, P., J. Leblon and P.H.E. Meijer. 2007. Gelation and transitions in biopolymers. pp. 22–27. *In*: The Physics of Phase Transitions. Springer, Berlin.

Park, H.R., S.H. Chough, Y.H. Yun and S.D. Yoon. 2005. Properties of starch/PVA blend films containing citric acid as additive. J. Polym. Environ. 13: 375–382.

Park, H.Y., S. Kim, K.M. Kim, Y. You, S.Y. Kim and J. Han. 2012. Development of antioxidant packaging material by applying corn-zein to LLDPE film in combination with phenolic compounds. J. Food Sci. 77: 273–279.

Pereda M., M.I. Aranguren and N.E. Marcovich. 2010. Caseinate films modified with tung oil. Food Hydrocolloid. 24: 800–808.

Pereda M., A.G. Ponce, N.E. Marcovich, R.A. Ruseckaite and J.F. Martucci. 2011. Chitosan–gelatin composites and bi-layer films with potential antimicrobial activity. Food Hydrocolloid. 25: 1372–1381.

Pérez, L.M., C.E. Balagué and A.C. Verdini. 2011. Evaluation of the biocide properties of whey-protein edible films with potassium sorbate to control non-O157 shiga toxin producing Escherichia coli. Procedia Food Science 1: 203–209.

Perumalla, A.V.S. and N.S. Hettiarachchy. 2011. Green tea and grape seed extracts-potential applications in food safety and quality. Food Res. Int. 44: 827–839.

Pommet, M., A. Redl, S. Guilbrt and M.H. Morel. 2005. Intrinsic influence of various plasticizers on functional properties and reactivity of wheat gluten thermoplastic materials. J. Cereal Sci. 42: 81–91.

Priyadarshini, S., V. Gopinath, N. Meera Priyadharsshini, D. Mubarak Ali and P. Velusamy. 2013. Synthesis of anisotropic silver nanoparticles using novel strain, Bacillus flexus and its biomedical application. Colloid Surf. B: Biointerfaces. 102: 232–237.

Realini, C.E. and B. Marcos. 2014. Active and intelligent packaging systems for a modern society. Meat Sci. 98: 404–419.

Restuccia, D., U.G. Spizzirri, O.I. Parisi, G. Cirillo, M. Cursio, F. Iemma, et al. 2010. New EU regulations aspects and global market of active and intelligent packaging for food industry applications. Food Control 21: 1425–1435.

Robertson, G.L. 2012. Food packaging: Principles and practice (3rd ed.). CRC Press, Boca Ratón, Florida, USA.

Robinson, D.K.R. and M.J. Morrison. 2010. Nanotechnologies for food packaging: Reporting the science and technology research trends. Report for the Observatory NANO. Phytotherapy Res. 15: 476–480.

Rouhi, J., S. Mahmud, N. Naderi, C.H.R. Ooi and M.R. Mahmood. 2013. Physical properties of fish gelatin-based bio-nanocomposite films incorporated with ZnO nanorods. Nanoscale Res. Lett. 8: 364–368.

Saenghirunwattana P., A. Noomhorm and V. Rungsardthong. 2014. Mechanical properties of soy protein based green composites reinforced with surface modified cornhusk fiber. Ind. Crop Prod. 60: 144–150.

Sanchez-Moreno, C., A. Jimenez-Escrig and F. Saura-Calixto. 2000. Study of low density lipoprotein oxidizability indexes to measure the antioxidant activity of dietary polyphenol. Nutr. Res. 20: 941–953.

Santos, A.O., T. Ueda-Nakamura, B.P. Dias Filho, Jr., V.F. Veiga, A.C. Pinto and C.V. Nakamura. 2008. Antimicrobial activity of Brazilian copaiba oils obtained from different species of the Copaifera genus. Memórias do Instituto Oswaldo Cruz, Rio de Janeiro 103(3): 277–281.

Sayanjali, S., B. Ghanbarzadeh and S. Ghiassifar. 2011. Evaluation of antimicrobial and physical properties of edible film based on carboxymethyl cellulose containing potassium sorbate on some mycotoxigenic Aspergillus species in fresh pistachios. LWT-Food Sci. Technol. 44(4): 1133–1138.

Seydim, A.C. and G. Sarikus. 2006. Antimicrobial activity of whey protein based edible films incorporated with oregano, rosemary and garlic essential oils. Food Res. Int. 39: 639–644.

Shakeri, M.-S., F. Shahidi, S. Beiraghi-Toosi and A. Bahrami. 2011. Antimicrobial activity of Zataria multiflora Boiss. essential oil incorporated with whey protein based films on pathogenic and probiotic bacteria. Int. J. Food Sci. Tech. 46: 549–554.

Shankar, S., S.V. More and R.S. Laxman. 2010. Recovery of silver from waste X-ray film by alkaline protease from Conidiobolus coronatus. Kathmandu Univ. J. Sci. Eng. Technol. 6: 60–69.

Shankar, S., X. Teng and J.W. Rhim. 2014a. Properties and characterization of agar/CuNP bionanocomposite films prepared with different copper salts and reducing agents. Carbohdr. Polym. 114: 484–492.

Shankar, S., X. Teng and J.W. Rhim. 2014b. Effects of concentration of ZnO nanoparticles on mechanical, optical, thermal, and antimicrobial properties of gelatin/ZnO nanocomposite films. Korean J. Packag. Sci. Technol. 20: 41–49.

Shankar, S., J.P. Reddy, J.W. Rhim and H.Y. Kim. 2015a. Preparation, characterization, and antimicrobial activity of chitin nanofibrils reinforced carrageenan nanocomposite films. Carbohdr. Polym. 117: 468–475.

Shankar, S., X. Teng, G. Li and J.W. Rhim. 2015b. Preparation, characterization, and antimicrobial activity of gelatin/ZnO nanocomposite films. Food Hydrocoll. 45: 264–271.

Shen, X.L., J.M. Wu, Y.H. Chen and G.H. Zhao. 2010. Antimicrobial and physical properties of sweet potato starch films incorporated with potassium sorbate or chitosan. Food Hydrocolloid. 24(4): 285–290.

Shiku, Y., P.Y. Hamaguchi, S. Benjakul, W. Visessanguan and M. Tanaka. 2004. Effect of surimi quality on properties of edible films based on Alaska Pollack. Food Chem. 86: 493–499.

Shimada, K. and J.C. Cheftel. 1998. Sulfhydryl group disulfide bond interchange during heat induced gelation of whey protein isolate. J. Agric. Food Chem. 37: 161–168.

Silva, N.H.C.S., C. Vilela, I.M. Marrucho, C.S.R. Freire, C. Pascoal Neto and A.J.D. Silvestre. 2014. Protein-based materials: from sources to innovative sustainable materials for biomedical applications. J. Mater. Chem. B. 2(24): 3715–3740.

Smith, C.R. 1921. Osmosis and swelling of gelatin. J. Am. Chem. Soc. 43: 1350–1366.

Sobrino-López, A. and O. Martin-Belloso. 2008. Enhancing the lethal effect of high-intensity pulsed electric field in milk by antimicrobial compounds as combined hurdles. J. Diary Sci. 91: 1759–1768.

Sothornvit, R. and J.M. Krochta. 2005. Plasticizers in edible films and coatings. pp. 403–433. In: Jung, H.H. (ed.). Innovations in Food Packaging. Academic Press, London.

Sperber, W.H. 2010. Introduction to the microbial spoilage of foods and beverages. pp. 1–40. In: Sperber, W.H. and M.P. Doyle (eds.). Compendium of the Microbial Spoilage of Foods and Beverages. Springer, N.Y.

Sudağidan, M. and A. Yemenicioğlu. 2012. Effects of nisin and lysozyme on growth inhibition and biofilm formation capacity of Staphylococcus aureus strains isolated from raw milk and cheese samples. J. Food Prot. 75: 1627–1633.

Sun, S., Y. Song and Q. Zheng. 2008. Thermo-molded wheat gluten plastics plasticized with glycerol: Effect of molding temperature. Food Hydrocolloid. 22: 1006–1013.

Suppakul, P., J. Miltz, K. Sonneveld and S.W. Bigger. 2003. Active packaging technologies with an emphasis on antimicrobial packaging and its applications. J. Food Sci. 68: 408–420.

Takahashi, H., M. Keshimura, S. Miya, S. Kuramoto, H. Koiso, T. Kuda and B. Kimura. 2012. Effect of paired antimicrobial combinations of Listeria monocytogenes growth inhibition in ready-to-eat seafood products. Food Control 26: 397–400.

Tayel, A.A., W.F. El-Tras, S. Moussa, A.F. El-Baz, H. Mahrous, M.F. Salem, et al. 2011. Antibacterial action of zinc oxide nanoparticles against food borne pathogens. J. Food Saf. 31: 211–218.

Taylor, P., Y. Song and Q. Zheng. 2014. Ecomaterials based on food proteins and polysaccharides. Polym. Rev. 54: 514–571.

Teixeira, B., A. Marques, C. Pires, C. Ramos, I. Batista, J.A. Saraiva, et al. 2014. Characterization of fish protein films incorporated with essential oils of clove, garlic and origanum: Physical, antioxidant and antibacterial properties. LWT-Food Sci. Technol. 59: 533–539.

Tharanathan, R.N. 2003. Biodegradable films and composite coatings: Past, present and future. Trends Food Sci. Technol. 14: 71–80.

Tongnuanchan, P., S. Benjakul, T. Prodpran and P. Songtipya. 2011. Characteristics of film based on protein isolate from red tilapia muscle with negligible yellow discoloration. Int. J. Biol. Macromol. 48: 758–767.

Türe, H., M. Gällstedt and M.S. Hedenqvist. 2012. Antimicrobial compression-molded wheat gluten films containing potassium sorbate. Food Res. Int. 45: 109–115.

U.S. Food And Drug Administration – FDA. Center for Food Safety and Applied Nutrition. Office of Premarket Approval. GRAS Notices. http://vm.cfsan.fda.gov. (2000).

Ünalan, İ.U., F. Korel and A. Yemenicioğlu. 2011. Active packaging of ground beef patties by edible zein films incorporated with partially purified lysozyme and Na₂EDTA. Int. J. Food Sci. Technol. 46: 1289–1295.

Ünalan, İ.U., I. Arcan, F. Korel and A. Yemenicioğlu. 2013. Application of active zein based films with controlled release properties to control Listeria monocytogenes growth and lipid oxidation in fresh Kashar cheese. Innovative Food Sci. Emerg. Technol. 20: 208–214.

Van de Velde, K. and P. Kiekens. 2002. Biopolymers: Overview of several properties and consequences on their applications. Polym. Test. 21(4): 433–442.

Vasconcellos, F.C.S., A.L. Woiciechowski, V.T. Soccol, D. Mantovani and C.R. Soccol. 2014. Antimicrobial and antioxidant properties of conglycinin and glycinin from soy protein isolate. Int. J. Current Microbiol. Appl. Sci. 3: 144–157.

Veiga, Jr. V.F., L. Zunino, J.B. Calixto, M.L. Patitucci and A.C. Pinto. 2001. Phytochemical and anti-oedematogenic studies of commercial copaíba oils available in Brazil. Phytother Res. 15(6): 476–480.

Veiga, Jr. V.F. and C.A. Pinto. 2002. The Copaifera L. genus (Gênero, L. Copaifera). Quimica Nova 25(2): 273–286.

Vermeiren, L., F. Devlieghere, M. van Beest, N. de Kruijf and J. Debevere. 1999. Developments in active packaging of foods. Trends Food Sci. Technol. 10: 77–86.

Vonasek, E., P. Le and N. Nitin. 2014. Encapsulation of bacteriophages in whey protein films for extended storage and release. Food Hydrocolloid. 37: 7–13.

Wang, H., D. Hu, Q. Ma and L. Wang. 2016. Physical and antioxidant properties of flexible soy protein isolate films by incorporating chestnut (Castanea mollissima) bur extracts. LWT-Food Sci. Technol. (in press).

Wang, L.F. and J.W. Rhim. 2015. Preparation and application of agar/alginate/collagen ternary blend functional food packaging films. Int. J. Biol. Macromol. 80: 460–468.

Wang, Q., L.L. Yin and G.W. Padua. 2008. Effect of hydrophilic and lipophilic compounds on zein microstructures. Food Biophys. 3: 174–181.

Wu, J., S. Chen, S. Ge, J. Miao and Q. Zhang. 2013. Preparation, properties and antioxidant activity of an active film from sliver carp (Hypophthalmichthys molitrix) skin gelatin incorporated with green tea extract. Food Hydrocolloid. 32(1): 42–51.

Wu, J., S. Ge, H. Liu, S. Wang, S. Chen, J. Wang, et al. 2014. Properties and antimicrobial activity of silver carp (Hypophthalmichthys molitrix) skin gelatin-chitosan films incorporated with oregano essential oil for fish preservation. Food Packag. Shelf life 2: 7–16.

Wu, J., H. Liu, S. Ge, S. Wang, Z. Qin, L. Chen, et al. 2015. The preparation, characterization, antimicrobial stability and in vitro release evaluation of fish gelatin films incorporated with cinnamon essential oil nanoliposomes. Food Hydrocolloid. 43: 427–435.

Yilmaz, Y. 2006. Novel uses of catechins in foods. Trends Food Sci. Technol. 17: 6471.

Yuan Y., Y.E. Sun, Z.L. Wan, X.Q. Yang, J.F. Wu, S.W. Yin, et al. 2014. Chitin microfibers reinforce soy protein gels cross-linked by transglutaminase. J. Agric. Food Chem. 62: 4434–4442.

Zhang, L., Y. Jiang, Y. Ding, N. Daskalakis, L. Jeuken, M. Povey, et al. 2010. Mechanistic investigation into antibacterial behavior of suspensions of ZnO nanoparticles against E. coli. J. Nanopart. Res. 12: 1625–1636.

Zhao, S., J. Yao, X. Fei, Z. Shao and X. Chen. 2013. An antimicrobial film by embedding in situ synthesized silver nanoparticles in soy protein isolate. Mater. Lett. 95: 142–144.

Zinoviadou, K.G., K.P. Koutsoumanis and C.G. Biliaderis. 2009. Physico-chemical properties of whey protein isolate films containing oregano oil and their antimicrobial action against spoilage flora of fresh beef. Meat Sci. 82: 338–345.

Zubeldía, F., M.R. Ansorena and N.E. Marcovich. 2015. Wheat gluten films obtained by compression molding. Polym. Test. 43: 68–77.

Book Conclusions

The ebook conclusions greatly exalts the results that all the researchers have put on the development of *"Biopackaging"* as well as techniques and methodologies for characterizations and determination of transport properties of film-forming biopolymers from different sources and industrial applications. For that, researchers and engineers have adopted a strategy based on the understanding of the relationship between structure and macroscopic properties of film-forming biopolymers.

Authors in this ebook have shown, through the diversity of the countries they belong, that biopolymers from animal, plant, microbial and seaweed sources can be extracted through eco-friendly processes and be used for biopackaging applications.

The design and engineering of innovative biopackaging deserve a special attention because it forms a fascinating interdisciplinary area that brings together food, biochemistry & biotechnology, materials science and nanotechnology.

The revolution in biopolymers applications comes from the use of proteins and polysaccharides in structured food and non-food materials with the elaboration of biocompatible and degradable films for active biopackaging applications.

In Chapter 1, *Edible Active Packaging for Food Application: Materials and Technology*, written by Mengxing Li and Ran Ye, it is stated that edible active packaging revolutionizes the concept of food packaging and extends the application of food packaging. As one of the most interesting applications, edible active packaging is used as a useful tool for bioactive delivery from agricultural and industrial waste, to consumers and their effectiveness in packaging, or to their economic and environmental impact. Edible packaging materials have been considered as attractive alternatives because of their unique properties, including the ability to protect foods with their barrier and mechanical properties, enhance sensory characteristics, control-release active ingredients, and control mass transfer between components of heterogeneous foods.

In Chapter 2, *Active Bio-Packaging in the Food Industry* written by Ricardo Stefani, it is concluded that the development of bio-based food packaging has been very active in the last decade. Bio-based packaging for food is mostly made from naturally occurring renewable polymers or natural polymers blended with artificial biodegradable polymers that have a good film and coating-forming properties. The use of these polymers in food packaging has a clear environmental advantage, since they are renewable and/or biodegradable, reducing environmental pollution. The biodegradable polymeric film into a low cost active or smart polymer. The food bio-packaging can be a versatile and low-cost approach to control and preserve the quality of a food product as it makes its way from the manufacturer to the final consumer.

In Chapter 3, *Antimicrobial Active Packaging* written by Cintia B. Contreras, German Charles, Ricardo Toselli and Miriam C. Strumia, it is presented that food packaging innovations need to decrease environmental pressure by taking into account a broad range of sustainability issues such as: waste prevention, efficient use of resources, process optimization and recycling or reuse. The use of renewable and environmentally friendly materials in future packaging technologies, with a specific interest in finding better cost/benefit ratios, represents the major challenge. Nanotechnology has demonstrated a great potential to expand the use of biodegradable polymers, since the preparation of bionanocomposites improves the overall performance of biopolymers, making them more competitive in a market dominated by non-biodegradable materials. The "ideal" packaging with active and intelligent properties still needs further research and development to address several challenges. The antimicrobial function of a container has been a need in the area of food. It is known that the addition of antimicrobial compounds to films can effectively inhibit or delay the growth of microorganisms that may be present on the surface of food products. Some antibacterial compounds including antibiotics, essential oils and inorganic nanoparticles could be directly incorporated or covalently bonded into the packaging material.

In Chapter 4, *Transport Phenomena in Biodegradable and Edible Films*, written by Tomy J. Gutiérrez and Kelvia Álvarez, it is concluded that transport properties of edible films and coatings play a key role in reducing, decelerating and improving the exchange of substances between the coated food and its surrounding environment. The fundamental aspects of mass transfer phenomena and the applicable methods for characterization and evaluation are presented. Similarly, it provides an introduction to the mathematical methods. Standard methods for mass transfer characterization were developed for synthetic polymer films; but applied to all types of films and coatings. However, they do not always lead to consistent results. The difficulty arises from changes in behavior with boundary conditions for the use of predictive models are needed to describe the actual behavior of the data obtained under ideal conditions.

In Chapter 5, *Formulation, Properties and Performance of Edible Films and Coatings from Marine Sources in Vegetable and Fruits*, written by Armida Rodríguez-Félix, Tomás J. Madera-Santana and Yolanda Freile-Pelegrín, it is stated that vegetables and fruits are highly consumable foods. The harvest performed in fruits and vegetables produces significant changes in the balance of gases such as the consumption of oxygen (O_2), water content and the production of carbon dioxide (CO_2). The increase in the rate of gas transfer produces in the fruit a metabolic loss, a gradual maturation and eventual senescence. However, internal or external factors are responsible for the gas transfer rate, such as the species, cultivar and maturity or grow state (internal), the atmospheric composition (CO_2, O_2 and ethylene), temperature, relative humidity and light exposure among other external factors. The cuticle or skin of fruits and vegetables prevents the contamination by pathogenic microorganisms and biochemical deterioration (off-flavour development, browning and loss of firmness or texture breakdown). In order to extend the post-harvest life of fresh fruits and vegetables, the edible films and coatings are suitable alternatives to storing them under a modified atmosphere that reduces the changes in quality and quantity losses. The main contribution from edible films and coatings is the reduction in moisture release, respiration, gas exchange and the oxidative reaction rate. The formulations, main physicochemical properties and performance of water-soluble edible films/coatings based on agar or chitosan in fresh produce (fruits and vegetables) includes recent advances by the incorporation of antimicrobials, texture enhancers and bioactive compounds (nutraceuticals) to improve the quality of fruits and vegetables.

Edible films and coatings have increased their applications in the food industry due to the wide range of properties that can show and their use for solving several problems in the storage and transport of fresh produce. Chitosan as edible film and coating is widely used in fruits and vegetables by antimicrobial characteristics.

Chapter 6, *Agroindustrial Biomass: Potential Materials for Production of Biopolymeric Films*, written by Delia R. Tapia-Blácido, Bianca C. Maniglia, Milena Martelli-Tosi and Vinícius F. Passos, has proposed the use of bio-polymers obtained from agroindustrial biomass for application in the preparation of biodegradable films as a polymer source or as fillers for biopolymers. The biopolymeric films produced from the remaining residue after extraction of turmeric dye and the use of pretreated soybean straw as filler in cassava starch are some examples of the use of agroindustrial biomass as a resource of biopolymers or filler to polymer films. The application of agroindustrial biomass as raw material to produce biopolymeric film is advantageous over the use of petroleum-based materials because they are biodegradable, inexpensive, abundant and renewable, contributing to the reduction of environmental pollution. However, the improvement of water vapour barrier and mechanical properties of films based on agroindustrial biomass is necessary to make the use of these residues viable in the packaging industry.

In Chapter 7, *Vegetable Nanocellulose in Food Packaging*, written by C. Gómez H., A. Serpa, J. Velásquez-Cock, C. Castro, B. Gómez H., L. Vélez, P. Gañán and R. Zuluaga, they study about the use of nanocellulose as a bio-based food packaging material. Nanocellulose obtained from plants or vegetable sources can be used to produce films, aerogels/foams and nanocomposites with barrier, mechanical and optical properties which are useful for the development of new materials for food packaging, because nanocellulose films exhibit barrier, optical and mechanical properties comparable with synthetic polymers. However, due to the hydrophilic nature of nanocellulose, there is an increasing interest in its use for biocomposites with other bio-polymers, such as PLA and starch, with the goal to develop nanostructured films, foams or edible coatings, as the addition of nanocellulose improves the mechanical and barrier properties. In spite of its advantages, there are still challenges to its commercialization, for instance, the lack of standardized methods for its characterization, the safety and regulations that shall be solved to achieve a successful commercial product in the food packaging industry.

In Chapter 8, *Xylan Polysaccharide Fabricated into Biopackaging Films*, written by Xiaofeng Chen, Junli Ren, Chuanfu Liu, Feng Peng and Runcang Sun, it is reported that Xylan, as one of the important polysaccharides, has received an increasing interest in last decades especially due to its abundance, degradability and renewability for the production of biodegradable films. Xylans have promising application in packaging due to their good intrinsic barrier properties against non-polar migrants. The chemical compositions and the structure of xylan especially Arx/Xyl ratio, have significant impacts on the properties of xylan-based films with potential application in agriculture and food packaging.

In Chapter 9, *Reactive Extrusion for the Production of Starch-based Biopackaging*, written by Tomy J. Gutiérrez, M. Paula Guarás and Vera A. Alvarez, thermoplastic starch (TPS) properties such as economics, readily available, biodegradable, edible and safe, are reported. Nevertheless, the high hydrophilic nature of starch makes it difficult to use as a biopackaging material. Efforts have been made in order to reduce this hydrophilic tendency. Reactive extrusion (REx) is a technology that has been widely studied as it can improve the properties of thermoplastic starch, whilst maintaining current production processes at an industrial level. Reactive extrusion (REx) is currently the most efficient and economical way for producing eco-friendly plastics from starch, as it combines the versatility of today's industrial processes with polymer science and technology. In addition, the chemical reactions such as polymerization, grafting, branching and functionalization that occur during REx do not need solvents for their processing. This gives REx the advantage over other processes since energy consumption required for the evaporation of solvents is eliminated.

In Chapter 10, *Polyhydroxyalkanoates – A Prospective Food Packaging Material: Overview of the State of the Art, Recent Developments and Potentials*, written by Katrin Jammernegg, Franz Stelzer and Stefan Spirk, the replacement of petrochemical-based plastics for biodegradable and renewable alternatives have been proposed. Besides their usage in the medical and pharmaceutical sector, the food packaging industry is the most promising scope of application for biodegradable materials. The state of the art and progress in research related to polyhydroxyalkanoates (PHA) with the focus on create sustainable materials for food packaging is presented.

In Chapter 11, *Active Biopackaging Based on Proteins*, written by Mariana Pereda, María R. Ansorena and Norma E. Marcovich, active films based on biopolymers are presented as a way to overcome the limitations of synthetic plastics, since they are biodegradable and thus, could limit the accumulation of residues, reducing, therefore, the environmental impact of conventional packaging materials. Films, casings and coatings made from proteins are effective barriers against oxygen (O_2), carbon dioxide (CO_2) and aroma. An additional advantage is that proteins can be processed by diverse methods such as dissolution-solvent evaporation or thermo-mechanical processing to produce films with suitable mechanical properties. The protein coatings or films require relatively lower amounts of added antimicrobial agents to reach the desired effect as compared to the synthetic polymer or to other biopolymer films. Therefore, numerous proteins such as corn zein, wheat gluten, soy, gelatin fish and milk proteins, etc. have been studied as potential film-forming agents. The antimicrobial packaging mainly targets the food surface on which microbiological changes occur most intensively. However, a sufficient antimicrobial effect could not be achieved unless the release rate of antimicrobial compounds from the packaging materials to the food surface is adjusted considering several complex factors. Some of these complex factors include the physical and chemical properties of the food, the growth kinetics of target pathogenic or spoilage microorganisms and the expected shelf-life of the food.

Martin A. Masuelli

Index

Editor's Biography

Martin A. Masuelli is a scientist at Instituto de Física Aplicada – CONICET – UNSL, San Luis, Argentina. He holds a master's degree and a PhD thesis in Membrane Technology from National University of San Luis (UNSL). He is Coordinator of Physics Chemistry Area and Director of Physics Chemistry Service Laboratory, UNSL. He is expert in polysaccharides and physics chemistry of macromolecules. He is author or co-author of more than 20 peer-reviewed international publications, 5 book chapters, 65 communications in national and international congresses, book editor ("Fiber Reinforced Polymers – The Technology Applied for Concrete Repair", INTECH, Croatia, 2013). Ebook editor of "Advances in Physicochemical Properties of Biopolymers Part 1 and 2" (Bentham Publishing, 2017, Dubai, UAE, 2017), and Author/Editor of book "Food Packaging and Biopackaging" (ICS Morebooks, Moldova, 2017). He is a member of the Sociedad Argentina de Ciencia y Tecnología Ambiental and Asociación Argentina de Fisicoquímica y Química Inorgánica. He is Editor in Chief and Founder in July 2013 of the Journal of Polymer and Biopolymers Physics Chemistry, Science and Education Publishing. He is on the editorial board of numerous journals. Email: masuelli@unsl.edu.ar

Printed and bound by CPI Group (UK) Ltd, Croydon, CR0 4YY

01/11/2024

01782624-0016